SYMBOLS

Maps use special signs or symbols to represent location and to give information of interest.

Map symbols can be points, lines or areas and vary in size, shape and colour. This allows a great range of different symbols to be created. These have to be carefully selected to make maps easy to understand. Usually the same symbols are used to represent features on maps of the same type and scale within an atlas.

An important part of any map is the key which explains what the symbols represent. Each map in this atlas has its own key. Shown below are typical examples of the keys found on each reference map in the atlas. The first is found on all of the British Isles 1:1 200 000 series of maps. The second is found on the smaller scale maps of the rest of the world.

TYPE STYLES

Various type styles are used to show the difference between features on the maps in this atlas. Physical features are shown in italic and a distinction is made between land and water features.

Mountain Peaks are shown in small italics.
eg. *Ben Nevis* *Mt Kenya* *Fuji-san*

Large mountain ranges are shown in bold italic capitals.
eg. ***HIMALAYA   ALPS***
ROCKY MOUNTAINS

Rivers are also shown in small italics but in a different typeface from mountain peaks.
eg. *Thames Euphrates Rhine Amazon*

Oceans are shown in large bold italic capitals.
eg. *ATLANTIC OCEAN*
PACIFIC OCEAN
INDIAN OCEAN

When a feature covers a large area the type is letterspaced and sometimes curved to follow the shape of the feature.
eg. *S A H A R A*
B E A U F O R T   S E A

Settlements are shown in upright type. Country capitals are shown in capitals.
eg. **LONDON**
PARIS
TOKYO
MOSCOW

The size and weight of the type increases with the population of a settlement.
eg. Westbury
Chippenham
Bristol
Birmingham

Administrative names are shown in capitals.
eg. EAST SUSSEX
RONDONIA
KERALA
CALIFORNIA

Country names are shown in large bold capitals.
eg. **CHINA**
KENYA
MEXICO

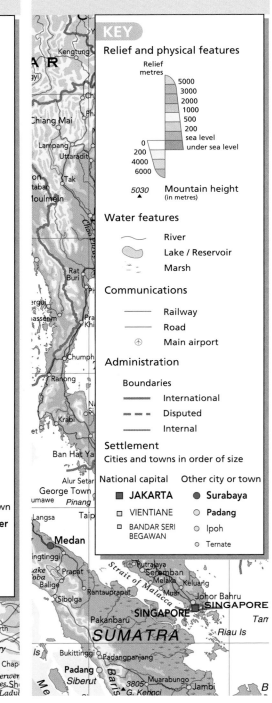

KEY

Relief and physical features

Relief metres
1000
500
200
100
0
sea level
under sea level
50
200

977 ▲ Mountain height (in metres)

Water features

〰 River
〰 Canal
⬤ Lake / Reservoir

Communications

── Railway
══ Motorway
── Road
···· Car ferry
⊕ Main airport
✈ Local airport

Administration

Boundaries

── International
── Internal

Settlement

▦ Urban area

Cities and towns in order of size

National capital | Other city or town
□ DUBLIN | ⬤ Manchester
 | ◉ Liverpool
 | ◦ Belfast
 | ◦ Carlisle
 | ◦ Keswick

KEY

Relief and physical features

Relief metres
5000
3000
2000
1000
500
200
0
sea level
under sea level
200
4000
6000

5030 ▲ Mountain height (in metres)

Water features

〰 River
⬤ Lake / Reservoir
〰 Marsh

Communications

── Railway
── Road
⊕ Main airport

Administration

Boundaries

── International
--- Disputed
── Internal

Settlement

Cities and towns in order of size

National capital | Other city or town
■ JAKARTA | ⬤ Surabaya
□ VIENTIANE | ◉ Padang
□ BANDAR SERI BEGAWAN | ◦ Ipoh
 | ◦ Ternate

An atlas map of the world shows the whole world on a flat surface of the page. yet in reality the earth is actually a sphere. This means that a system has to be used to turn the round surface of the earth into a flat map of the world, or part of the world. This cannot be done without some distortion - on a map some parts of the world have been stretched, other parts have been compressed.

A system for turning the globe into a flat map is called a **projection.**

There are many different projections, each of which distort different things to achieve a flat map. Correct area, correct shape, correct distances or correct directions can be achieved by a projection; but by achieving any one of these things the others have to be distorted. When choosing the projection to use for a particular map it is important to decide which of these things is the most important to have correct.

The projections below illustrate the main types of projections, and include some of those used in this atlas.

Cylindrical projection

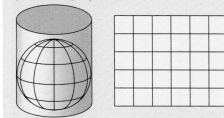

Cylindrical projections are constructed by projecting the surface of the globe on to a cylinder just touching the globe.

Conic projection

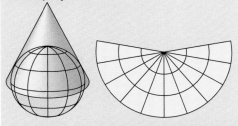

Conic projections are constructed by projecting part of the globe on to a cone which just touches a circle on the globe.

Azimuthal projection

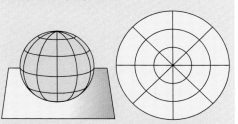

Azimuthal projections are constructed by projecting part of a globe on to a plane which touches the globe only at one point

Examples of projections

Mercator
Southeast Asia pp104-105

Mercator is a cylindrical projection. It is a useful projection for areas 15° N or S of the equator where distortion of shape is minimal. The projection is useful for navigation as directions can be plotted as straight lines.

Eckert IV
World pp 114-115

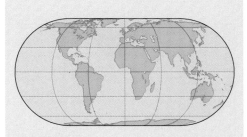

Eckert IV is an equal area projection. Equal area projections are useful for world thematic maps where it is important to show the correct relative sizes of continental areas. Ecker IV has a straight central meridian but all others are curved which help suggest the spherical nature of the earth.

Albers Equal Area Conic
Europe pp 34-35

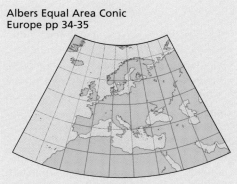

Conic projections are best suited for areas between 30° and 60° N and S with longer east-west extent than north-south. Such an area would be Europe. Meridians are straight and equally spaced.

Chamberlin Trimetric
Canada pp 62-63

Chamberlin trimetric is an equidistant projection. It shows correct distances from approximately three points. It is used for areas with a greater north-south than east-west extent, such as North America.

Lambert Azimuthal Equal Area
Australia p 110

Lambert's projection is uselful for areas which have similar east-west, north-south dimensions such as Australia.

Polar stereographic
Antarctica p 112

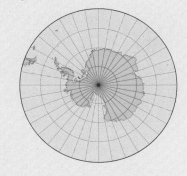

This projection shows no angular or shape distortion over small areas. All points on the map are in constant relative position and distance from the centre.

LATITUDE

Lines of latitude are imaginary lines which run in an east-west direction around the world. They are also called **parallels** of latitude because they run parallel to each other. Latitude is measured in **degrees** (°).

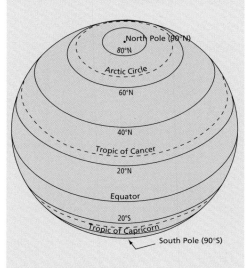

The most important line of latitude is the **Equator** (0°). The North Pole is 90° North (90°N) and the South Pole is 90° South (90°S). All other lines of latitude are given a number between 0° and 90°, either North (N) or South (S) of the Equator. Some other important lines of latitude are the Tropic of Cancer (23$\frac{1}{2}$°N), Tropic of Capricorn (23$\frac{1}{2}$°S), Arctic Circle (66$\frac{1}{2}$°N) and Antarctic Circle (66$\frac{1}{2}$°S).

The Equator can also be used as a line to divide the Earth into two halves. The northern half, north of the Equator, is the **Northern Hemisphere**. The southern half, south of the Equator, is the **Southern Hemisphere**.

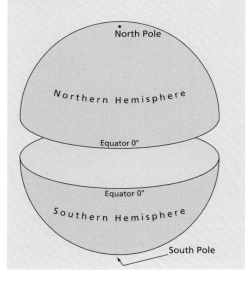

LONGITUDE

Lines of longitude are imaginary lines which run in a north-south direction, from the North Pole to the South Pole. These lines are also called **meridians** of longitude. They are also measured in **degrees** (°).

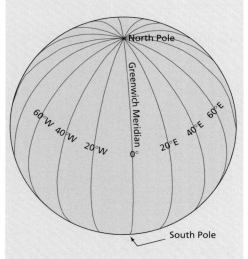

The most important line of longitude is the prime meridian (0°). This line runs through the Greenwich Observatory in London and is therefore known as the Greenwich Meridian. Exactly opposite the Greenwich Meridian on the other side of the world is the 180° line of longitude known as the International Date Line. All the other lines of longitude are given a number between 0° and 180°, either East (E) or West (W) of the Greenwich Meridian.

The Greenwich Meridian (0°) and the International Date Line (180°) can also be used to divide the world into two halves. The half to the west of the Greenwich Meridian is the Western Hemisphere. The half to the east of the Greenwich Meridian is the Eastern Hemisphere.

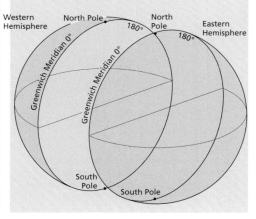

FINDING PLACES USING LATITUDE AND LONGITUDE

When lines of latitude and longitude are drawn on a map they form a grid pattern, very much like a pattern of squares.

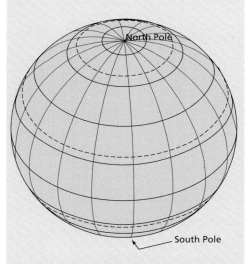

By stating the **latitude** and then the **longitude** of a place, it becomes much easier to find. On the map (below) Point A is very easy to find because it is exactly latitude 58° North of the Equator and longitude 4° West of the Greenwich Meridian (58°N,4°W).

To be even more accurate in locating a place, each degree of latitude and longitude can also be divided into smaller units called **minutes** ('). There are 60 minutes in each degree. On the map (below) Halkirk is one half (or 30/60ths) of the way past latitude 58°N, and one-half (or 30/60ths) of the way past longitude 3°W. Its latitude is therefore 58 degrees 30 minutes North and its longitude is 3 degrees 30 minutes West. This can be shortened to 58°30'N, 3°30'W.

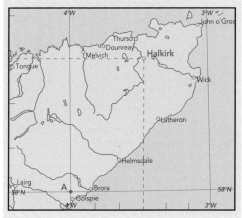

SCALE

To draw a map of any part of the world, the area must be reduced in size, or scaled down so that it will fit on to a page. The scale of a map tells us by how much the area has been reduced in size.

The scale of a map can also be used to work out distance and area. The scale of a map will show the relationship between distances on the map and distances on the ground.

Scale can be shown on a map in a number of ways:

(a) **in words**
e.g. 'one cm. to one km.' (one cm. on the map represents one km. on the ground). 'one cm. to one m.' (one cm. on the map represents one m. on the ground).

(b) **in numbers**
e.g. '1 : 100 000' or '1/100 000' (one cm. on the map represents 100 000 cm., or one km., on the ground). '1 : 25 000' or '1/25 000' (one cm. on the map represents 25 000 cm, or 250 m., on the ground). '1 : 100' or '1/100' (one cm. on the map represents 100 cm, or one m., on the ground).

(c) **as a line scale**
e.g.

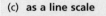

MEASURING DISTANCE ON A MAP

When a map does not have distances printed on it, we can use the scale of the map to work out how far it is from one place to another. The easiest scale to use is a line scale. You must find out how far the places are apart on the map and then see what this distance represents on the line scale. To measure the straight line distance between two points:

a) Place a piece of paper between the two points on the map,
(b) Mark off the distance between the two points along the edge of the paper,
(c) Place the paper along the line scale,
(d) Read off the distance on the scale.

Step 1

Line up the paper and mark off the distance from A to B.

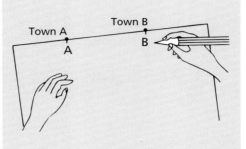

Step 2

Compare this distance with the line scale at the bottom of the map. The distance between A and B is 1.5 km on the line scale.

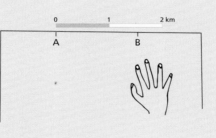

To measure the distance between two points where there are bends or curves:
(a) Place a sheet of paper on the map and mark off the start point on the edge of the paper,
(b) Now move the paper so that its edge follows the bends and curves on the map (Hint: Use the tip of your pencil to pin the edge of the paper to the curve as you pivot the paper around the curve),
(c) Mark off the end point on your sheet of paper,
(d) Place the paper along the line scale,
(e) Read off the distance on the scale.

Using a sheet of paper around a curve : Mark off the start point then twist the paper to follow the curve.

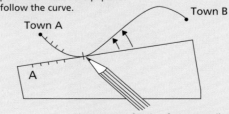

You can use the tip of your pencil to pin the paper to the curve. This stops the paper jumping off course.

MAP SCALE AND MAP INFORMATION

The scale of a map also determines how much information can be shown on it. As the area shown on a map becomes larger and larger, the amount of detail and the accuracy of the map becomes less and less.

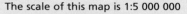
The scale of this map is 1:5 000 000

The scale of this map is 1:10 000 000

The scale of this map is 1:20 000 000

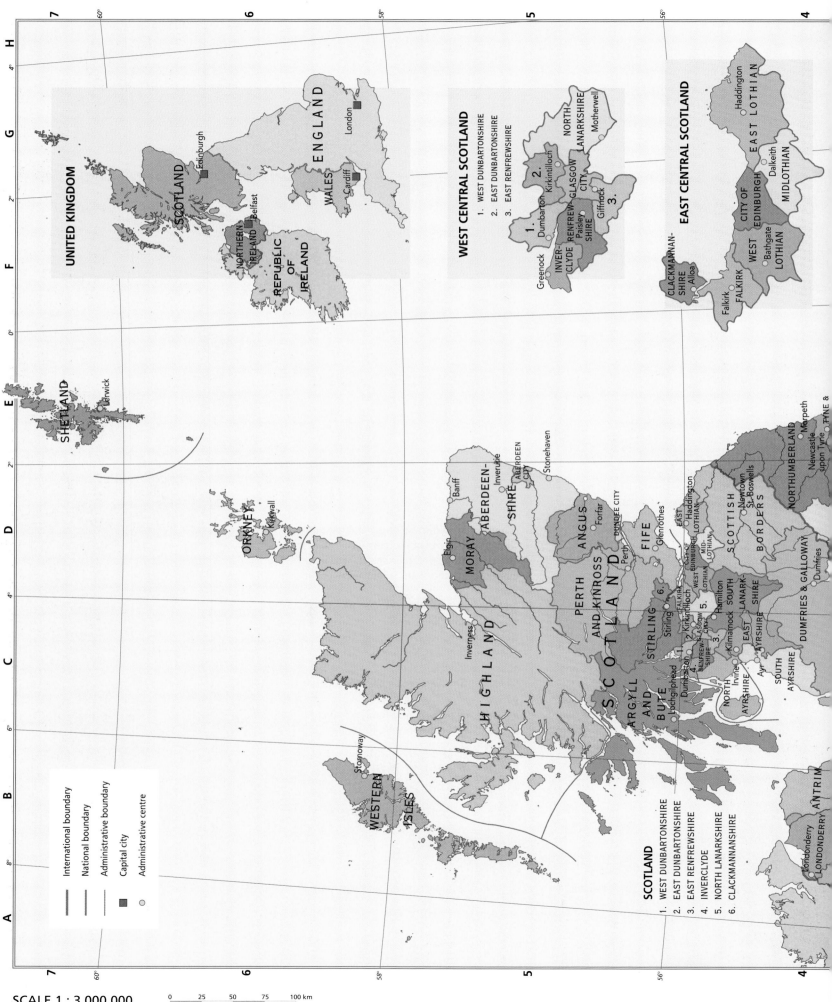

SCOTLAND
1. WEST DUNBARTONSHIRE
2. EAST DUNBARTONSHIRE
3. EAST RENFREWSHIRE
4. INVERCLYDE
5. NORTH LANARKSHIRE
6. CLACKMANNANSHIRE

UNITED KINGDOM

SCOTLAND
ENGLAND
WALES
NORTHERN IRELAND
REPUBLIC OF IRELAND

Edinburgh
London
Cardiff
Belfast

WEST CENTRAL SCOTLAND
1. WEST DUNBARTONSHIRE
2. EAST DUNBARTONSHIRE
3. EAST RENFREWSHIRE

WEST DUNBARTONSHIRE
EAST DUNBARTONSHIRE
NORTH LANARKSHIRE
GLASGOW CITY
RENFREWSHIRE
INVER-CLYDE
EAST RENFREWSHIRE

Greenock
Dumbarton
Kirkintilloch
Paisley
Giffnock
Motherwell

EAST CENTRAL SCOTLAND

EAST LOTHIAN
CITY OF EDINBURGH
MIDLOTHIAN
WEST LOTHIAN
CLACKMANNAN-SHIRE
FALKIRK

Haddington
Dalkeith
Bathgate
Alloa
Falkirk

SHETLAND
Lerwick

ORKNEY
Kirkwall

HIGHLAND
Inverness

MORAY
Elgin

ABERDEEN-SHIRE
Banff
Inverurie
ABERDEEN CITY
Stonehaven

PERTH AND KINROSS
Perth

ANGUS
Forfar

DUNDEE CITY

FIFE
Glenrothes

SCOTLAND

STIRLING
Stirling

ARGYLL AND BUTE
Lochgilphead

EAST LOTHIAN
Haddington
WEST LOTHIAN
MID-LOTHIAN

SCOTTISH BORDERS
Newtown
St. Boswells

NORTHUMBERLAND
Morpeth
Newcastle upon Tyne &
TYNE &

Dumbarton
Kirkintilloch
RENFREW-SHIRE
GLASGOW CITY

Hamilton
SOUTH LANARK-SHIRE
Kilmarnock
EAST AYRSHIRE

NORTH AYRSHIRE
Irvine
Ayr
SOUTH AYRSHIRE

DUMFRIES & GALLOWAY
Dumfries

WESTERN ISLES
Stornoway

ANTRIM
Londonderry
LONDONDERRY

International boundary
National boundary
Administrative boundary
Capital city
Administrative centre

SCALE 1 : 3 000 000

0   25   50   75   100 km

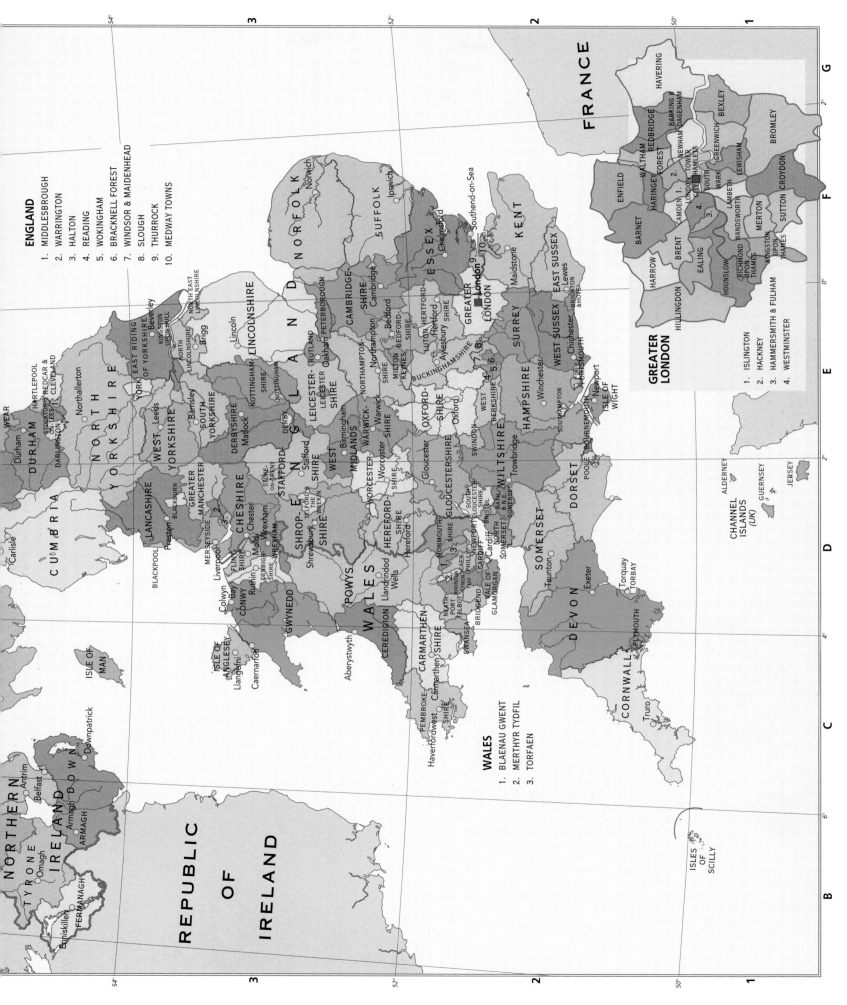

ENGLAND
1. MIDDLESBROUGH
2. WARRINGTON
3. HALTON
4. READING
5. WOKINGHAM
6. BRACKNELL FOREST
7. WINDSOR & MAIDENHEAD
8. SLOUGH
9. THURROCK
10. MEDWAY TOWNS

WALES
1. BLAENAU GWENT
2. MERTHYR TYDFIL
3. TORFAEN

GREATER LONDON
1. ISLINGTON
2. HACKNEY
3. HAMMERSMITH & FULHAM
4. WESTMINSTER

Conic projection

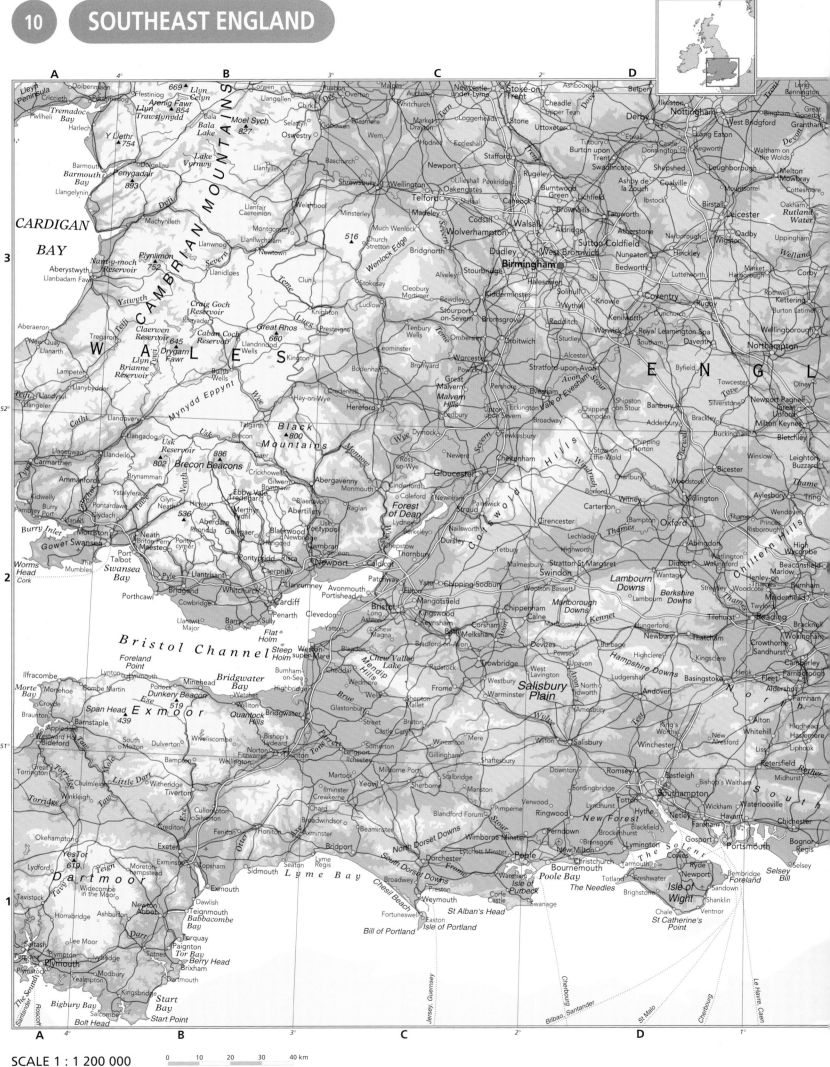

SCALE 1 : 1 200 000

0   10   20   30   40 km

Conic projection

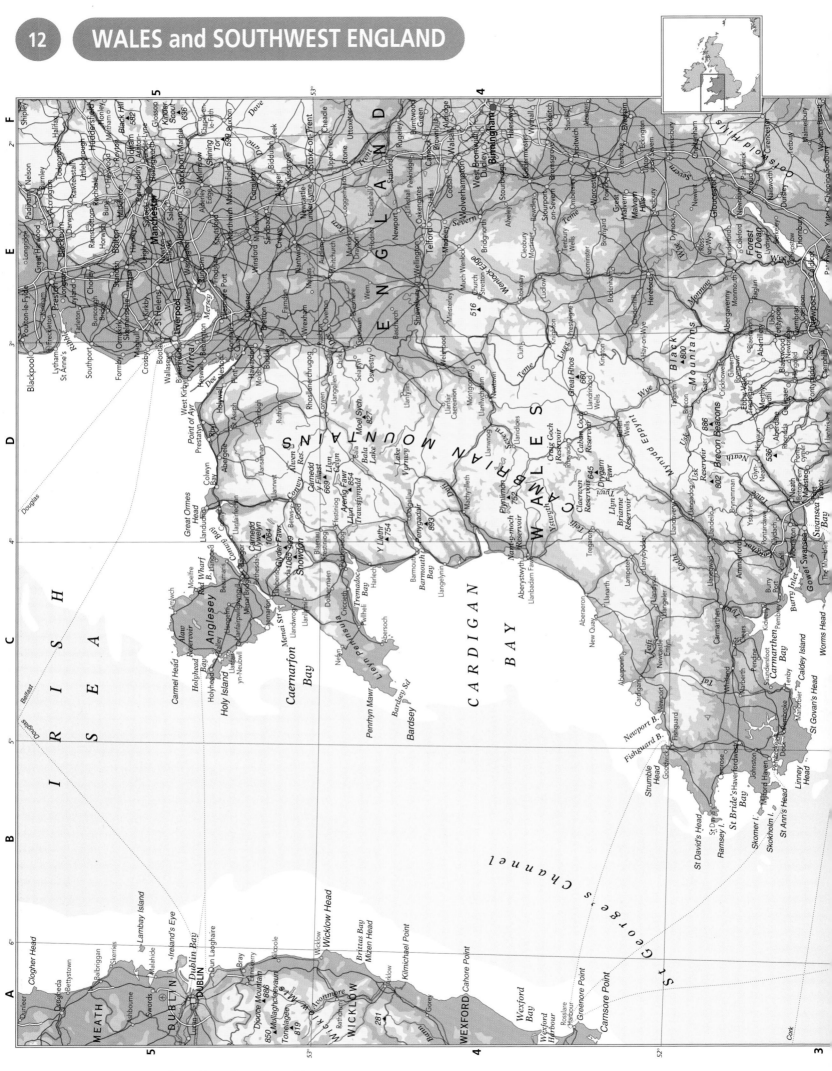

SCALE 1 : 1 200 000

0   10   20   30   40 km

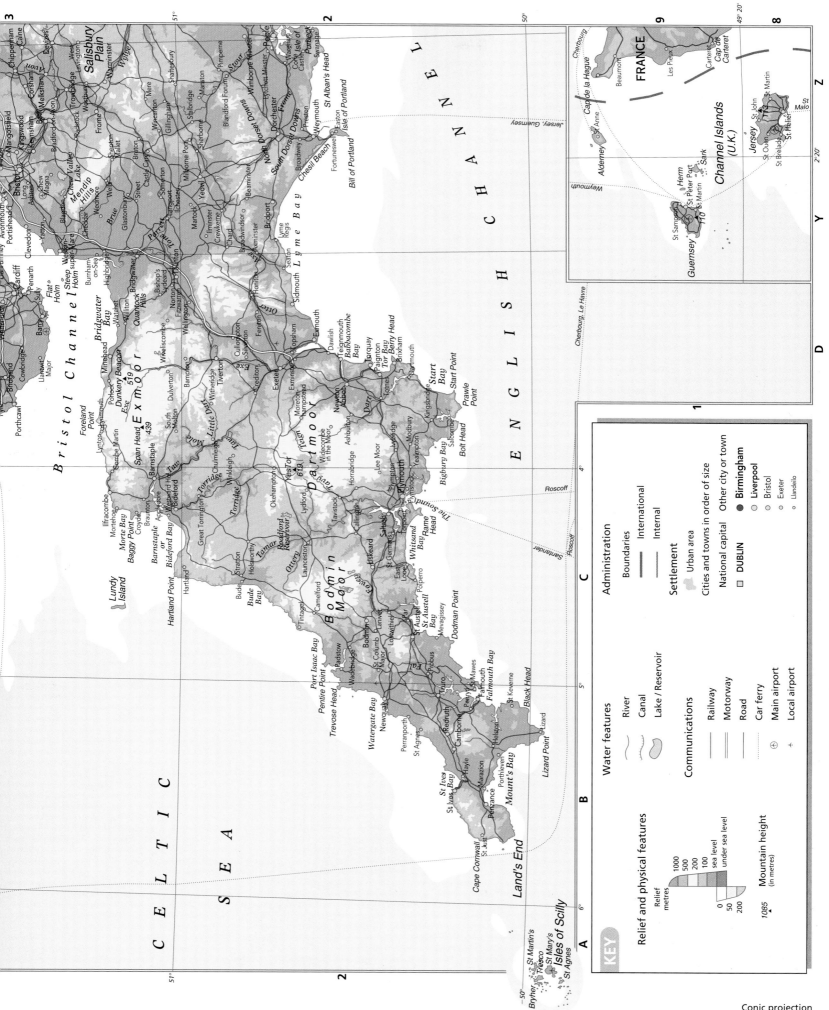

Conic projection

SCALE 1 : 1 200 000

0   10   20   30   40 km

KEY

Relief and physical features

Relief
metres
1000
500
200
100
0 sea level
50
200 under sea level

977 ▲ Mountain height
(in metres)

Water features

～ River

～ Canal

Lake / Reservoir

Communications

— Railway

= Motorway

— Road

····· Car ferry

⊕ Main airport

✈ Local airport

Administration

Boundaries

━━ International

── Internal

Settlement

Urban area

Cities and towns in order of size

National capital | Other city or town

▢ DUBLIN | ● Manchester

○ Liverpool

◉ Belfast

◎ Carlisle

○ Keswick

NORTH
SEA

E   2°   F   1°   G   0°   H

ENGLAND

North York Moors

Conic projection

SCALE 1 : 1 200 000

0   10   20   30   40 km

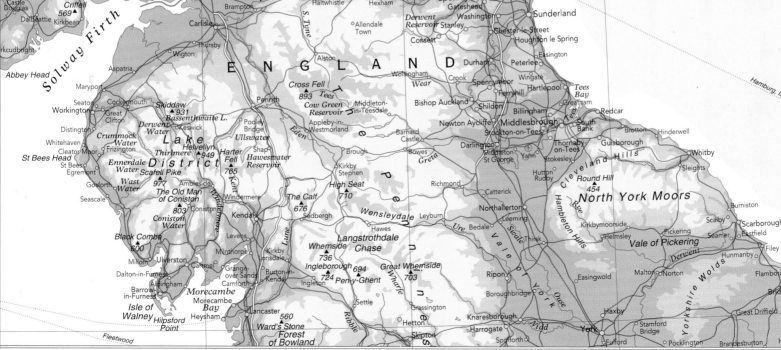

Conic projection

KEY

Relief and physical features

Relief
metres
1000
500
200
100
sea level
0
50
under sea level
200

1344 ▲ Mountain height (in metres)

Water features

~~~ River
~~~ Canal
⬭ Lake / Reservoir

Communications

——— Railway
——— Road
········· Car ferry
⊕ Main airport
⟟ Local airport

Settlement

Cities and towns in order of size
◉ Aberdeen
○ Inverness
∘ Kirkwall

SCALE 1 : 1 200 000

0 10 20 30 40 km

E 4° F 3° G 2° X 2° Y 1° Z

Herma Ness

Unst

Noup Head   Mull Head
Papa
Westray   North
Ronaldsay   Torshavn

Westray   The
North Sound   N. Ronaldsay Firth

Point of
Fethaland   Fetlar

Westray Firth   Eday   Sanday

Ronas Hill
450   Yell

Brough Head   Rousay   Sanday
Sound

Egilsay   Stronsay

Esha Ness   Toft

St. Magnus
Bay   Voe

Orkney
Islands   Stronsay
Firth

Loch of
Hurray   Shapinsay   Auskerry

Papa
Stour   Muckle
Roe

Loch of
Stenness   Wide Firth
Kirkwall   Melby

Stromness   Mainland   Skaill

Ward Hill
479   Scapa
Flow   Copinsay   Shetland
Islands   West
Burra   Lerwick

Hoy   Flotta   Burray   Isle of
Noss

Foula   Bressay

St Margaret's Hope   Bergen (& Hanstholm)
(summer only)

South
Ronaldsay   Mousa

Pentland   South
Walls   Burwick
Brough Ness   Pentland Skerries   Sumburgh

Firth   Sumburgh
Head

Dunnet Head   Island of
Stroma   Duncansby
Head

Strathy
Point   Thurso   Dunnet
Bay
B.   John o'Groats   Stromness

Thurso   Dunnet
Dounreay   Bay   Fair Isle

Melvich   Thurso   Loch
Watten   Sinclair's
Bay   Aberdeen

Loch
Hope   Halkirk

Ben Hope   Tongue   Wick
927   764
Ben Loyal Loch   CAITHNESS   Wick
Loyal

SUTHERLAND   Loch
Rimsdale

L. Naver   Thurso

Ben
Klibreck
961   Latheron

Helmsdale

Loch Shin   Lairg   Brora   Helmsdale

Brora

Golspie

Bonar Bridge   Dornoch

Dornoch Firth   Tarbat Ness

Tain

TER
SS   Loch Glass   Balintore

Ben Wyvis   Nigg
1046   Invergordon   Bay   Cromarty

Dingwall   Cromarty Firth   Moray Firth   Lossiemouth   Portknockie   Cullen   Troup Head   Fraserburgh
Burghead   Buckie   Portsoy   Macduff

Black   Conon Bridge   Forres   Elgin   Banff   Loch of Strathbeg
Isle   Fochabers   Rattray Head

Beauly   Portrose   Nairn   Lossie   Knock Hill   New   N. Ugie   Crimond
Moray   Isla   430   Pitsligo
Firth   Rothes   Keith   Aberchirder   Turriff
Beauly Firth   Inverness   Dufftown   Peterhead

Findhorn   Spey   Deveron   Huntly   Mintlaw   Boddam
STRATHBOGIE   Cruden Bay

Glen More   Ness   Nairn   Strathspey   Bogie   Insch   Oldmeldrum   Elon   Ythan
Loch Ness   Grantown   Don   Inverurie
-on-Spey   Urie
Hills of   Kemnay   Kintore   Dyce
Cromdale

Geal Charn   Avon   Westhill   Aberdeen
821
Aviemore   Dee   NORTH
Monadhliath Mountains   Cairn Gorm   Portlethen
1245   Aboyne   Banchory
Kingussie   Ben Macdui   SEA
Newtonmore   1291   1309   Ballater   Newtonhill
Cairn Toul

Creag   Braemar   Dee   Mount Keen   Stonehaven
Meagaidh   1155   939
1130   Lochnagar   N. Esk
L. Laggan   Water of Saughs   Inverbervie

Ben Alder   Carn nan   Mayar   Laurencekirk
1148   Beinn Dearg Gabhar   928   S. Esk
1008   Backwater   Hillside
Loch   Forest of Atholl 1121   Reservoir   Brechin
Garry   Glenshee   Montrose
Loch Ericht   Blair Atholl   Isla   Kirriemuir
Loch   Schiehallion   Pitlochry   Forfar   Lunan Bay
Laidon   1083   Loch Tummel

E   F   3°   G   2°   H   1°   I

Conic projection

KEY

Relief and physical features

Relief
metres
1000
500
200
100
0 sea level
under sea level
200
4000

1041 ▲ Mountain height
(in metres)

Water features

〰 River
≋ Canal
▢ Lake / Reservoir
Marsh

Communications

Railway
Motorway
Road
⊕ Main airport

Administration

Boundaries
International
Internal

Settlement
Cities and towns in order of size

National capital   Other city or town
◻ DUBLIN   ○ Cork
                  ○ Killarney

SCALE 1 : 2 000 000

0   20   40   60   80 km

Conic projection

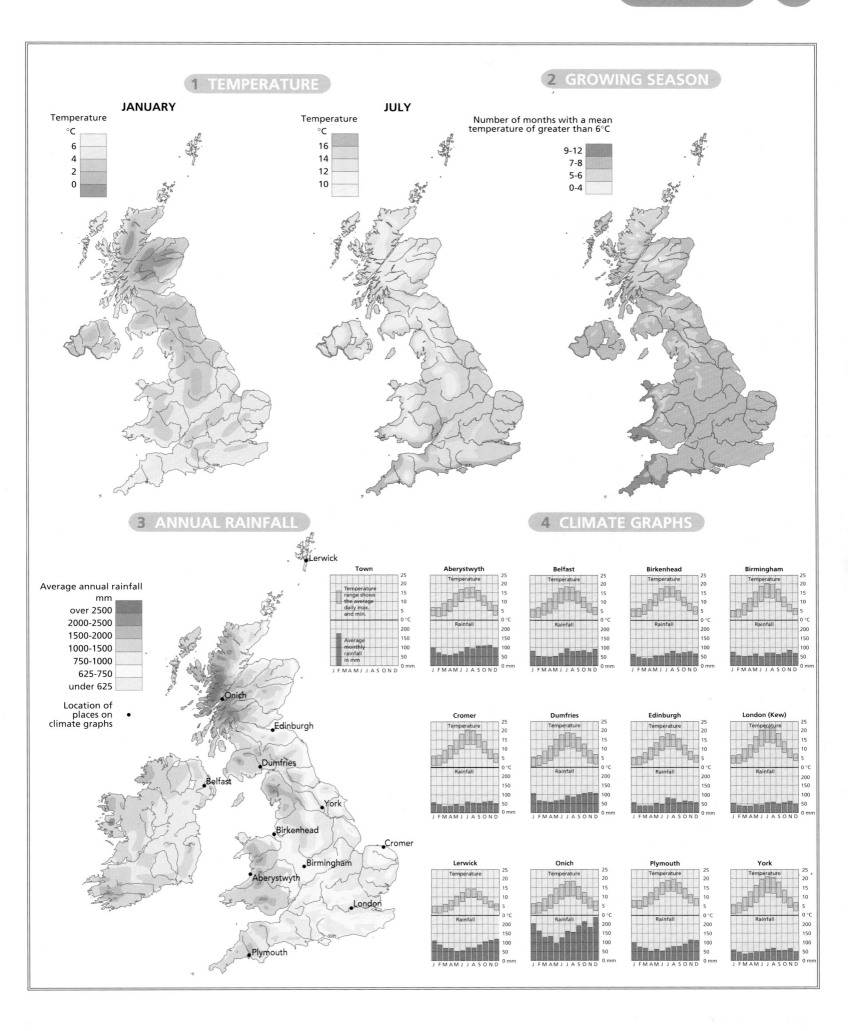

1 TEMPERATURE

JANUARY

Temperature
°C
6
4
2
0

JULY

Temperature
°C
16
14
12
10

2 GROWING SEASON

Number of months with a mean
temperature of greater than 6°C

9-12
7-8
5-6
0-4

3 ANNUAL RAINFALL

Average annual rainfall
mm
over 2500
2000-2500
1500-2000
1000-1500
750-1000
625-750
under 625

Location of
places on
climate graphs •

Lerwick
Onich
Edinburgh
Dumfries
Belfast
York
Birkenhead
Cromer
Birmingham
Aberystwyth
London
Plymouth

4 CLIMATE GRAPHS

Town
Temperature
range shows
the average
daily max.
and min.

Average
monthly
rainfall
in mm

Aberystwyth
Temperature
Rainfall

Belfast
Temperature
Rainfall

Birkenhead
Temperature
Rainfall

Birmingham
Temperature
Rainfall

Cromer
Temperature
Rainfall

Dumfries
Temperature
Rainfall

Edinburgh
Temperature
Rainfall

London (Kew)
Temperature
Rainfall

Lerwick
Temperature
Rainfall

Onich
Temperature
Rainfall

Plymouth
Temperature
Rainfall

York
Temperature
Rainfall

Shetland Islands

Relief metres
1000
500
200
100
0 sea level
200

SCALE 1 : 4 000 000

0   50   100   150 km

Conic projection

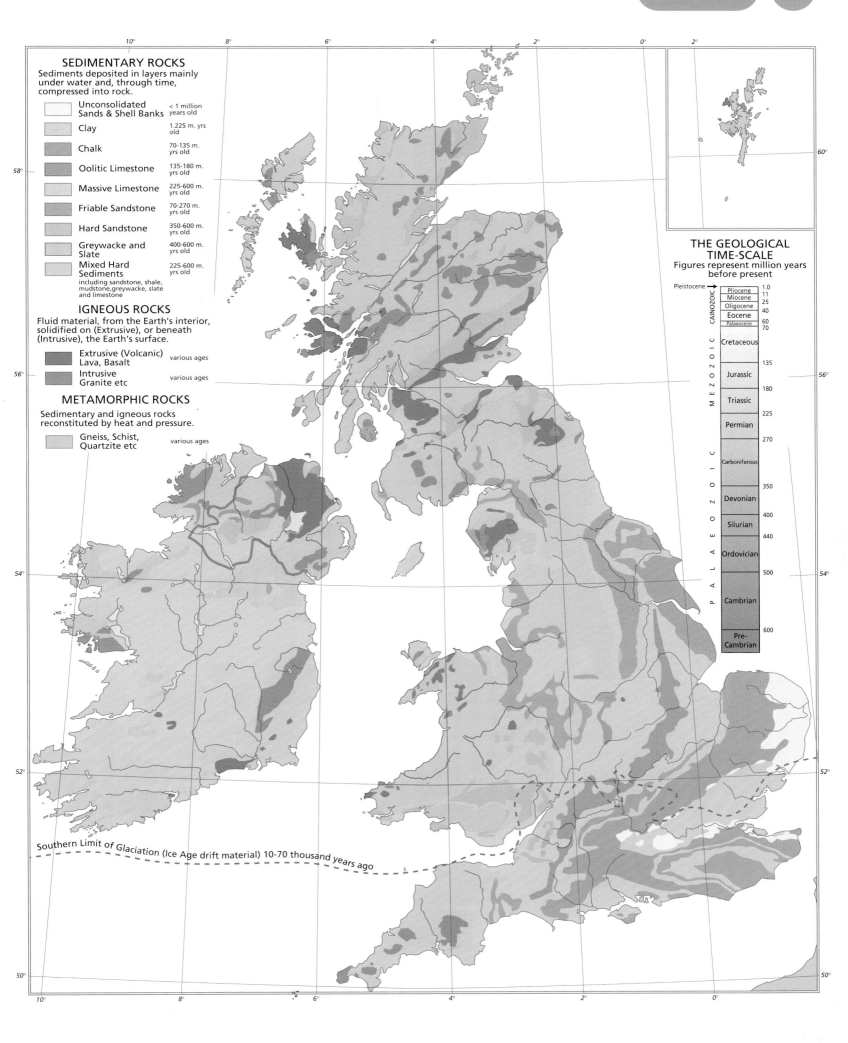

SEDIMENTARY ROCKS
Sediments deposited in layers mainly under water and, through time, compressed into rock.

	Unconsolidated Sands & Shell Banks	< 1 million years old
	Clay	1.225 m. yrs old
	Chalk	70-135 m. yrs old
	Oolitic Limestone	135-180 m. yrs old
	Massive Limestone	225-600 m. yrs old
	Friable Sandstone	70-270 m. yrs old
	Hard Sandstone	350-600 m. yrs old
	Greywacke and Slate	400-600 m. yrs old
	Mixed Hard Sediments including sandstone, shale, mudstone, greywacke, slate and limestone	225-600 m. yrs old

IGNEOUS ROCKS
Fluid material, from the Earth's interior, solidified on (Extrusive), or beneath (Intrusive), the Earth's surface.

| | Extrusive (Volcanic) Lava, Basalt | various ages |
| | Intrusive Granite etc | various ages |

METAMORPHIC ROCKS
Sedimentary and igneous rocks reconstituted by heat and pressure.

| | Gneiss, Schist, Quartzite etc | various ages |

THE GEOLOGICAL TIME-SCALE
Figures represent million years before present

Pleistocene →

CAINOZOIC	Pliocene	1.0
	Miocene	11
	Oligocene	25
	Eocene	40
		60
	Palaeocene	70
MESOZOIC	Cretaceous	
		135
	Jurassic	
		180
	Triassic	
		225
	Permian	
		270
PALAEOZOIC	Carboniferous	
		350
	Devonian	
		400
	Silurian	
		440
	Ordovician	
		500
	Cambrian	
		600
	Pre-Cambrian	

Southern Limit of Glaciation (Ice Age drift material) 10-70 thousand years ago

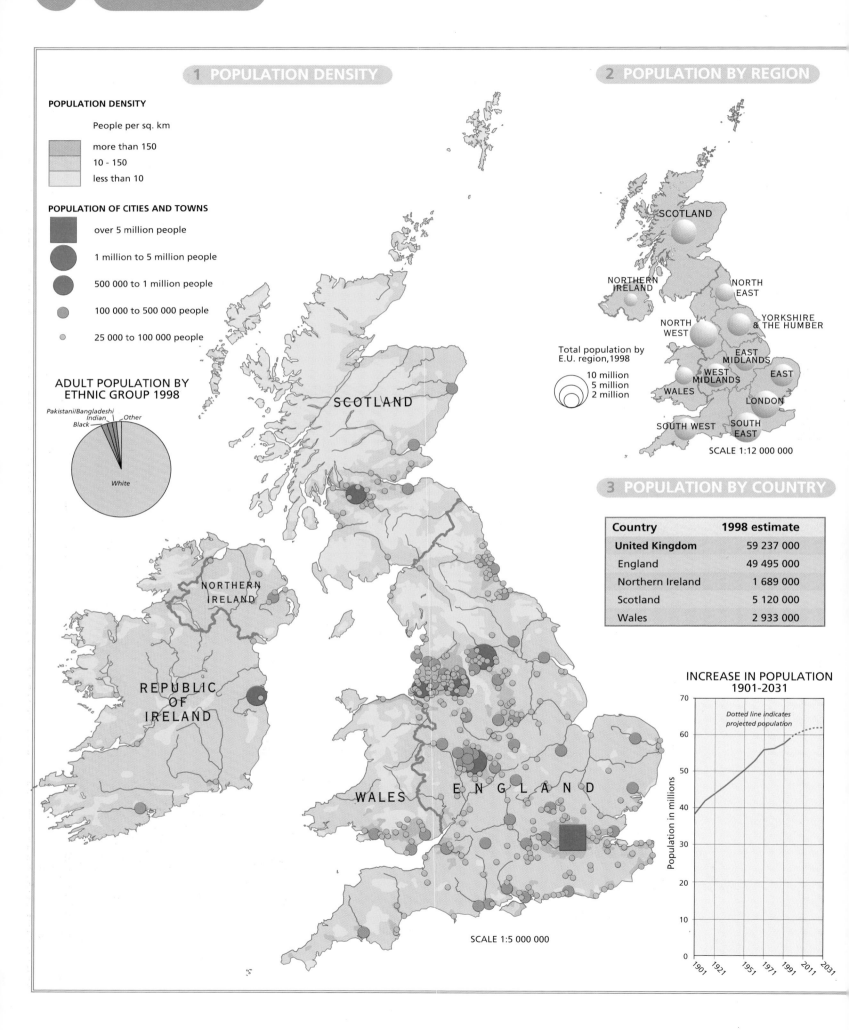

1 POPULATION DENSITY

POPULATION DENSITY

People per sq. km

more than 150
10 - 150
less than 10

POPULATION OF CITIES AND TOWNS

over 5 million people

1 million to 5 million people

500 000 to 1 million people

100 000 to 500 000 people

25 000 to 100 000 people

ADULT POPULATION BY ETHNIC GROUP 1998

Pakistani/Bangladeshi
Indian
Black
Other

White

SCOTLAND

NORTHERN IRELAND

REPUBLIC OF IRELAND

WALES

ENGLAND

SCALE 1:5 000 000

2 POPULATION BY REGION

SCOTLAND

NORTHERN IRELAND

NORTH EAST

NORTH WEST

YORKSHIRE & THE HUMBER

EAST MIDLANDS

WEST MIDLANDS

WALES

EAST

LONDON

SOUTH WEST

SOUTH EAST

Total population by E.U. region, 1998

10 million
5 million
2 million

SCALE 1:12 000 000

3 POPULATION BY COUNTRY

Country	1998 estimate
United Kingdom	59 237 000
England	49 495 000
Northern Ireland	1 689 000
Scotland	5 120 000
Wales	2 933 000

INCREASE IN POPULATION 1901-2031

Dotted line indicates projected population

Population in millions

70
60
50
40
30
20
10
0

1901 1921 1951 1971 1991 2011 2031

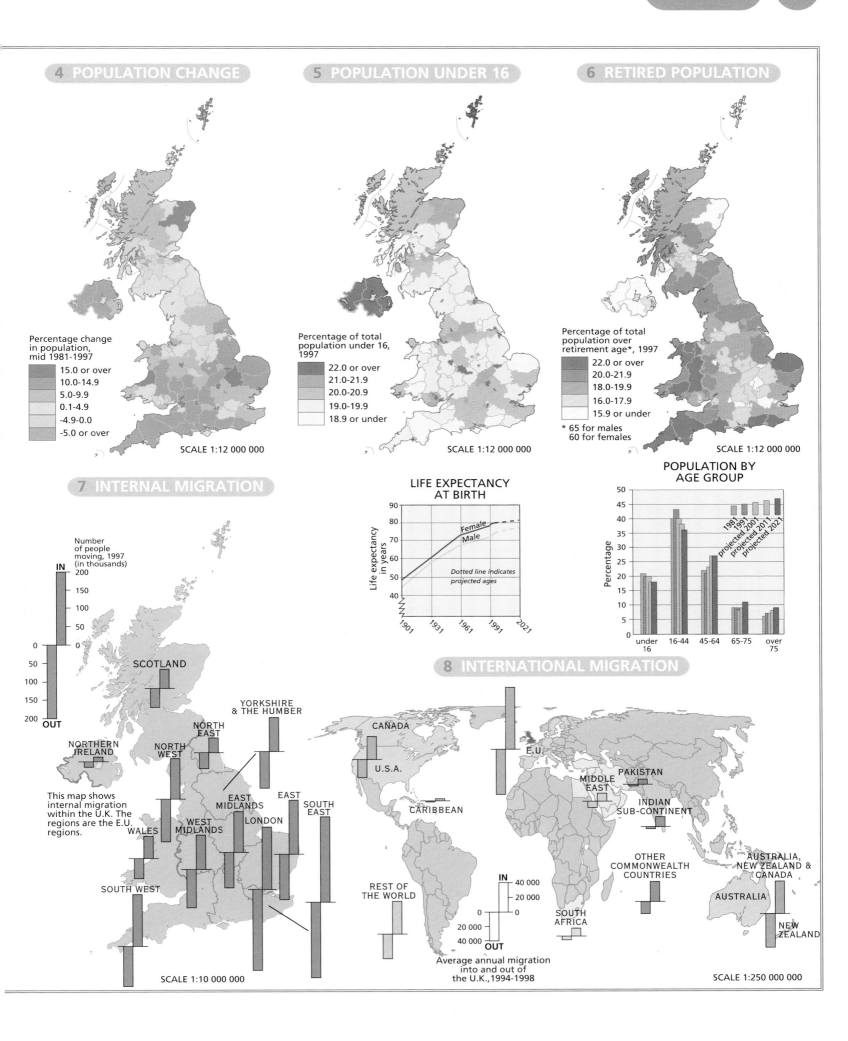

4 POPULATION CHANGE

Percentage change
in population,
mid 1981-1997
- 15.0 or over
- 10.0-14.9
- 5.0-9.9
- 0.1-4.9
- -4.9-0.0
- -5.0 or over

SCALE 1:12 000 000

5 POPULATION UNDER 16

Percentage of total
population under 16,
1997
- 22.0 or over
- 21.0-21.9
- 20.0-20.9
- 19.0-19.9
- 18.9 or under

SCALE 1:12 000 000

6 RETIRED POPULATION

Percentage of total
population over
retirement age*, 1997
- 22.0 or over
- 20.0-21.9
- 18.0-19.9
- 16.0-17.9
- 15.9 or under

* 65 for males
60 for females

SCALE 1:12 000 000

7 INTERNAL MIGRATION

Number
of people
moving, 1997
(in thousands)

IN
200
150
100
50
0
0
50
100
150
200
OUT

SCOTLAND

NORTHERN
IRELAND

NORTH
EAST

NORTH
WEST

YORKSHIRE
& THE HUMBER

EAST
MIDLANDS

EAST

WEST
MIDLANDS

LONDON

SOUTH
EAST

WALES

SOUTH WEST

This map shows
internal migration
within the U.K. The
regions are the E.U.
regions.

SCALE 1:10 000 000

**LIFE EXPECTANCY
AT BIRTH**

90
80
70
60
50
40

Life expectancy
in years

Female
Male

Dotted line indicates
projected ages

1901   1931   1961   1991   2021

**POPULATION BY
AGE GROUP**

50
45
40
35
30
25
20
15
10
5
0

Percentage

1981
1991
projected 2001
projected 2011
projected 2021

under 16   16-44   45-64   65-75   over 75

8 INTERNATIONAL MIGRATION

CANADA

U.S.A.

CARIBBEAN

E.U

MIDDLE
EAST

PAKISTAN

INDIAN
SUB-CONTINENT

OTHER
COMMONWEALTH
COUNTRIES

AUSTRALIA,
NEW ZEALAND &
CANADA

AUSTRALIA

NEW
ZEALAND

SOUTH
AFRICA

REST OF
THE WORLD

IN
40 000
20 000
0
0
20 000
40 000
OUT

Average annual migration
into and out of
the U.K.,1994-1998

SCALE 1:250 000 000

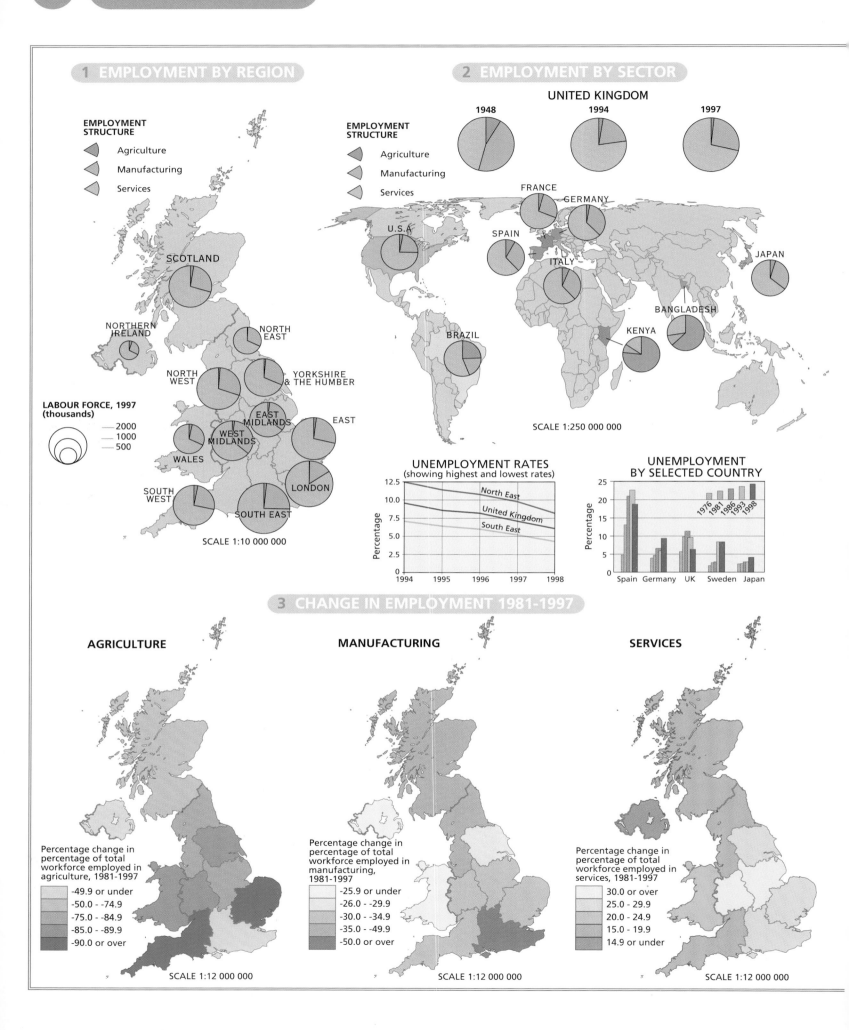

1 EMPLOYMENT BY REGION

EMPLOYMENT
STRUCTURE

◄ Agriculture
◄ Manufacturing
◄ Services

LABOUR FORCE, 1997
(thousands)

2000
1000
500

SCOTLAND

NORTHERN
IRELAND

NORTH
EAST

NORTH
WEST

YORKSHIRE
& THE HUMBER

EAST
MIDLANDS

WEST
MIDLANDS

EAST

WALES

LONDON

SOUTH
WEST

SOUTH EAST

SCALE 1:10 000 000

2 EMPLOYMENT BY SECTOR

UNITED KINGDOM

1948   1994   1997

EMPLOYMENT
STRUCTURE

◄ Agriculture
◄ Manufacturing
◄ Services

FRANCE

GERMANY

U.S.A

SPAIN

ITALY

JAPAN

BRAZIL

BANGLADESH

KENYA

SCALE 1:250 000 000

UNEMPLOYMENT RATES
(showing highest and lowest rates)

North East
United Kingdom
South East

Percentage

12.5
10.0
7.5
5.0
2.5
0

1994   1995   1996   1997   1998

UNEMPLOYMENT
BY SELECTED COUNTRY

Percentage

25
20
15
10
5
0

1976
1981
1986
1993
1998

Spain   Germany   UK   Sweden   Japan

3 CHANGE IN EMPLOYMENT 1981-1997

AGRICULTURE

Percentage change in
percentage of total
workforce employed in
agriculture, 1981-1997

-49.9 or under
-50.0 - -74.9
-75.0 - -84.9
-85.0 - -89.9
-90.0 or over

SCALE 1:12 000 000

MANUFACTURING

Percentage change in
percentage of total
workforce employed in
manufacturing,
1981-1997

-25.9 or under
-26.0 - -29.9
-30.0 - -34.9
-35.0 - -49.9
-50.0 or over

SCALE 1:12 000 000

SERVICES

Percentage change in
percentage of total
workforce employed in
services, 1981-1997

30.0 or over
25.0 - 29.9
20.0 - 24.9
15.0 - 19.9
14.9 or under

SCALE 1:12 000 000

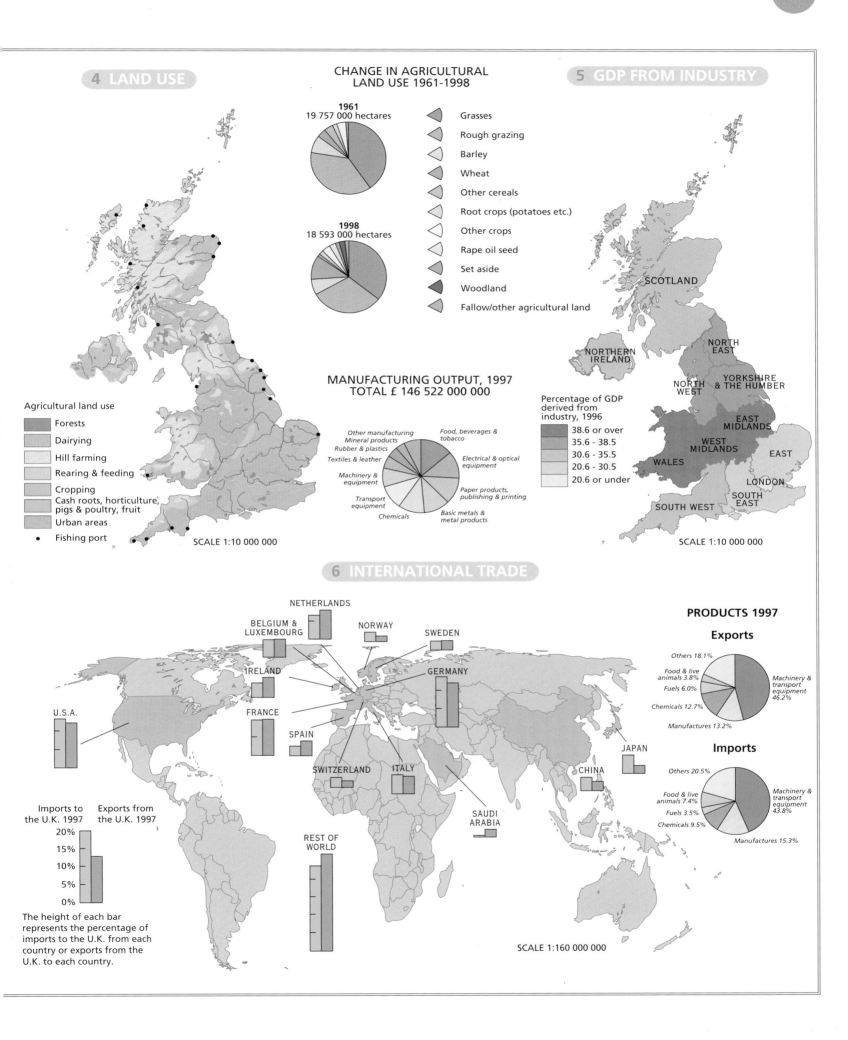

4 LAND USE

Agricultural land use
- Forests
- Dairying
- Hill farming
- Rearing & feeding
- Cropping
- Cash roots, horticulture, pigs & poultry, fruit
- Urban areas
- • Fishing port

SCALE 1:10 000 000

CHANGE IN AGRICULTURAL LAND USE 1961-1998

1961
19 757 000 hectares

1998
18 593 000 hectares

- Grasses
- Rough grazing
- Barley
- Wheat
- Other cereals
- Root crops (potatoes etc.)
- Other crops
- Rape oil seed
- Set aside
- Woodland
- Fallow/other agricultural land

MANUFACTURING OUTPUT, 1997
TOTAL £ 146 522 000 000

Other manufacturing
Mineral products
Rubber & plastics
Textiles & leather
Machinery & equipment
Transport equipment
Chemicals
Basic metals & metal products
Paper products, publishing & printing
Electrical & optical equipment
Food, beverages & tobacco

5 GDP FROM INDUSTRY

SCOTLAND

NORTHERN IRELAND

NORTH EAST

NORTH WEST

YORKSHIRE & THE HUMBER

EAST MIDLANDS

WEST MIDLANDS

WALES

EAST

LONDON

SOUTH EAST

SOUTH WEST

Percentage of GDP derived from industry, 1996
- 38.6 or over
- 35.6 - 38.5
- 30.6 - 35.5
- 20.6 - 30.5
- 20.6 or under

SCALE 1:10 000 000

6 INTERNATIONAL TRADE

NETHERLANDS
BELGIUM & LUXEMBOURG
NORWAY
SWEDEN
IRELAND
GERMANY
U.S.A.
FRANCE
SPAIN
SWITZERLAND
ITALY
JAPAN
CHINA
SAUDI ARABIA
REST OF WORLD

Imports to the U.K. 1997   Exports from the U.K. 1997
- 20%
- 15%
- 10%
- 5%
- 0%

The height of each bar represents the percentage of imports to the U.K. from each country or exports from the U.K. to each country.

SCALE 1:160 000 000

PRODUCTS 1997

Exports
- Others 18.1%
- Food & live animals 3.8%
- Fuels 6.0%
- Chemicals 12.7%
- Manufactures 13.2%
- Machinery & transport equipment 46.2%

Imports
- Others 20.5%
- Food & live animals 7.4%
- Fuels 3.5%
- Chemicals 9.5%
- Manufactures 15.3%
- Machinery & transport equipment 43.8%

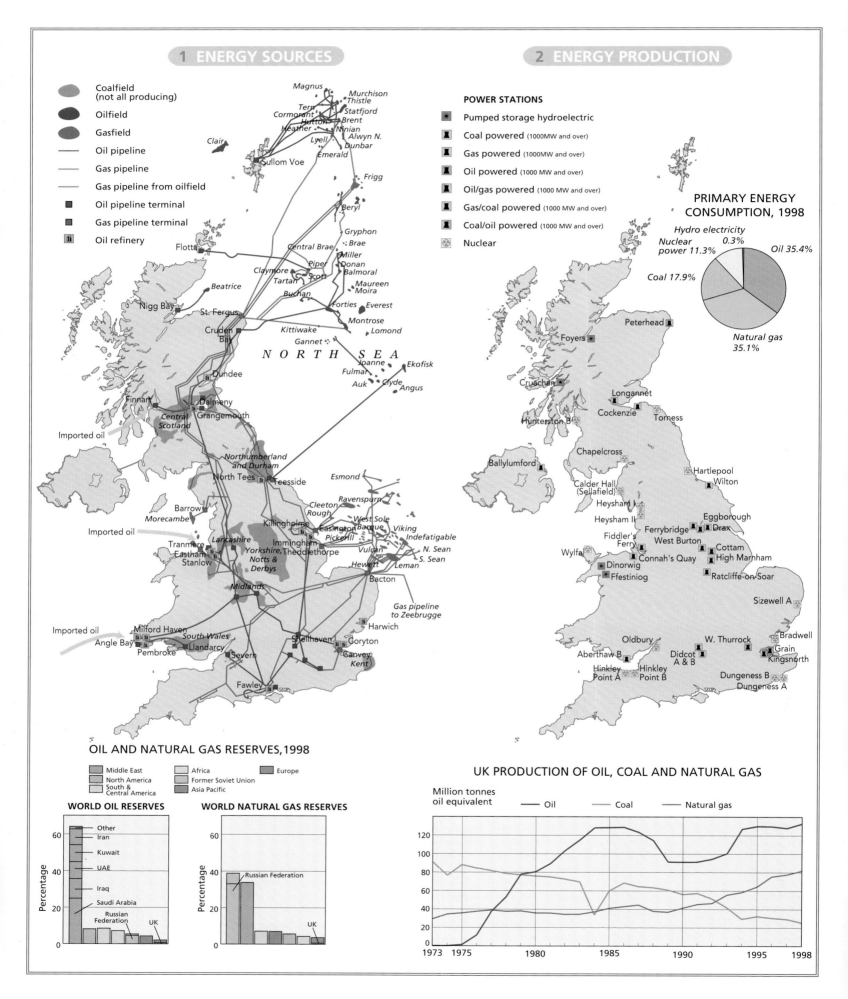

1 ENERGY SOURCES

Coalfield (not all producing)
Oilfield
Gasfield
Oil pipeline
Gas pipeline
Gas pipeline from oilfield
Oil pipeline terminal
Gas pipeline terminal
Oil refinery

Magnus, Murchison, Thistle, Tern, Cormorant, Statfjord, Hutton, Brent, Heather, Ninian, Lyell, Alwyn N., Dunbar, Emerald, Clair, Sullom Voe, Frigg, Beryl, Gryphon, Brae, Flotta, Central Brae, Miller, Claymore, Piper, Donan, Balmoral, Tartan, Scott, Buchan, Maureen, Moira, Beatrice, Forties, Everest, Nigg Bay, St. Fergus, Montrose, Cruden Bay, Kittiwake, Lomond, Gannet, Joanne, Ekofisk, Dundee, Fulmar, Auk, Clyde, Angus, Finnart, Dalmeny, Central Scotland, Grangemouth, Imported oil, Northumberland and Durham, North Tees, Teesside, Esmond, Barrow, Ravenspurn, Morecambe, Cleeton, Rough, West Sole, Killingholme, Easington, Barque, Viking, Immingham, Pickerill, Indefatigable, Tranmere, Lancashire, Theddlethorpe, Vulcan, N. Sean, Eastham, Yorkshire, Notts & Derbys, Hewett, S. Sean, Stanlow, Leman, Midlands, Bacton, Gas pipeline to Zeebrugge, Imported oil, Harwich, Milford Haven, South Wales, Shellhaven, Coryton, Angle Bay, Llandarcy, Pembroke, Severn, Canvey, Kent, Fawley

NORTH SEA

OIL AND NATURAL GAS RESERVES, 1998

Middle East
North America
South & Central America
Africa
Former Soviet Union
Asia Pacific
Europe

WORLD OIL RESERVES
Percentage
Other
Iran
Kuwait
UAE
Iraq
Saudi Arabia
Russian Federation
UK
60, 40, 20, 0

WORLD NATURAL GAS RESERVES
Percentage
Russian Federation
UK
60, 40, 20, 0

2 ENERGY PRODUCTION

POWER STATIONS

Pumped storage hydroelectric
Coal powered (1000MW and over)
Gas powered (1000MW and over)
Oil powered (1000 MW and over)
Oil/gas powered (1000 MW and over)
Gas/coal powered (1000 MW and over)
Coal/oil powered (1000 MW and over)
Nuclear

PRIMARY ENERGY CONSUMPTION, 1998

Hydro electricity 0.3%
Nuclear power 11.3%
Coal 17.9%
Oil 35.4%
Natural gas 35.1%

Peterhead, Foyers, Cruachan, Longannet, Cockenzie, Torness, Hunterston B, Chapelcross, Ballylumford, Calder Hall (Sellafield), Hartlepool, Wilton, Heysham I, Heysham II, Eggborough, Ferrybridge, Drax, Fiddler's Ferry, West Burton, Wylfa, Connah's Quay, Cottam, High Marnham, Dinorwig, Ratcliffe-on-Soar, Ffestiniog, Sizewell A, Oldbury, W. Thurrock, Bradwell, Aberthaw B, Didcot A & B, Grain, Kingsnorth, Hinkley Point A, Hinkley Point B, Dungeness B, Dungeness A

UK PRODUCTION OF OIL, COAL AND NATURAL GAS

Million tonnes oil equivalent
Oil
Coal
Natural gas
120, 100, 80, 60, 40, 20
1973, 1975, 1980, 1985, 1990, 1995, 1998

SCALE 1 : 8 000 000

Conic projection

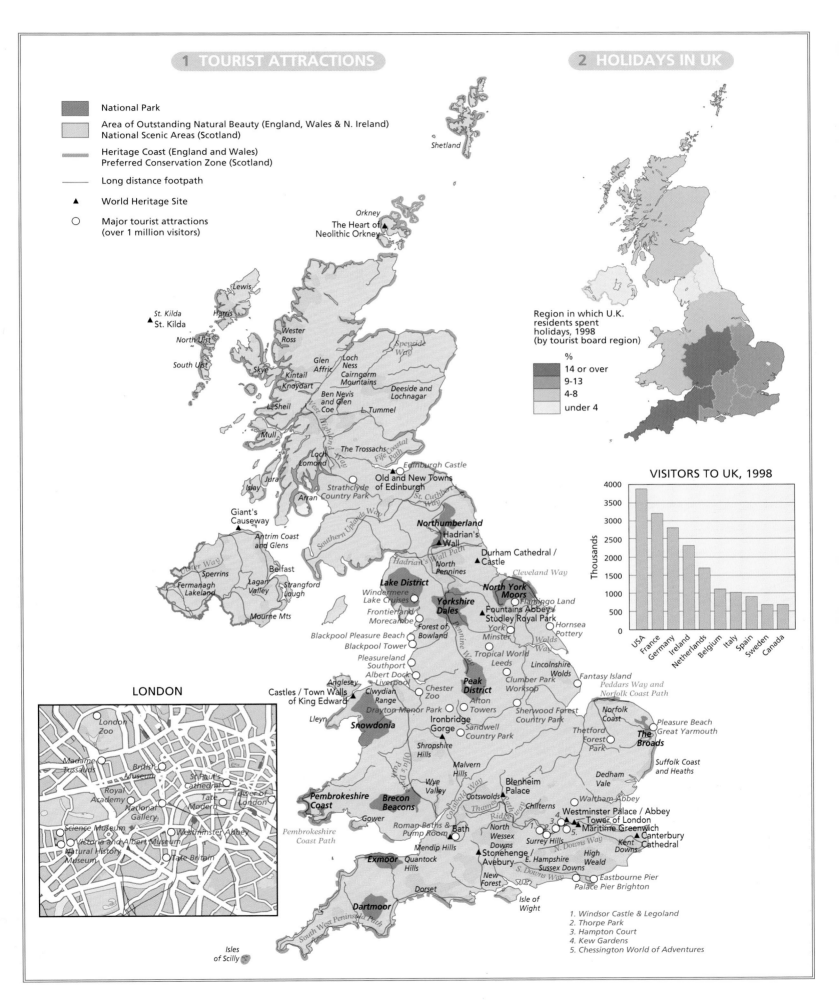

1 TOURIST ATTRACTIONS

National Park

Area of Outstanding Natural Beauty (England, Wales & N. Ireland)
National Scenic Areas (Scotland)

Heritage Coast (England and Wales)
Preferred Conservation Zone (Scotland)

— Long distance footpath

▲ World Heritage Site

○ Major tourist attractions
(over 1 million visitors)

Shetland

Orkney
The Heart of ▲
Neolithic Orkney

Lewis

St. Kilda ▲
St. Kilda

Harris

North Uist

South Uist

Wester Ross

Skye

Glen Affric

Loch Ness

Cairngorm Mountains

Speyside Way

Kintail

Knoydart

Ben Nevis and Glen Coe

Deeside and Lochnagar

L. Sheil

L. Tummel

Mull

The Trossachs

West Highland Way

Jura

Loch Lomond

Islay

Arran

Fife Coastal Path

Edinburgh Castle ▲
Old and New Towns of Edinburgh

Strathclyde Country Park

St. Cuthbert's Way

Southern Uplands Way

Giant's Causeway ▲

Antrim Coast and Glens

Belfast

Ulster Way

Sperrins

Fermanagh Lakeland

Lagan Valley

Strangford Lough

Mourne Mts

Northumberland
Hadrian's Wall ▲

Hadrian's Wall Path

North Pennines

Durham Cathedral / Castle ▲

Cleveland Way

Lake District

Windermere Lake Cruises

Yorkshire Dales

North York Moors

Flamingo Land

Fountains Abbey / Studley Royal Park ▲

Frontierland Morecambe

Forest of Bowland

York Minster

Hornsea Pottery

Blackpool Pleasure Beach
Blackpool Tower

Pennine Way

Tropical World Leeds

Wolds Way

Pleasureland Southport

Albert Dock Liverpool

Lincolnshire Wolds

Fantasy Island

Anglesey

Castles / Town Walls of King Edward

Clwydian Range

Chester Zoo

Peak District

Alton Towers

Clumber Park Worksop

Lleyn

Drayton Manor Park

Snowdonia

Ironbridge Gorge

Sandwell Country Park

Sherwood Forest Country Park

Norfolk Coast

Pleasure Beach Great Yarmouth

Shropshire Hills

Thetford Forest Park

The Broads

Offa's Dyke Path

Malvern Hills

Suffolk Coast and Heaths

Pembrokeshire Coast

Brecon Beacons

Wye Valley

Cotswolds

Blenheim Palace

Dedham Vale

Waltham Abbey

Chilterns

Westminster Palace / Abbey ▲
Tower of London ▲

Maritime Greenwich ▲

Canterbury Cathedral ▲

Gower

Thames Path

Ridgeway

Roman Baths & Pump Room

Bath

North Wessex Downs

Surrey Hills

N. Downs Way

Kent Downs

High Weald

Pembrokeshire Coast Path

Mendip Hills

Stonehenge / Avebury ▲

E. Hampshire

Sussex Downs

S. Downs Way

Exmoor

Quantock Hills

New Forest

Eastbourne Pier
Palace Pier Brighton

Dorset

Isle of Wight

Dartmoor

South West Peninsula Path

Isles of Scilly

1. Windsor Castle & Legoland
2. Thorpe Park
3. Hampton Court
4. Kew Gardens
5. Chessington World of Adventures

LONDON

London Zoo

Madame Tussauds

British Museum

St. Paul's Cathedral

Royal Academy

Tate Modern

Tower of London

National Gallery

Science Museum

Victoria and Albert Museum

Natural History Museum

Westminster Abbey

Tate Britain

2 HOLIDAYS IN UK

Region in which U.K. residents spent holidays, 1998 (by tourist board region)
%
14 or over
9-13
4-8
under 4

VISITORS TO UK, 1998

	Thousands
4000	
3500	
3000	
2500	
2000	
1500	
1000	
500	
0	

USA • France • Germany • Ireland • Netherlands • Belgium • Italy • Spain • Sweden • Canada

SCALE 1 : 5 000 000

Conic projection

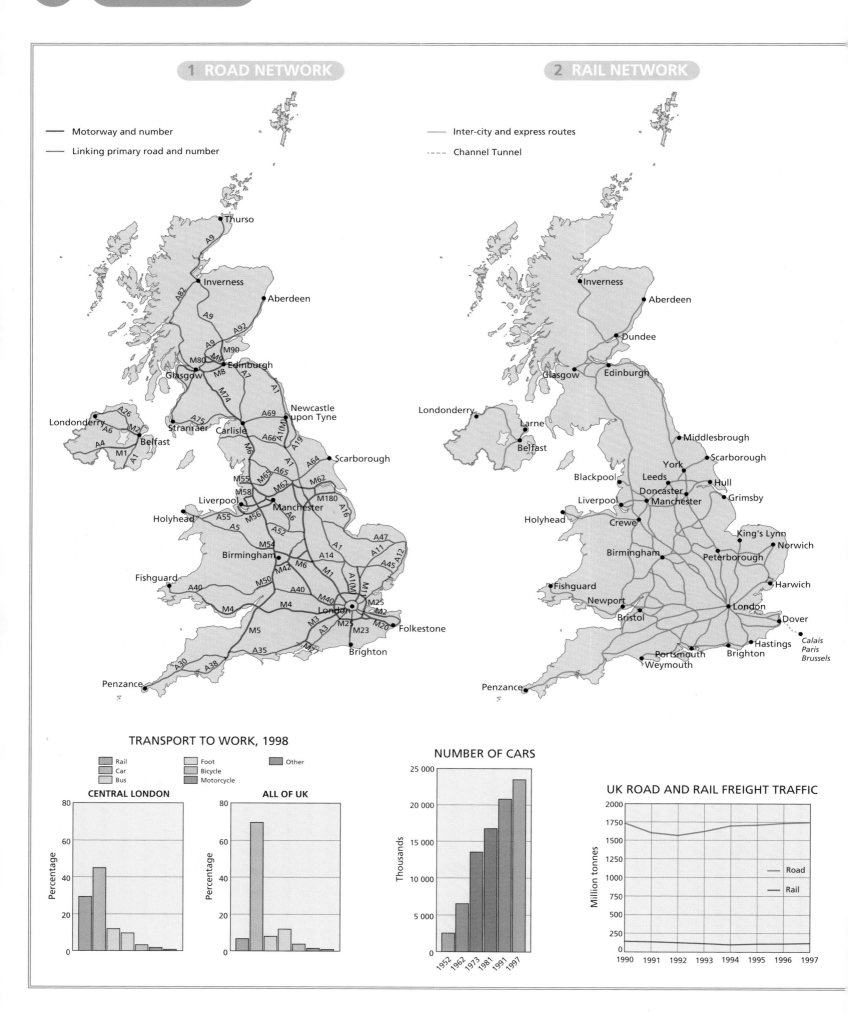

1 ROAD NETWORK

— Motorway and number

— Linking primary road and number

2 RAIL NETWORK

— Inter-city and express routes

- - - Channel Tunnel

TRANSPORT TO WORK, 1998

- Rail
- Car
- Bus
- Foot
- Bicycle
- Motorcycle
- Other

CENTRAL LONDON

ALL OF UK

NUMBER OF CARS

UK ROAD AND RAIL FREIGHT TRAFFIC

Road

Rail

SCALE 1 : 8 000 000

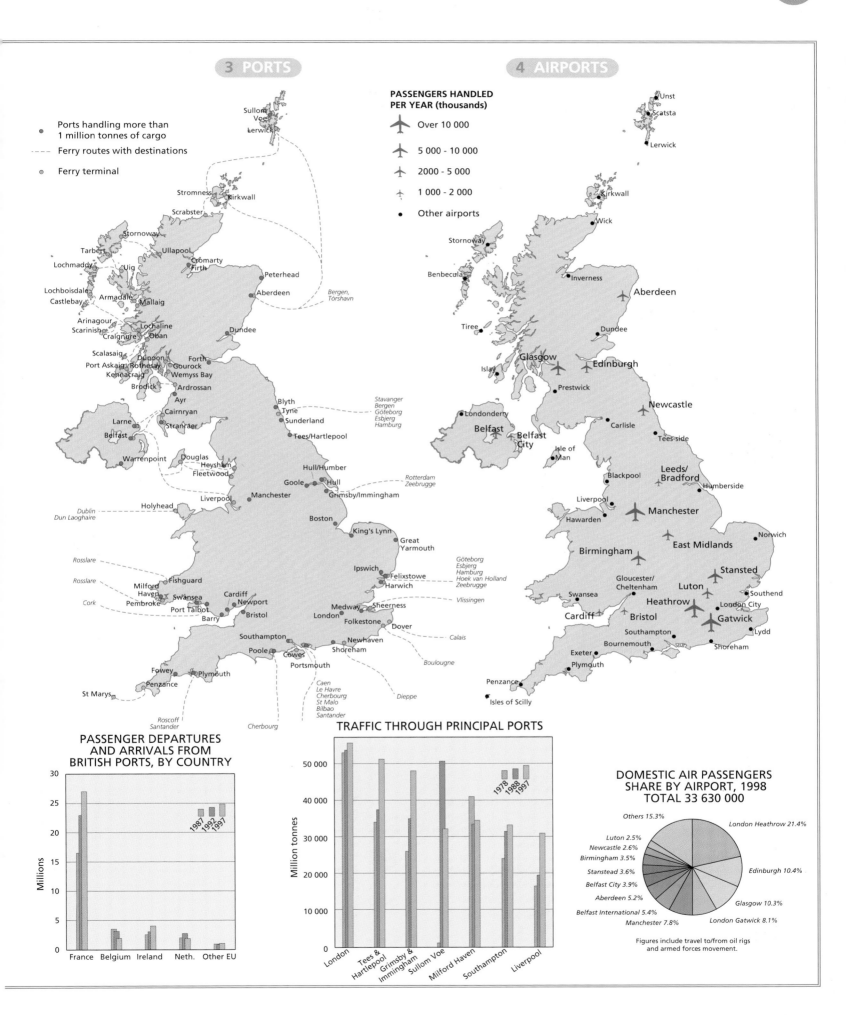

3 PORTS

4 AIRPORTS

Ports handling more than 1 million tonnes of cargo

Ferry routes with destinations

Ferry terminal

PASSENGERS HANDLED PER YEAR (thousands)

Over 10 000

5 000 - 10 000

2000 - 5 000

1 000 - 2 000

Other airports

PASSENGER DEPARTURES AND ARRIVALS FROM BRITISH PORTS, BY COUNTRY

1987 1992 1997

Millions

France Belgium Ireland Neth. Other EU

TRAFFIC THROUGH PRINCIPAL PORTS

1978 1988 1997

Million tonnes

London Tees & Hartlepool Grimsby & Immingham Sullom Voe Milford Haven Southampton Liverpool

DOMESTIC AIR PASSENGERS SHARE BY AIRPORT, 1998 TOTAL 33 630 000

Others 15.3%
Luton 2.5%
Newcastle 2.6%
Birmingham 3.5%
Stanstead 3.6%
Belfast City 3.9%
Aberdeen 5.2%
Belfast International 5.4%
Manchester 7.8%
London Heathrow 21.4%
Edinburgh 10.4%
Glasgow 10.3%
London Gatwick 8.1%

Figures include travel to/from oil rigs and armed forces movement.

Conic projection

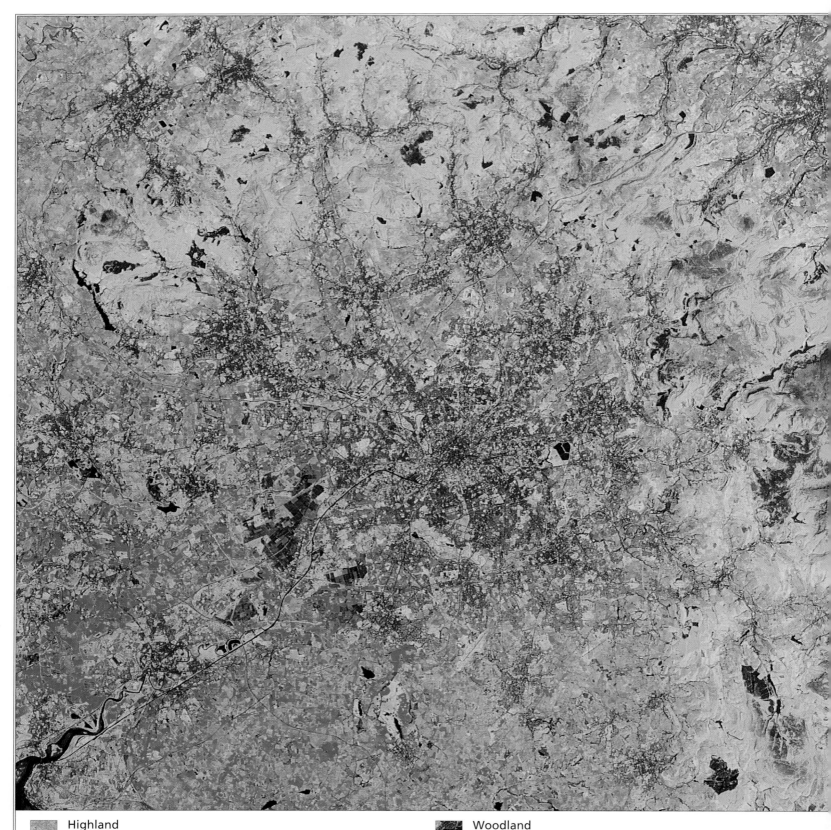

Highland

The blue/green colour corresponds to grassland over 300 metres above sea level on the map opposite. In the higher areas of the Pennines the colour becomes greener as grassland changes to moorland, for example around Shining Tor.

Lowland and arable land

The areas around Manchester appear as shades of orange and red. The cultivated areas near the river Mersey are redder.

Built up area

These areas are dark blue on the satellite image. The largest area is the Manchester urban sprawl. In the top left of the image the built up areas of Blackburn and Accrington stand out from the surrounding farmland.

Woodland

Some areas of woodland can be seen on the lower slopes of Shining Tor. There is also a small area near Alderley Edge.

Reservoir

The small distinctive shape of these can be seen in the Pennines area. Examples are Watergrove Reservoir near Whitworth and Errwood Reservoir south of Whaley Bridge.

Canal

The straight line of the Manchester Ship Canal can be seen running alongside the winding course of the river Mersey.

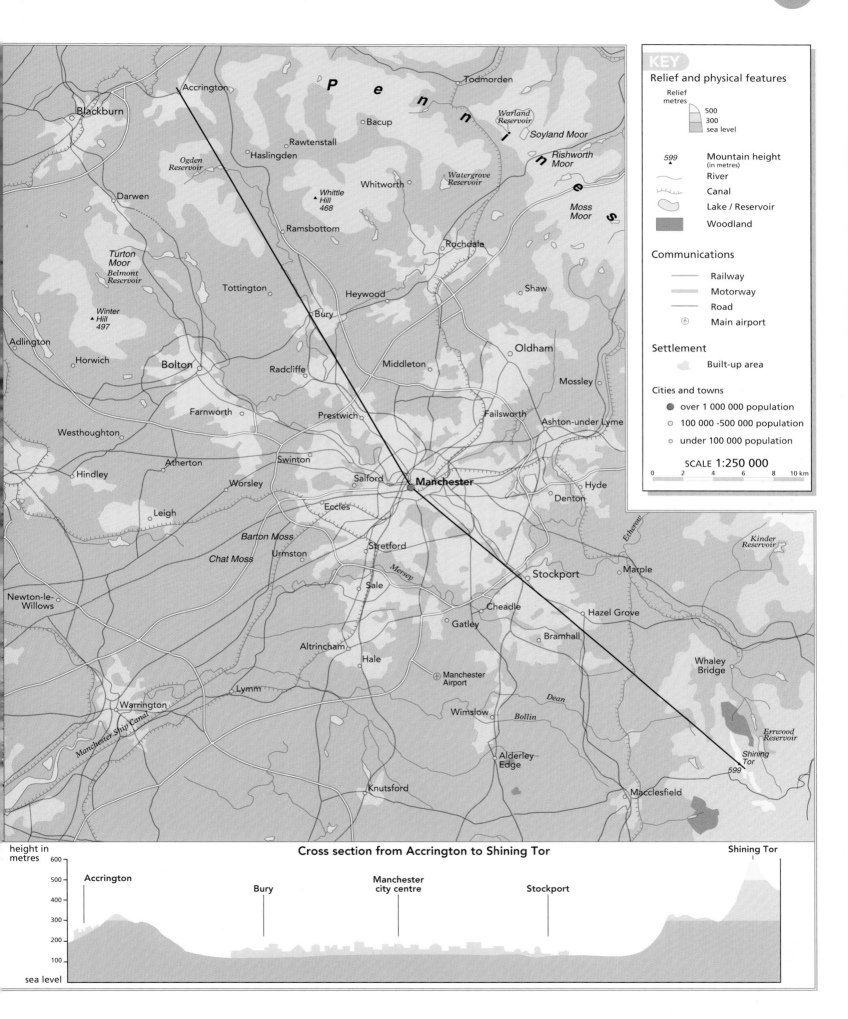

KEY

Relief and physical features

Relief
metres
500
300
sea level

599 ▲ Mountain height
(in metres)

River

Canal

Lake / Reservoir

Woodland

Communications

Railway

Motorway

Road

⊕ Main airport

Settlement

Built-up area

Cities and towns

● over 1 000 000 population

○ 100 000 -500 000 population

○ under 100 000 population

SCALE 1:250 000

0 2 4 6 8 10 km

Blackburn
Accrington
Pennines
Todmorden
Bacup
Rawtenstall
Haslingden
Ogden Reservoir
Warland Reservoir
Soyland Moor
Rishworth Moor
Darwen
Whitworth
Watergrove Reservoir
Moss Moor
Whittle Hill 468 ▲
Ramsbottom
Turton Moor
Belmont Reservoir
Rochdale
Shaw
Tottington
Heywood
Winter Hill 497 ▲
Bury
Adlington
Horwich
Bolton
Radcliffe
Middleton
Oldham
Farnworth
Prestwich
Mossley
Westhoughton
Failsworth
Ashton-under Lyme
Atherton
Swinton
Hindley
Worsley
Salford
Manchester
Hyde
Leigh
Eccles
Denton
Barton Moss
Stretford
Stockport
Marple
Chat Moss
Urmston
Mersey
Kinder Reservoir
Etherow
Newton-le-Willows
Sale
Cheadle
Hazel Grove
Gatley
Bramhall
Altrincham
Hale
Whaley Bridge
Lymm
Manchester Airport
Dean
Warrington
Wimslow
Bollin
Errwood Reservoir
Manchester Ship Canal
Shining Tor 599
Alderley Edge
Knutsford
Macclesfield

Cross section from Accrington to Shining Tor

height in metres

600

Shining Tor

500

Accrington

Manchester city centre

400

Bury

Stockport

300

200

100

sea level

SCALE 1 : 16 000 000

0   200   400   600   800 km

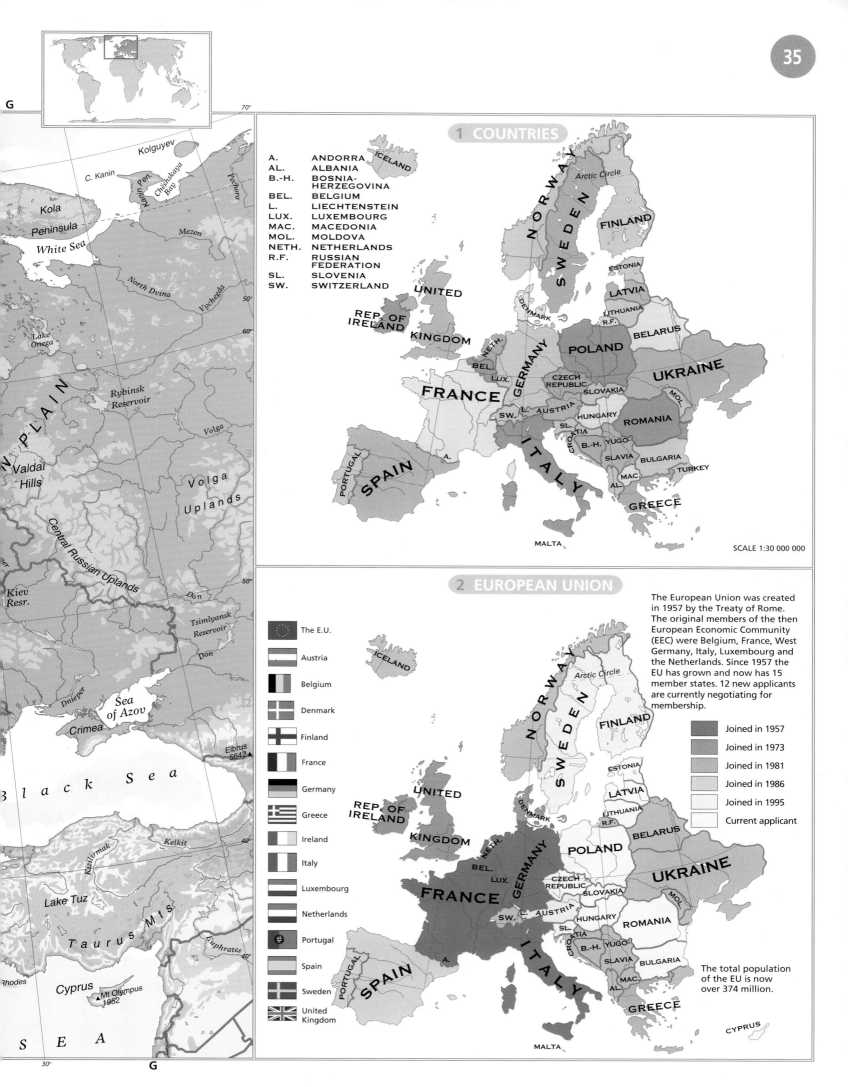

1 COUNTRIES

A.	ANDORRA
AL.	ALBANIA
B.-H.	BOSNIA-HERZEGOVINA
BEL.	BELGIUM
L.	LIECHTENSTEIN
LUX.	LUXEMBOURG
MAC.	MACEDONIA
MOL.	MOLDOVA
NETH.	NETHERLANDS
R.F.	RUSSIAN FEDERATION
SL.	SLOVENIA
SW.	SWITZERLAND

SCALE 1:30 000 000

2 EUROPEAN UNION

The European Union was created in 1957 by the Treaty of Rome. The original members of the then European Economic Community (EEC) were Belgium, France, West Germany, Italy, Luxembourg and the Netherlands. Since 1957 the EU has grown and now has 15 member states. 12 new applicants are currently negotiating for membership.

Joined in 1957
Joined in 1973
Joined in 1981
Joined in 1986
Joined in 1995
Current applicant

The total population of the EU is now over 374 million.

The E.U.
Austria
Belgium
Denmark
Finland
France
Germany
Greece
Ireland
Italy
Luxembourg
Netherlands
Portugal
Spain
Sweden
United Kingdom

Albers Equal Area Conic projection

Conic projection

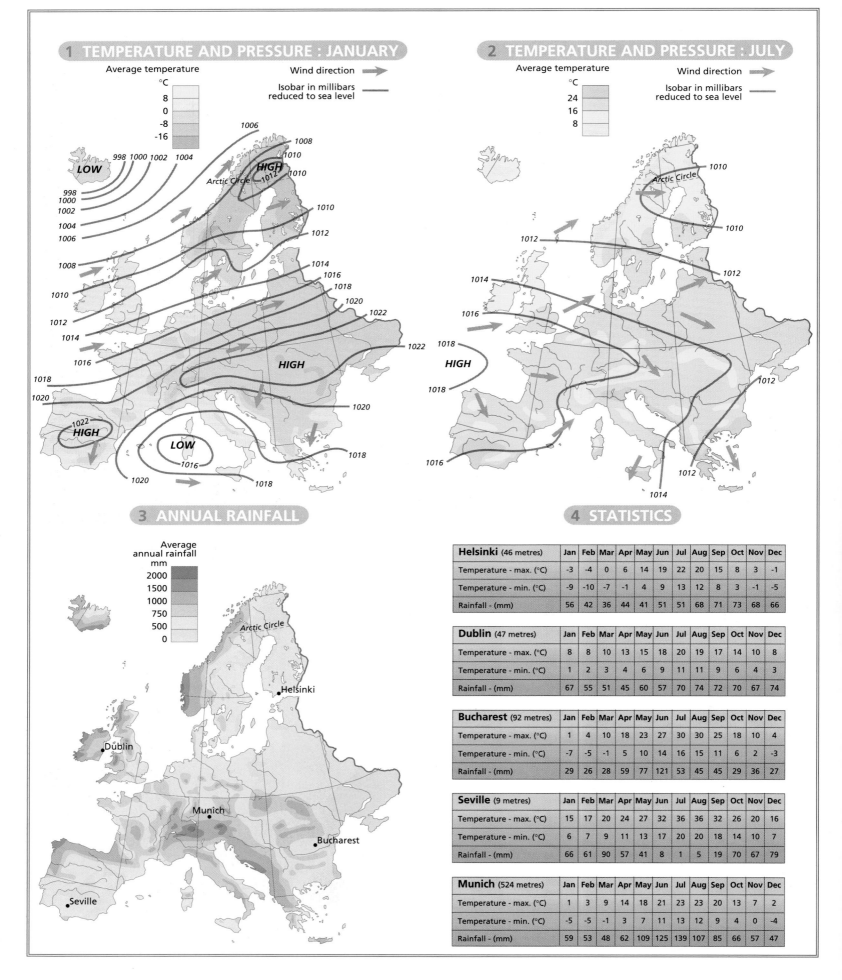

1 TEMPERATURE AND PRESSURE : JANUARY

Average temperature

°C
8
0
-8
-16

Wind direction

Isobar in millibars
reduced to sea level

998 1000 1002 1004

LOW

998
1000
1002
1004
1006
1008
1010
1012
1014
1016
1018
1020

1006
1008
1010
1012 1010

Arctic Circle

1010
1012
1014
1016
1018
1020
1022
1022

HIGH

1022
HIGH

LOW
1016

1020

1018

1018

2 TEMPERATURE AND PRESSURE : JULY

Average temperature

°C
24
16
8

Wind direction

Isobar in millibars
reduced to sea level

1010

Arctic Circle

1010

1012

1012

1014

1016

1018
HIGH
1018

1016

1012

1012

1014

3 ANNUAL RAINFALL

Average
annual rainfall
mm
2000
1500
1000
750
500
0

Arctic Circle

Helsinki

Dublin

Munich

Bucharest

Seville

4 STATISTICS

Helsinki (46 metres)	Jan	Feb	Mar	Apr	May	Jun	Jul	Aug	Sep	Oct	Nov	Dec
Temperature - max. (°C)	-3	-4	0	6	14	19	22	20	15	8	3	-1
Temperature - min. (°C)	-9	-10	-7	-1	4	9	13	12	8	3	-1	-5
Rainfall - (mm)	56	42	36	44	41	51	51	68	71	73	68	66

Dublin (47 metres)	Jan	Feb	Mar	Apr	May	Jun	Jul	Aug	Sep	Oct	Nov	Dec
Temperature - max. (°C)	8	8	10	13	15	18	20	19	17	14	10	8
Temperature - min. (°C)	1	2	3	4	6	9	11	11	9	6	4	3
Rainfall - (mm)	67	55	51	45	60	57	70	74	72	70	67	74

Bucharest (92 metres)	Jan	Feb	Mar	Apr	May	Jun	Jul	Aug	Sep	Oct	Nov	Dec
Temperature - max. (°C)	1	4	10	18	23	27	30	30	25	18	10	4
Temperature - min. (°C)	-7	-5	-1	5	10	14	16	15	11	6	2	-3
Rainfall - (mm)	29	26	28	59	77	121	53	45	45	29	36	27

Seville (9 metres)	Jan	Feb	Mar	Apr	May	Jun	Jul	Aug	Sep	Oct	Nov	Dec
Temperature - max. (°C)	15	17	20	24	27	32	36	36	32	26	20	16
Temperature - min. (°C)	6	7	9	11	13	17	20	20	18	14	10	7
Rainfall - (mm)	66	61	90	57	41	8	1	5	19	70	67	79

Munich (524 metres)	Jan	Feb	Mar	Apr	May	Jun	Jul	Aug	Sep	Oct	Nov	Dec
Temperature - max. (°C)	1	3	9	14	18	21	23	23	20	13	7	2
Temperature - min. (°C)	-5	-5	-1	3	7	11	13	12	9	4	0	-4
Rainfall - (mm)	59	53	48	62	109	125	139	107	85	66	57	47

SCALE 1 : 40 000 000

0 400 800 1200 1600 km

Conic projection

1 POPULATION DENSITY

POPULATION
Persons per sq. km

- over 200
- 100-200
- 50-100
- 10-50
- 1-10
- 0-1

Urban agglomerations
- over 5 000 000
- 1 000 000-5 000 000
- 500 000-1 000 000

Arctic Circle

London
Essen-Dortmund
Paris

SCALE 1:30 000 000

2 POPULATION TABLE

Country	% Change in annual growth rate 1995-2000	Life expectancy (years) 1995-2000
Albania	-0.4	73
Austria	0.5	77
Belarus	-0.3	68
Belgium	0.1	77
Bosnia-Herzegovina	3.0	73
Bulgaria	-0.7	71
Croatia	-0.1	73
Czech Republic	-0.2	74
Denmark	0.3	76
Estonia	-1.2	69
Finland	0.3	77
France	0.4	78
Germany	0.1	77
Greece	0.3	78
Hungary	-0.4	71
Iceland	0.9	79
Italy	0.0	78
Latvia	-1.5	68
Lithuania	-0.3	70
Luxembourg	1.1	77
Macedonia	0.6	73
Malta	0.7	77
Moldova	0.0	68
Netherlands	0.4	78
Norway	0.5	78
Poland	0.1	73
Portugal	0.0	75
Republic of Ireland	0.7	76
Romania	-0.4	70
Slovakia	0.1	73
Slovenia	0.0	74
Spain	0.0	78
Sweden	0.2	79
Switzerland	0.7	79
Ukraine	-0.4	69
United Kingdom	0.2	77
Yugoslavia	0.1	73

3 POPULATION UNDER 16

Arctic Circle

Percentage of total population in either category, 1998

- 27.5
- 25
- 22.5
- 20
- 17.5
- 15
- 7.5

SCALE 1:40 000 000

4 POPULATION OVER 60

Arctic Circle

SCALE 1:40 000 000

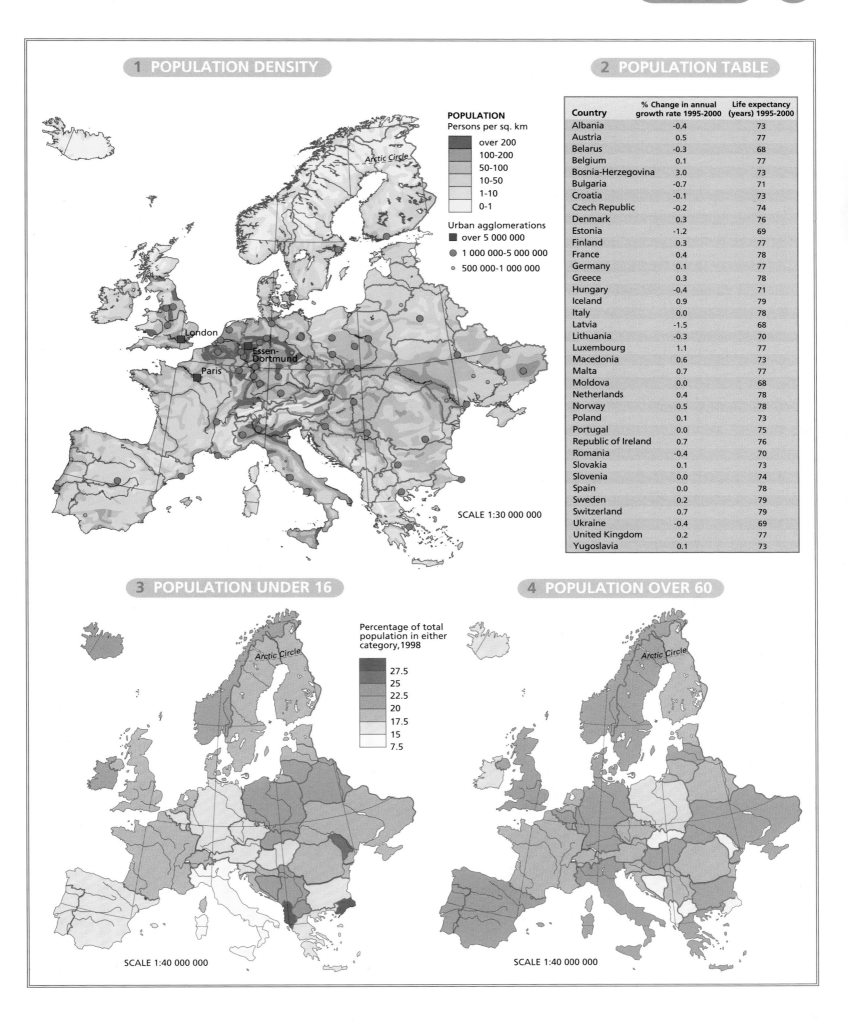

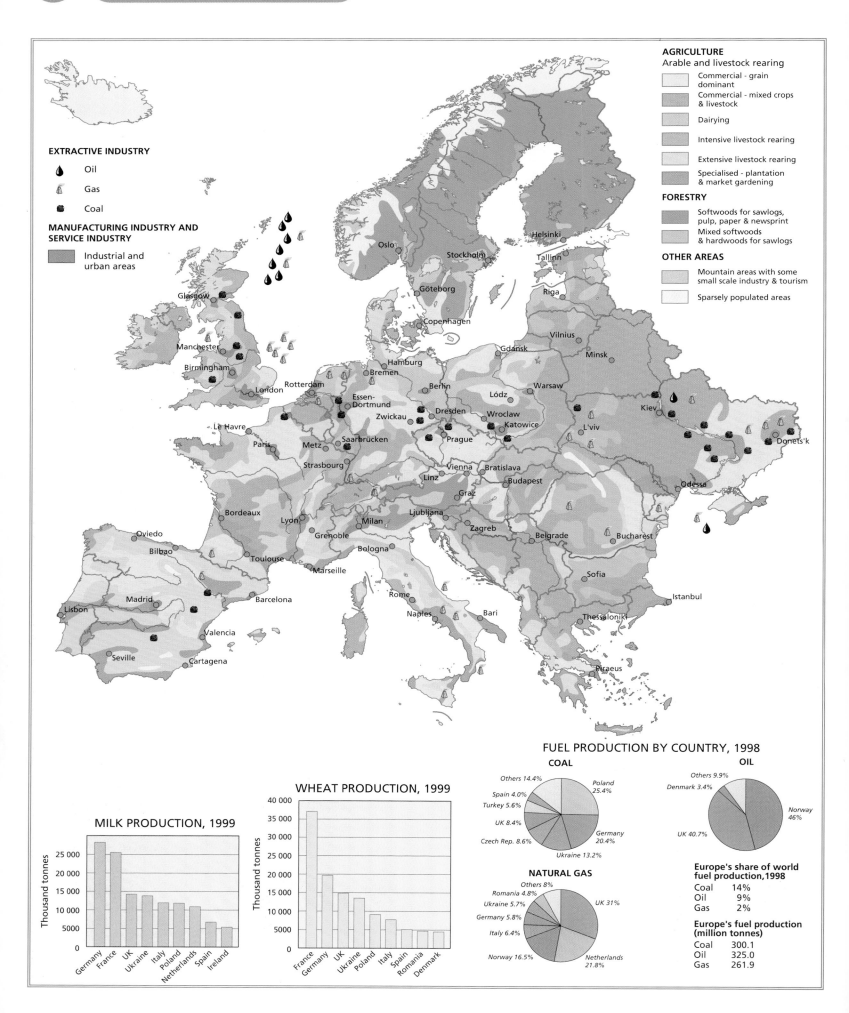

EXTRACTIVE INDUSTRY

- Oil
- Gas
- Coal

MANUFACTURING INDUSTRY AND SERVICE INDUSTRY

- Industrial and urban areas

AGRICULTURE
Arable and livestock rearing

- Commercial - grain dominant
- Commercial - mixed crops & livestock
- Dairying
- Intensive livestock rearing
- Extensive livestock rearing
- Specialised - plantation & market gardening

FORESTRY

- Softwoods for sawlogs, pulp, paper & newsprint
- Mixed softwoods & hardwoods for sawlogs

OTHER AREAS

- Mountain areas with some small scale industry & tourism
- Sparsely populated areas

MILK PRODUCTION, 1999

Thousand tonnes

Germany, France, UK, Ukraine, Italy, Poland, Netherlands, Spain, Ireland

WHEAT PRODUCTION, 1999

Thousand tonnes

France, Germany, UK, Ukraine, Poland, Italy, Spain, Romania, Denmark

FUEL PRODUCTION BY COUNTRY, 1998

COAL

- Others 14.4%
- Spain 4.0%
- Turkey 5.6%
- UK 8.4%
- Czech Rep. 8.6%
- Ukraine 13.2%
- Germany 20.4%
- Poland 25.4%

OIL

- Others 9.9%
- Denmark 3.4%
- Norway 46%
- UK 40.7%

NATURAL GAS

- Others 8%
- Romania 4.8%
- Ukraine 5.7%
- Germany 5.8%
- Italy 6.4%
- Norway 16.5%
- Netherlands 21.8%
- UK 31%

Europe's share of world fuel production, 1998

Coal	14%
Oil	9%
Gas	2%

Europe's fuel production (million tonnes)

Coal	300.1
Oil	325.0
Gas	261.9

SCALE 1 : 20 000 000

Albers equal area conic projection

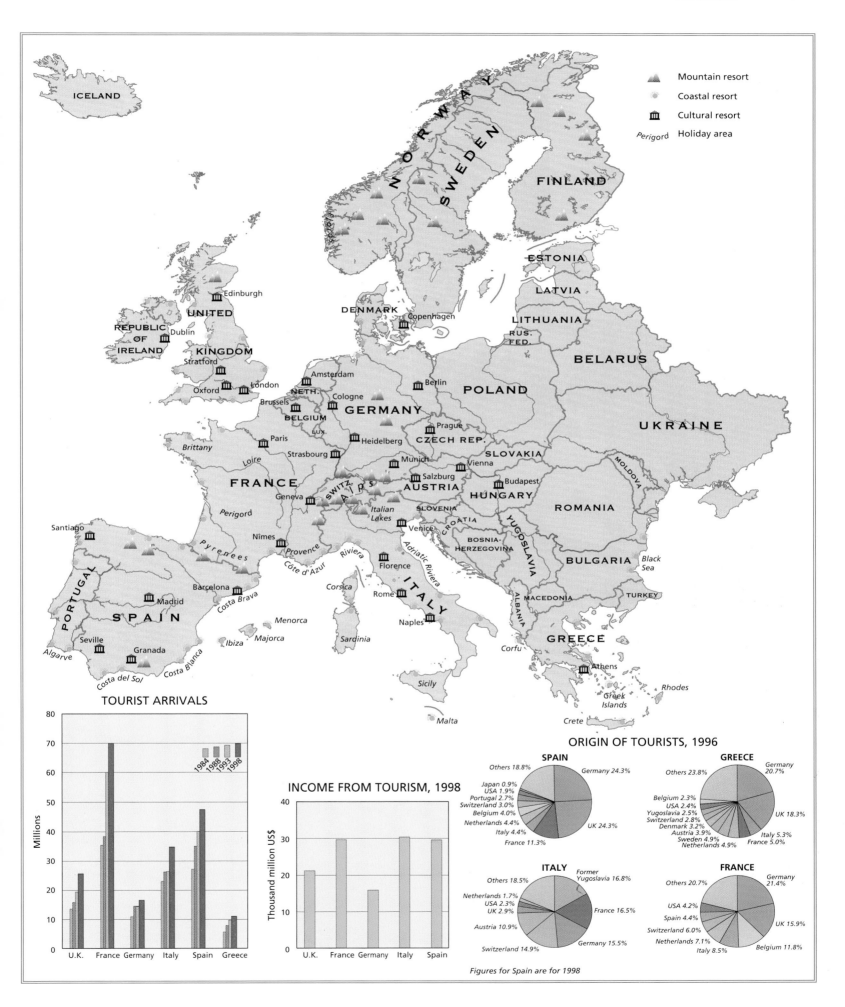

ICELAND

NORWAY SWEDEN FINLAND

ESTONIA
LATVIA
LITHUANIA
RUS. FED.

BELARUS

UKRAINE

DENMARK
Copenhagen

Edinburgh

UNITED
KINGDOM
Stratford
Oxford London
Brussels
BELGIUM
Lux.

REPUBLIC
OF
IRELAND
Dublin

Amsterdam
NETH.
Cologne
Berlin

POLAND

Prague
CZECH REP.
SLOVAKIA
MOLDOVA

Paris
Brittany
Loire
Strasbourg
Heidelberg

GERMANY

Munich
Salzburg
Vienna
Budapest
HUNGARY
ROMANIA

FRANCE

Geneva
SWITZ. ALPS
AUSTRIA
SLOVENIA
Italian
Lakes
Venice
CROATIA
YUGOSLAVIA
BOSNIA-
HERZEGOVINA
BULGARIA
Black
Sea

Perigord

Nîmes
Provence
Riviera
Côte d'Azur
Adriatic Riviera
Florence
Corsica
Rome
ITALY
ALBANIA
MACEDONIA
TURKEY

Santiago

Pyrenees

Barcelona
Costa Brava
Madrid
Costa Blanca

PORTUGAL

SPAIN

Seville
Granada
Algarve
Costa del Sol

Ibiza
Menorca
Majorca
Sardinia

Naples

GREECE

Corfu
Athens
Rhodes

Sicily
Malta
Greek
Islands
Crete

Mountain resort
Coastal resort
Cultural resort
Perigord Holiday area

TOURIST ARRIVALS

Millions

80
70
60
50
40
30
20
10
0

U.K. France Germany Italy Spain Greece

1984
1988
1993
1998

INCOME FROM TOURISM, 1998

Thousand million US$

40
30
20
10
0

U.K. France Germany Italy Spain

ORIGIN OF TOURISTS, 1996

SPAIN
Others 18.8%
Germany 24.3%
Japan 0.9%
USA 1.9%
Portugal 2.7%
Switzerland 3.0%
Belgium 4.0%
Netherlands 4.4%
Italy 4.4%
France 11.3%
UK 24.3%

GREECE
Others 23.8%
Germany 20.7%
Belgium 2.3%
USA 2.4%
Yugoslavia 2.5%
Switzerland 2.8%
Denmark 3.2%
Austria 3.9%
Sweden 4.9%
Netherlands 4.9%
France 5.0%
Italy 5.3%
UK 18.3%

ITALY
Others 18.5%
Former
Yugoslavia 16.8%
Netherlands 1.7%
USA 2.3%
UK 2.9%
Austria 10.9%
France 16.5%
Germany 15.5%
Switzerland 14.9%

FRANCE
Others 20.7%
Germany 21.4%
USA 4.2%
Spain 4.4%
Switzerland 6.0%
Netherlands 7.1%
Italy 8.5%
UK 15.9%
Belgium 11.8%

Figures for Spain are for 1998

SCALE 1 : 20 000 000

Albers equal area conic projection

Built-up area
The main built up areas, which can be identified on the satellite image, are Rotterdam, Dordrecht and Antwerpen.

Farmland
These areas appear as a greenish yellow pattern in the top right of the satellite image.

Woodland
Patchy areas of darkbrown/red lying north of Antwerpen are areas of woodland.

Canal
The pattern of dark thin lines is the canal system which cuts across islands and peninsulas to link the cities of Rotterdam and Antwerpen.

Dunes
Dunes appear as white linear features along most of the coast. Extensive areas of dunes are also found in the Schelde estuary.

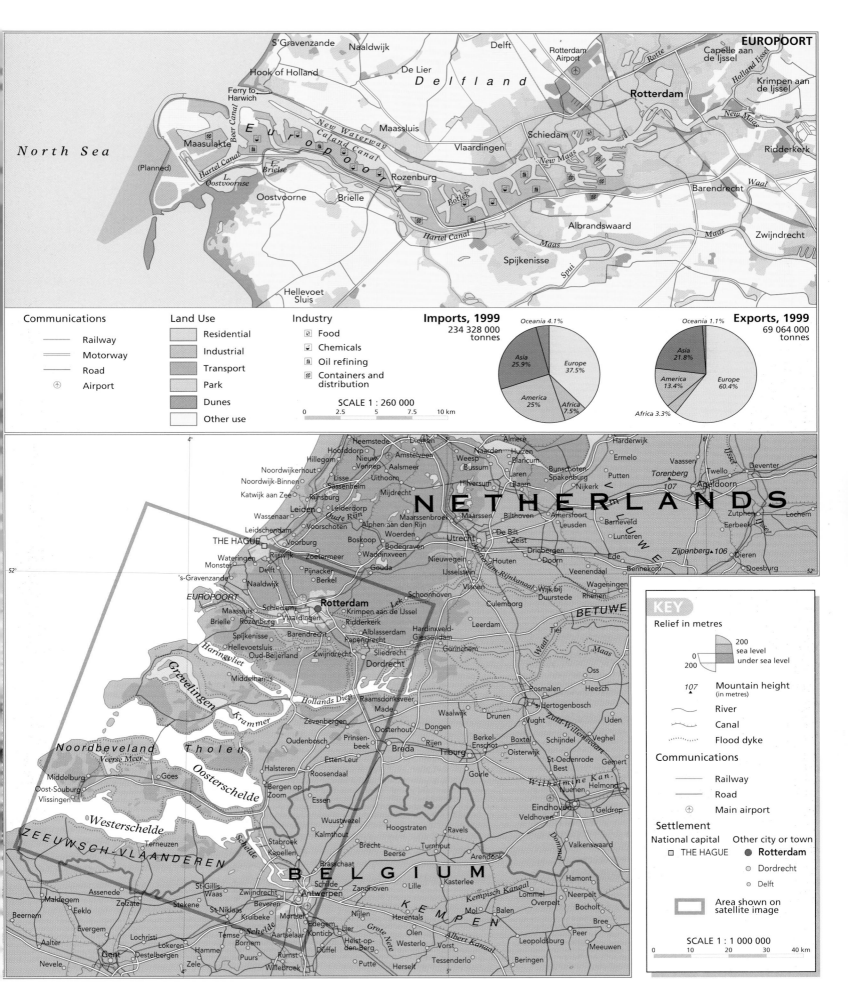

EUROPOORT

North Sea

S'Gravenzande   Naaldwijk   Delft   Rotterdam Airport   Capelle aan de Ijssel

Hook of Holland   De Lier   *Delfland*   **Rotterdam**   Holland Ijssel   Krimpen aan de Ijssel

Ferry to Harwich   Maassluis   Schiedam   New Maas

Maasulakte   Beer Canal   New Waterway Caland Canal   Vlaardingen   Ridderkerk

(Planned)   Hartel Canal   L. Brielse   Europoort   New Maas   Barendrecht   Waal

L. Oostvoornse   Rozenburg   Botlek   Albrandswaard   Zwijndrecht

Oostvoorne   Brielle   Hartel Canal   Maas   Maas

Spijkenisse   Spui

Hellevoet Sluis

Communications
— Railway
═ Motorway
— Road
⊕ Airport

Land Use
Residential
Industrial
Transport
Park
Dunes
Other use

Industry
Food
Chemicals
Oil refining
Containers and distribution

SCALE 1 : 260 000
0   2.5   5   7.5   10 km

Imports, 1999
234 328 000 tonnes
Oceania 4.1%
Asia 25.9%
Europe 37.5%
America 25%
Africa 7.5%

Exports, 1999
69 064 000 tonnes
Oceania 1.1%
Asia 21.8%
America 13.4%
Europe 60.4%
Africa 3.3%

NETHERLANDS

Heemstede   Diemen   Almere   Harderwijk
Hoofddorp   Amstelveen   Naarden   Huizen   Ermelo
Hillegom   Nieuw Vennep   Aalsmeer   Weesp   Blaricum   Bunschoten Spakenburg   Putten   Vaassen   Deventer
Noordwijkerhout   Uithoorn   Bussum   Laren   Nijkerk   Torenberg   107   Apeldoorn   Twello
Noordwijk-Binnen   Lisse   Sassenheim   Mijdrecht   Hilversum   Baarn   <Vm   Zutphen   Lochem
Katwijk aan Zee   Rijnsburg   Leiderdorp   Maarssenbroek   Maarssen   Bilthoven   Amersfoort   Barneveld   Veluwe   Eerbeek
Wassenaar   Leiden   Oude Rijn   Alphen aan den Rijn   Woerden   De Bilt   Leusden   Lunteren   Zijpenberg 106   Dieren
Leidschendam   Voorschoten   Boskoop   Bodegraven   Utrecht   Zeist   Driebergen   Ede   Doorn   Veenendaal   Bennekom   Doesburg
THE HAGUE   Voorburg   Zoetermeer   Waddinxveen   Nieuwegein   Houten   Rhenen   Wageningen
Wateringen   Rijswijk   Gouda   IJsselstein   Vianen   Wijk bij Duurstede
Monster   Delft   Pijnacker   Berkel   Schoonhoven   Culemborg   **BETUWE**
's-Gravenzande   Naaldwijk   **Rotterdam**   Lek   Leerdam   Tiel
EUROPOORT   Schiedam   Krimpen aan de IJssel   Hardinxveld-Giessendam   Gorinchem   Waal   Oss
Maassluis   Rozenburg   Vlaardingen   Ridderkerk   Alblasserdam   Papendrecht   Rosmalen   Heesch
Brielle   Barendrecht   Sliedrecht   Zwijndrecht   Dordrecht   Hertogenbosch   Uden   Veghel
Spijkenisse   *Haringvliet*   Oud-Beijerland   Zwijndrecht   Waalwijk   Drunen   Vught   Berkel-Enschot   Schijndel
Hellevoetsluis   Dordrecht   Hollands Diep   Raamsdonksveer   Made   Waalwijk   Dongen   Boxtel   Oisterwijk   Best   Gemert
Grevelingen   Middelharnis   Zevenbergen   Oosterhout   Breda   Rijen   Tilburg   St-Oedenrode   Helmond
Krammer   Oudenbosch   Prinsenbeek   Gorle   Best   Nuenen   Geldrop
Noordbeveland   Veerse Meer   *Tholen*   Etten-Leur   Roosendaal   Halsteren   Eindhoven   Veldhoven
Middelburg   Goes   *Oosterschelde*   Bergen op Zoom   Essen   Wuustwezel   Hoogstraten   Ravels   Valkenswaard
Oost-Souburg   Vlissingen   Kalmthout   Brecht   Beerse   Arendonk
Westerschelde   Terneuzen   Stabroek   Kapellen   Brasschaat   Turnhout
ZEEUWSCH-VLAANDEREN   Schelde   St-Gillis-Waas   Zwijndrecht   Schilde   Zandhoven   Lille   Kasterlee   Hamont
Maldegem   Zelzate   Stekene   Beveren   **BELGIUM**   **KEMPEN**   Neerpelt   Bocholt
Assenede   Eeklo   Antwerpen   Herentals   Olen   Mol   Balen   Bree
Beernem   Evergem   Lochristi   Temse   Schelde   Edegem   Lier   Heist-op-den-Berg   Westerlo   Vorst   Leopoldsburg   Peer   Meeuwen
Aalter   Nevele   Gent   Destelbergen   Hamme   Bornem   Kontich   Duffel   Putte   Tessenderlo   Beringen
Lokeren   Rumst   Willebroek   Herselt

KEY

Relief in metres
200 sea level
0 under sea level
200

107 ▲ Mountain height (in metres)

~ River
Canal
····· Flood dyke

Communications
— Railway
— Road
⊕ Main airport

Settlement
National capital   Other city or town
□ THE HAGUE   ● **Rotterdam**
○ Dordrecht
○ Delft

□ Area shown on satellite image

SCALE 1 : 1 000 000
0   10   20   30   40 km

Lambert Azimuthal Equal Area projection

KEY

Relief and physical features

Relief
metres

5000
3000
2000
1000
500
200
sea level
0
under sea level
200
4000
6000

818 ▲ Mountain height
(in metres)

Permanent ice

Water features

River

Canal

Lake / Reservoir

Marsh

Communications

Railway

Motorway

Road

⊕ Main airport

Administration

Boundaries

International

Internal

Settlement
Cities and towns in order of size

National capital   Other city or town

■ **AMSTERDAM**   ● **Rotterdam**

□ **THE HAGUE**   ○ Dortmund

□ BONN   ○ Maastricht

□ LUXEMBOURG   ○ Oostende

SCALE 1 : 2 000 000

0   20   40   60   80 km

Conic projection

SCALE 1 : 7 500 000

0   100   200   300 km

Conic projection

Lambert Conformal Conic projection

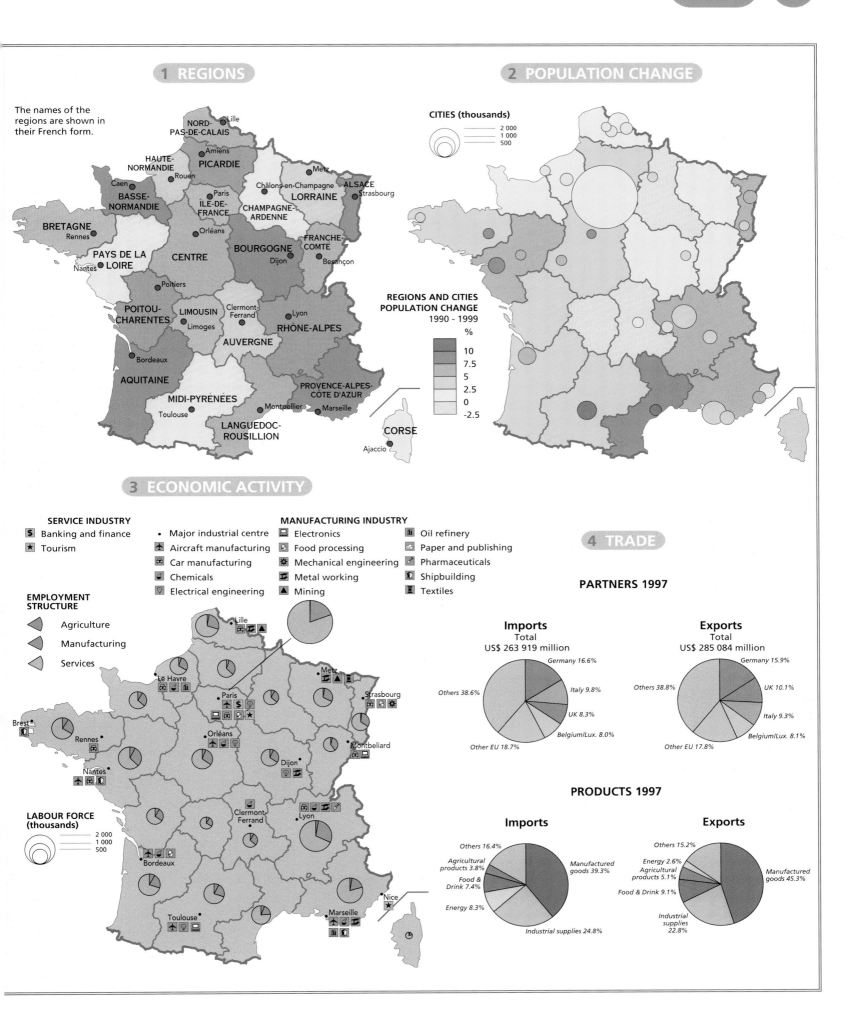

1 REGIONS

The names of the regions are shown in their French form.

NORD-PAS-DE-CALAIS
Lille

HAUTE-NORMANDIE
Amiens
PICARDIE
Rouen
Caen
Metz
Châlons-en-Champagne
ALSACE
Strasbourg
BASSE-NORMANDIE
Paris
ÎLE-DE-FRANCE
LORRAINE
CHAMPAGNE-ARDENNE
BRETAGNE
Rennes
Orléans
FRANCHE-COMTÉ
PAYS DE LA LOIRE
Nantes
CENTRE
BOURGOGNE
Dijon
Besançon
Poitiers
POITOU-CHARENTES
LIMOUSIN
Limoges
Clermont-Ferrand
Lyon
RHÔNE-ALPES
AUVERGNE
Bordeaux
AQUITAINE
PROVENCE-ALPES-CÔTE D'AZUR
MIDI-PYRÉNÉES
Montpellier
Marseille
Toulouse
LANGUEDOC-ROUSILLION
CORSE
Ajaccio

2 POPULATION CHANGE

CITIES (thousands)
2 000
1 000
500

REGIONS AND CITIES
POPULATION CHANGE
1990 - 1999
%
10
7.5
5
2.5
0
-2.5

3 ECONOMIC ACTIVITY

SERVICE INDUSTRY
$ Banking and finance
★ Tourism

MANUFACTURING INDUSTRY
• Major industrial centre
✈ Aircraft manufacturing
🚗 Car manufacturing
Chemicals
Electrical engineering
Electronics
Food processing
Mechanical engineering
Metal working
▲ Mining
Oil refinery
Paper and publishing
Pharmaceuticals
Shipbuilding
Textiles

EMPLOYMENT STRUCTURE
Agriculture
Manufacturing
Services

Lille
Le Havre
Metz
Paris
Strasbourg
Brest
Rennes
Orléans
Montbeliard
Nantes
Dijon
Clermont-Ferrand
Lyon
Bordeaux
Nice
Toulouse
Marseille

LABOUR FORCE (thousands)
2 000
1 000
500

4 TRADE

PARTNERS 1997

Imports
Total
US$ 263 919 million
Germany 16.6%
Italy 9.8%
UK 8.3%
Belgium/Lux. 8.0%
Other EU 18.7%
Others 38.6%

Exports
Total
US$ 285 084 million
Germany 15.9%
UK 10.1%
Italy 9.3%
Belgium/Lux. 8.1%
Other EU 17.8%
Others 38.8%

PRODUCTS 1997

Imports
Others 16.4%
Agricultural products 3.8%
Food & Drink 7.4%
Energy 8.3%
Industrial supplies 24.8%
Manufactured goods 39.3%

Exports
Others 15.2%
Energy 2.6%
Agricultural products 5.1%
Food & Drink 9.1%
Industrial supplies 22.8%
Manufactured goods 45.3%

SCALE 1 : 10 000 000
0   100   200   300 km

KEY

Relief and physical features

Relief
metres
5000
3000
2000
1000
500
200
0 sea level
  under sea level
200
4000
6000

3482 ▲ Mountain height
(in metres)

Water features

～～ River
～～～ Intermittent river
Canal
Lake / Reservoir
Marsh

Communications

Railway
Motorway
Road
⊕ Main airport

Administration

Boundaries

International

Settlement

Cities and towns in order of size

National capital
■ MADRID
□ ANDORRA
   LA VELLA

Other city or town
● Barcelona
● Málaga
○ Pamplona
○ Benidorm

SCALE 1 : 5 000 000

0   50   100   150   200 km

Lambert Conformal Conic projection

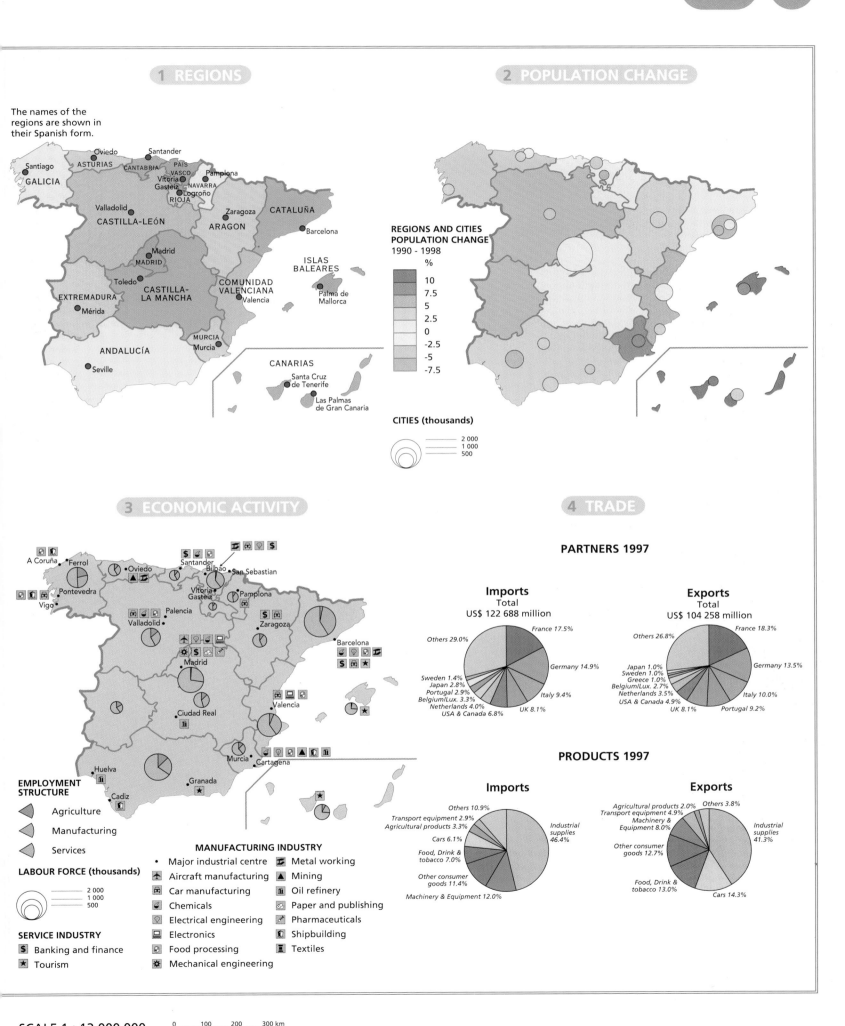

1 REGIONS

The names of the regions are shown in their Spanish form.

Santiago
GALICIA
Oviedo
ASTURIAS
Santander
CANTABRIA
PAÍS VASCO
Vitoria Gasteiz
Pamplona
NAVARRA
RIOJA
Logroño
Valladolid
CASTILLA-LEÓN
ARAGON
Zaragoza
CATALUÑA
Barcelona
Madrid
MADRID
ISLAS BALEARES
Toledo
CASTILLA-LA MANCHA
COMUNIDAD VALENCIANA
Valencia
Palma de Mallorca
EXTREMADURA
Mérida
MURCIA
Murcia
ANDALUCÍA
Seville
CANARIAS
Santa Cruz de Tenerife
Las Palmas de Gran Canaria

2 POPULATION CHANGE

REGIONS AND CITIES POPULATION CHANGE
1990 - 1998
%
10
7.5
5
2.5
0
-2.5
-5
-7.5

CITIES (thousands)
2 000
1 000
500

3 ECONOMIC ACTIVITY

A Coruña
Ferrol
Oviedo
Santander
Bilbao
San Sebastian
Pontevedra
Vigo
Vitoria Gasteiz
Pamplona
Palencia
Valladolid
Zaragoza
Barcelona
Madrid
Valencia
Ciudad Real
Murcia
Cartagena
Huelva
Cadiz
Granada

EMPLOYMENT STRUCTURE

Agriculture

Manufacturing

Services

LABOUR FORCE (thousands)
2 000
1 000
500

SERVICE INDUSTRY
$ Banking and finance
★ Tourism

MANUFACTURING INDUSTRY
• Major industrial centre
✈ Aircraft manufacturing
🚗 Car manufacturing
Chemicals
Electrical engineering
Electronics
Food processing
Mechanical engineering
Metal working
▲ Mining
Oil refinery
Paper and publishing
Pharmaceuticals
Shipbuilding
Textiles

4 TRADE

PARTNERS 1997

Imports
Total
US$ 122 688 million

Others 29.0%
France 17.5%
Germany 14.9%
Italy 9.4%
UK 8.1%
USA & Canada 6.8%
Netherlands 4.0%
Belgium/Lux. 3.3%
Portugal 2.9%
Japan 2.8%
Sweden 1.4%

Exports
Total
US$ 104 258 million

Others 26.8%
France 18.3%
Germany 13.5%
Italy 10.0%
Portugal 9.2%
UK 8.1%
USA & Canada 4.9%
Netherlands 3.5%
Belgium/Lux. 2.7%
Greece 1.0%
Sweden 1.0%
Japan 1.0%

PRODUCTS 1997

Imports
Others 10.9%
Transport equipment 2.9%
Agricultural products 3.3%
Cars 6.1%
Food, Drink & tobacco 7.0%
Other consumer goods 11.4%
Machinery & Equipment 12.0%
Industrial supplies 46.4%

Exports
Agricultural products 2.0%
Others 3.8%
Transport equipment 4.9%
Machinery & Equipment 8.0%
Other consumer goods 12.7%
Food, Drink & tobacco 13.0%
Cars 14.3%
Industrial supplies 41.3%

SCALE 1 : 12 000 000

0   100   200   300 km

KEY

Relief and physical features

Relief
metres
5000
3000
2000
1000
500
200
sea level
under sea level
0
200
4000
6000

1142 ▲ Mountain height
(in metres)

Permanent ice

Water features

River

Intermittent river

Canal

Lake / Reservoir

Marsh

Communications

Railway

Motorway

Road

⊕ Main airport

Administration

Boundaries

International

Settlement

Cities and towns in order of size

National capital

■ BERLIN   ● Munich

□ ZAGREB   ○ Dortmund

□ LJUBLJANA   ○ Ulm

□ LUXEMBOURG   ○ Tuttlingen

Other city or town

SCALE 1 : 4 500 000

0   50   100   150   200 km

Lambert Conformal Conic projection

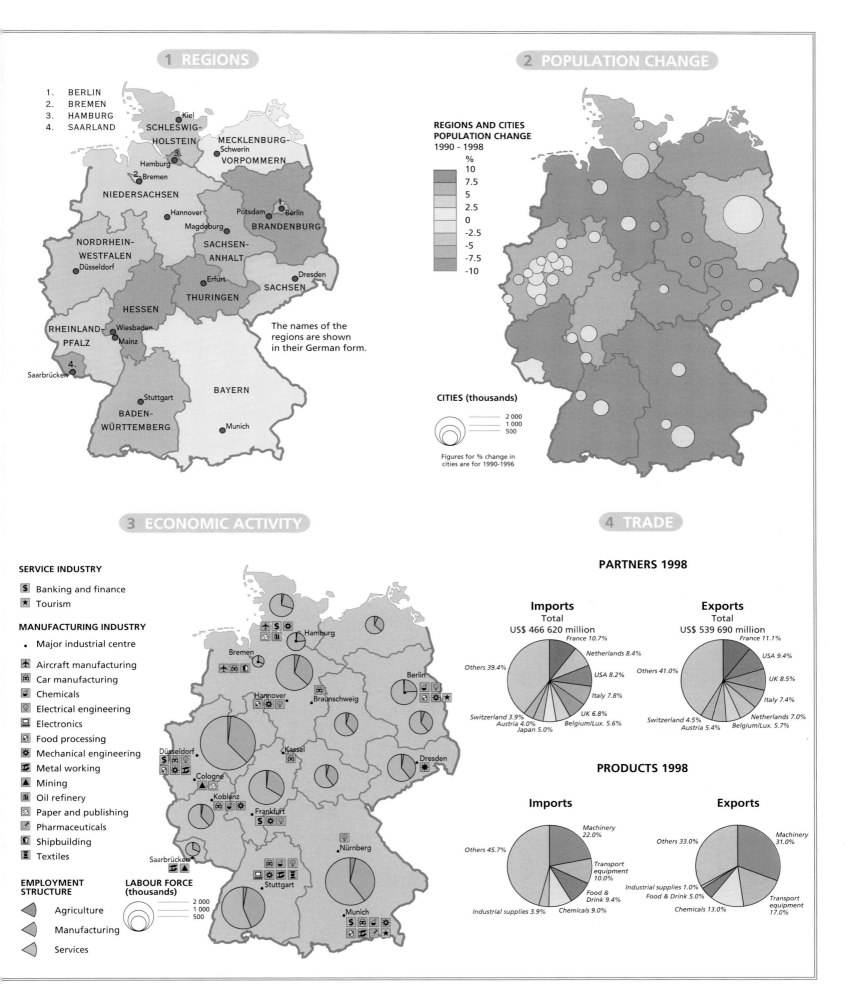

1 REGIONS

1. BERLIN
2. BREMEN
3. HAMBURG
4. SAARLAND

Kiel
SCHLESWIG-HOLSTEIN
3.
Hamburg
MECKLENBURG-VORPOMMERN
Schwerin
2. Bremen
NIEDERSACHSEN
Hannover
Potsdam 1. Berlin
Magdeburg
BRANDENBURG
NORDRHEIN-WESTFALEN
Düsseldorf
SACHSEN-ANHALT
Erfurt
Dresden
THÜRINGEN
SACHSEN
HESSEN
RHEINLAND-PFALZ
Wiesbaden
Mainz
4.
Saarbrücken
Stuttgart
BAYERN
BADEN-WÜRTTEMBERG
Munich

The names of the
regions are shown
in their German form.

2 POPULATION CHANGE

REGIONS AND CITIES
POPULATION CHANGE
1990 - 1998
%
10
7.5
5
2.5
0
-2.5
-5
-7.5
-10

CITIES (thousands)

2 000
1 000
500

Figures for % change in
cities are for 1990-1996

3 ECONOMIC ACTIVITY

SERVICE INDUSTRY

$ Banking and finance
★ Tourism

MANUFACTURING INDUSTRY

• Major industrial centre

✈ Aircraft manufacturing
🚗 Car manufacturing
⚗ Chemicals
💡 Electrical engineering
🖥 Electronics
🍽 Food processing
✳ Mechanical engineering
⚒ Metal working
▲ Mining
🛢 Oil refinery
📰 Paper and publishing
💊 Pharmaceuticals
🚢 Shipbuilding
📏 Textiles

**EMPLOYMENT
STRUCTURE**

◁ Agriculture
◁ Manufacturing
◁ Services

**LABOUR FORCE
(thousands)**

2 000
1 000
500

Hamburg
Bremen
Hannover
Braunschweig
Berlin
Düsseldorf
Cologne
Kassel
Dresden
Koblenz
Frankfurt
Nürnberg
Saarbrücken
Stuttgart
Munich

4 TRADE

PARTNERS 1998

Imports
Total
US$ 466 620 million

France 10.7%
Netherlands 8.4%
Others 39.4%
USA 8.2%
Italy 7.8%
Switzerland 3.9%
Austria 4.0%
UK 6.8%
Japan 5.0%
Belgium/Lux. 5.6%

Exports
Total
US$ 539 690 million

France 11.1%
USA 9.4%
Others 41.0%
UK 8.5%
Italy 7.4%
Switzerland 4.5%
Netherlands 7.0%
Austria 5.4%
Belgium/Lux. 5.7%

PRODUCTS 1998

Imports

Machinery
22.0%
Others 45.7%
Transport
equipment
10.0%
Food &
Drink 9.4%
Industrial supplies 3.9%
Chemicals 9.0%

Exports

Machinery
31.0%
Others 33.0%
Industrial supplies 1.0%
Food & Drink 5.0%
Transport
equipment
17.0%
Chemicals 13.0%

SCALE 1 : 7 500 000

0   100   200   300 km

Administration

Boundaries
International

Settlement
Cities and towns in order of size

National capital	Other city or town
■ ROME	● Milan
□ SARAJEVO	○ Genoa
□ SAN MARINO	○ Venice
	○ Ragusa

KEY

Relief and physical features

Relief metres
5000
3000
2000
1000
500
200
sea level
0
under sea level
200
4000
6000

4634 ▲ Mountain height (in metres)

Permanent ice

Water features

River

Canal

Lake / Reservoir

Communications

Railway

Motorway

Road

⊕ Main airport

SCALE 1 : 5 000 000

0   50   100   150   200 km

Lambert Conformal Conic projection

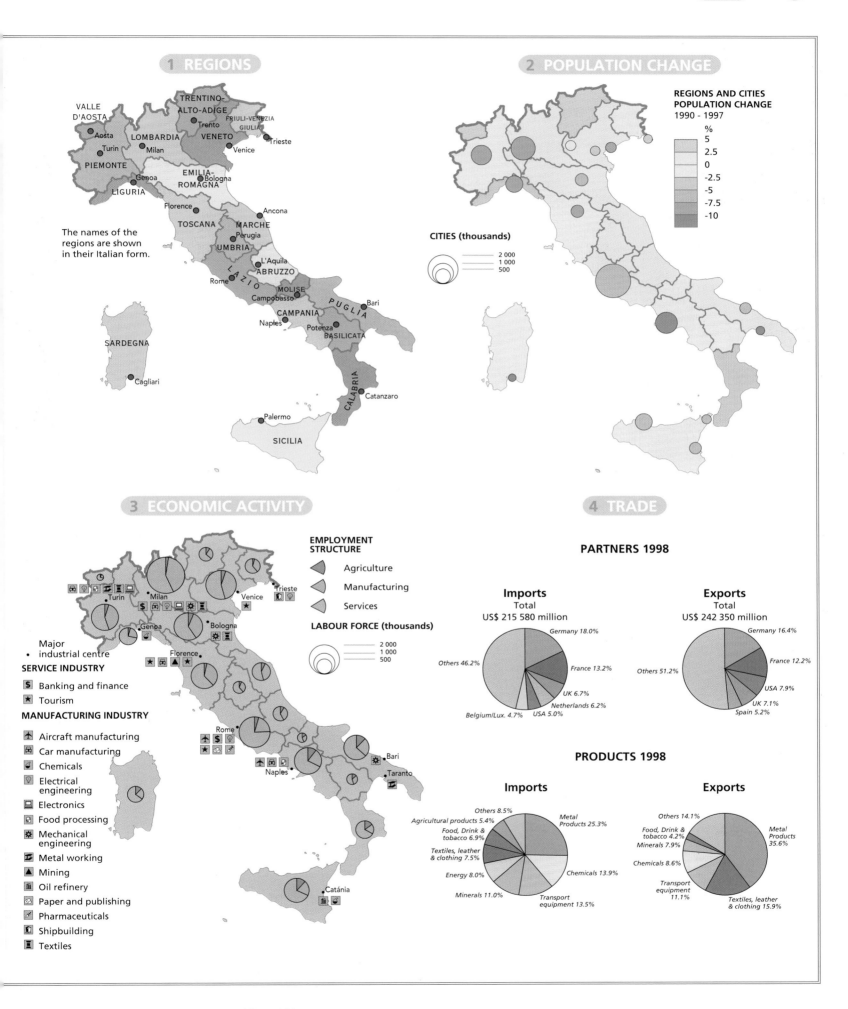

1 REGIONS

VALLE D'AOSTA
Aosta
Turin
PIEMONTE
LOMBARDIA
Milan
TRENTINO-ALTO-ADIGE
Trento
FRIULI-VENEZIA GIULIA
VENETO
Venice
Trieste
Genoa
LIGURIA
EMILIA-ROMAGNA
Bologna
Florence
TOSCANA
Ancona
MARCHE
Perugia
UMBRIA
L'Aquila
ABRUZZO
Rome
LAZIO
MOLISE
Campobasso
PUGLIA
Bari
CAMPANIA
Naples
Potenza
BASILICATA
SARDEGNA
Cagliari
CALABRIA
Catanzaro
Palermo
SICILIA

The names of the regions are shown in their Italian form.

2 POPULATION CHANGE

REGIONS AND CITIES POPULATION CHANGE
1990 - 1997

%
5
2.5
0
-2.5
-5
-7.5
-10

CITIES (thousands)
2 000
1 000
500

3 ECONOMIC ACTIVITY

Turin
Milan
Venice
Trieste
Genoa
Bologna
Florence
Rome
Bari
Naples
Taranto
Catánia

EMPLOYMENT STRUCTURE
Agriculture
Manufacturing
Services

LABOUR FORCE (thousands)
2 000
1 000
500

Major
• industrial centre

SERVICE INDUSTRY
$ Banking and finance
★ Tourism

MANUFACTURING INDUSTRY
Aircraft manufacturing
Car manufacturing
Chemicals
Electrical engineering
Electronics
Food processing
Mechanical engineering
Metal working
Mining
Oil refinery
Paper and publishing
Pharmaceuticals
Shipbuilding
Textiles

4 TRADE

PARTNERS 1998

Imports
Total
US$ 215 580 million

Germany 18.0%
France 13.2%
UK 6.7%
Netherlands 6.2%
USA 5.0%
Belgium/Lux. 4.7%
Others 46.2%

Exports
Total
US$ 242 350 million

Germany 16.4%
France 12.2%
USA 7.9%
UK 7.1%
Spain 5.2%
Others 51.2%

PRODUCTS 1998

Imports

Others 8.5%
Agricultural products 5.4%
Food, Drink & tobacco 6.9%
Textiles, leather & clothing 7.5%
Energy 8.0%
Minerals 11.0%
Transport equipment 13.5%
Chemicals 13.9%
Metal Products 25.3%

Exports

Others 14.1%
Food, Drink & tobacco 4.2%
Minerals 7.9%
Chemicals 8.6%
Transport equipment 11.1%
Textiles, leather & clothing 15.9%
Metal Products 35.6%

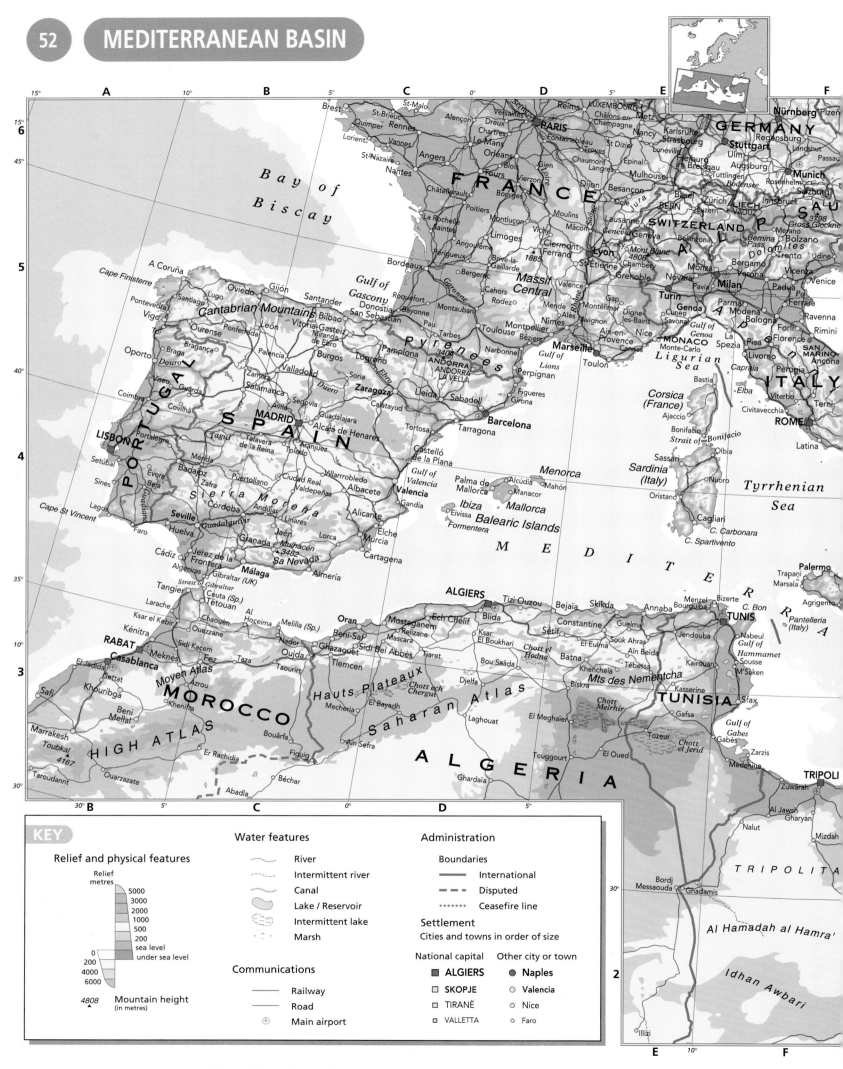

SCALE 1 : 10 000 000

0   100   200   300   400 km

CZECH REPUBLIC

Svitavy
Ostrava
Olomouc
Bielsko Biała
Khmel'nyts'kyy
Zhmerynka
Uman'
Kirovohrad
Dnipropetrovs'k
Zaporizhzhya
Donets'k
Tagan-
Rostov-na-
Donu

Linz
Brno
České
Budějovice
Žilina
Poprad
Prešov
Košice
Stryj
Ivano-Frankivs'k
Kolomyya
Dniester
Mohyliv-
Podil's'kyy
Voz4esens'k
Kryvyy Rih
Nikopol'
Gulf of Taganrog
Yeysk
Sal'sk

SLOVAKIA
Uzhhorod
Chernivtsi
Soroca
Berezivka
Mykolayiv
Melitopol'
Berdyans'k
Pavlovskaya
Tikhoretsk

VIENNA
BRATISLAVA
Miskolc
Tisza
Satu
Mare
Botoşani
Suceava
Bălţi
MOLDOVA
Tighina
Kherson
Henichés'k
Dzhankoy
Kerch
Timashevsk
Kropotkin

Wiener
Neustadt
Leoben
Győr
BUDAPEST
Nyíregyháza
Debrecen
Dej
Iaşi
CHIŞINĂU
Tiraspol
Odessa
Dnieper
Krasnodar
Armavir

STRIA
Graz
Szombathely
Székesfehérvár
Kecskemét
Oradea
Cluj-Napoca
Târgu
Mureş
Tecuci
Bilhorod-
Dnistrovs'kyy
Yevpatoriya
Simferopol'
Feodosiya
Tuapse
Maykop

Klagenfurt
HUNGARY
Nagykanizsa
Szekszárd
Szeged
Arad
Deva
Sibiu
Braşov
Galaţi
Crimea
Novorossiysk
Sochi

Kranj
SLOVENIA
Maribor
Balaton
Hungarian Plain
ROMANIA
Râmnicu Vâlcea
Buzău
Brăila
Sevastopol'

LJUBLJANA
Trieste
ZAGREB
CROATIA
Osijek
Sombor
VOJVODINA
Zrenjanin
Târgu
Jiu
Piteşti
Ploieşti
BUCHAREST
Constanţa

Pula
Rijeka
Karlovac
Sisak
Banja Luka
BELGRADE
Drobeta-
Turnu Severin
Craiova
Danube

Zadar
Šibenik
Bihać
BOSNIA-
HERZEGOVINA
Tuzla
YUGOSLAVIA
Ruse
Pleven
Shumen

Split
Metković
Zenica
SARAJEVO
Titovo
Užice
SERBIA
Zaječar
Veliko Turnovo
Sliven
Varna

Mostar
Kragujevac
BULGARIA
Stara Zagora
Burgas

Dubrovnik
MONTE-
NEGRO
Podgorica
Kotor
Peć
Priština
KOSOVO
Prizren
Niš
SOFIA
Pernik
Plovdiv
Khaskovo

Pescara
Shkodër
Kumanovo
Vranje
Strúma
Rhodope Mts
Edirne
Zonguldak
Karabük

TIRANË
SKOPJE
Veles
MACEDONIA
Bitola
Drama
Xanthi
Komotini
İstanbul
Kartal
Adapazari
Bolu

Foggia
Barletta
Durrës
Elbasan
Korçë
Serres
Kavala
Tekirdağ
Sea of
Marmara
Körféz
ANKARA
Yozgat

Bari
Vesuvius
1281
ALBANIA
Vlorë
Thasos
Samothraki
Gökçeada
Gallipoli
Bandirma
Bursa
Kırıkkale

Naples
Salerno
Potenza
Brindisi
Lecce
Kozani
Mt Olympus
2917
Larisa
Limnos
Çanakkale
Balıkesir
Eskişehir
TURKEY

Gulf of
Salerno
Sapri
Taranto
Gallipoli
Ioannina
Pindus Mts
Trikala
Volos
Akhisar
Kütahya
Afyon
3916
Mt Erciyas
Malatya

Corfu
Corfu
Igoumenitsa
Karditsa
Aegean
Skyros
Lesvos
Manisa
Usak
Akşehir
Lake
Tuz
Kayseri
Kahraman
Maras

Cosenza
Crotone
Lefkada
GREECE
Evvoia
Sea
Chios
İzmir
Aydın
Denizli
Isparta
Nigde
Birecik
Gaziantep

Lipari
Islands
Stromboli
Catanzaro
Kefallonia
Mesolongion
Chalkida
Piraeus
ATHENS
Andros
Samos
Ikaria
Söke
Yatağan
Burdur
Beyşehir
Ereğli
Konya
Adana
Tarsus

Sicily
Messina
Reggio di
Calabria
Zakynthos
Patras
Corinth
Tripoli
Tinos
Naxos
Marmaris
Fethiye
Gulf of
Antalya
Antalya
Karaman
Mersin
İskenderun
Aleppo

Mt Etna
3323
Catania
Ionian
Kyparissia
Sparti
Paros
Kalamata
Ios
Amorgos
Dodecanese
Kos
Antakya
Latakia
Hamâh

Caltanissetta
Gela
Sea
Milos
Thira
Rodos
Rhodes
(Greece)
Karpathos
Keryneia
Famagusta
SYRIA
Homs

Ragusa
C. Matapan
Kythira
Sea of Crete
NICOSIA
Tripoli
LEBANON
Zahle

Chania
Iraklion
Mt Olympus
1952
Limassol
CYPRUS
BEIRUT
DAMASCUS

VALLETTA
MALTA
Crete
Saida
Tyre
L. Tiberias
Irbid

Haifa
Nazareth
Mafraq
Zarqa

ISRAEL
Tel Aviv-Yafo
Holon
JERUSALEM
AMMAN
WEST
BANK

Gaza
GAZA
Dead Sea

Beersheba
Port Said
Dumyât
El Mansûra
JORDAN
Ma'an

Al Khums
Misratah
Al Bayda'
Darnah
Tubruq
Marsa
Matrûh
Alexandria
El Mansûra
Ismâ'iliya
Beersheba

Bani
Walid
Sabkhat al
Hayshah
Al Marj
Al Jabal al Akhdar
Umm Sa'ad
Damanhûr
Tanta
Zagazig
Suez
Canal

Al Qaddahiyah
Banghazi
Gulf of Sirte
Shibîn el Kôm
El Giza
CAIRO
Suez
Sinai

LIBYA
An Nawfaliyah
As Sidrah
Ajdabiya
Libyan Plateau
El Faiyûm
G. Katherina
2637

Sirte
Marsa
al Burayqah
Qattara Depression
Siwa
Beni Suef
Gulf of Suez
Ras Muhammad

Al 'Uqaylah
Al Jaghbub
Maghâgha
El Minya
Hurghada
Red

Waddan
Maradah
Jalu
Great Sand Sea
Bawiti
Western Desert
Mallawi
Asyût
Bûr Safâga
Sea

Sabha
Calanscio Sand Sea
CYRENAICA
ASSARIR
EGYPT
Sohâg
Qena
Luxor
Quseir

Adriatic Sea
Dinaric Alps
Dalmatia
Transylvanian Alps
Carpathian Mts
Balkan Mts
Pontine Mountains
Samsun
Ordu
Trabzon
Giresun
Erzincan

BLACK SEA
Sea of Azov
İnebolu
Sinop
Bafra
Merzifon
Amasya
Çorum
Turhal
Sivas

Gulf of
Karkinitskiy
Kartal
Zonguldak
Karabük
Kızılırmak
Çankırı

Idfu
Aswan

Conic projection

KEY

Relief and physical features

Relief metres
5000
3000
2000
1000
500
200
0 sea level
200 under sea level
4000
6000

3798 ▲ Mountain height (in metres)

Permanent ice

Water features

~~~ River

~~~ Canal

Lake / Reservoir

Marsh

Communications

Railway

Motorway

Road

⊕ Main airport

Administration

Boundaries

International

Internal

Settlement

Cities and towns in order of size

National capital | Other city or town
■ WARSAW | ● Kharkiv
□ CHIŞINAU | ○ Kraków
□ BRATISLAVA | ○ Brno
□ VADUZ | ○ Chelm

SCALE 1 : 5 000 000

0 50 100 150 200 km

KEY

Relief and physical features

Relief
metres
5000
3000
2000
1000
500
200
sea level
0
under sea level
200
4000
6000

3971 ▲ Mountain height
(in metres)

Water features

~~~ River
- - - Intermittent river
~~~ Canal
Lake / Reservoir
Intermittent lake
Marsh

Communications

——— Railway
≡≡≡ Motorway
——— Road
⊕ Main airport

Administration

Boundaries
━━━ International
─── Internal
······· Ceasefire line

Settlement

Cities and towns in order of size

National capital | Other city or town
■ ATHENS | ● İstanbul
□ SKOPJE | ○ Konya
□ NICOSIA | ○ Split
| ○ Dubrovnik

SCALE 1 : 5 000 000

0   50   100   150   200 km

Conic projection

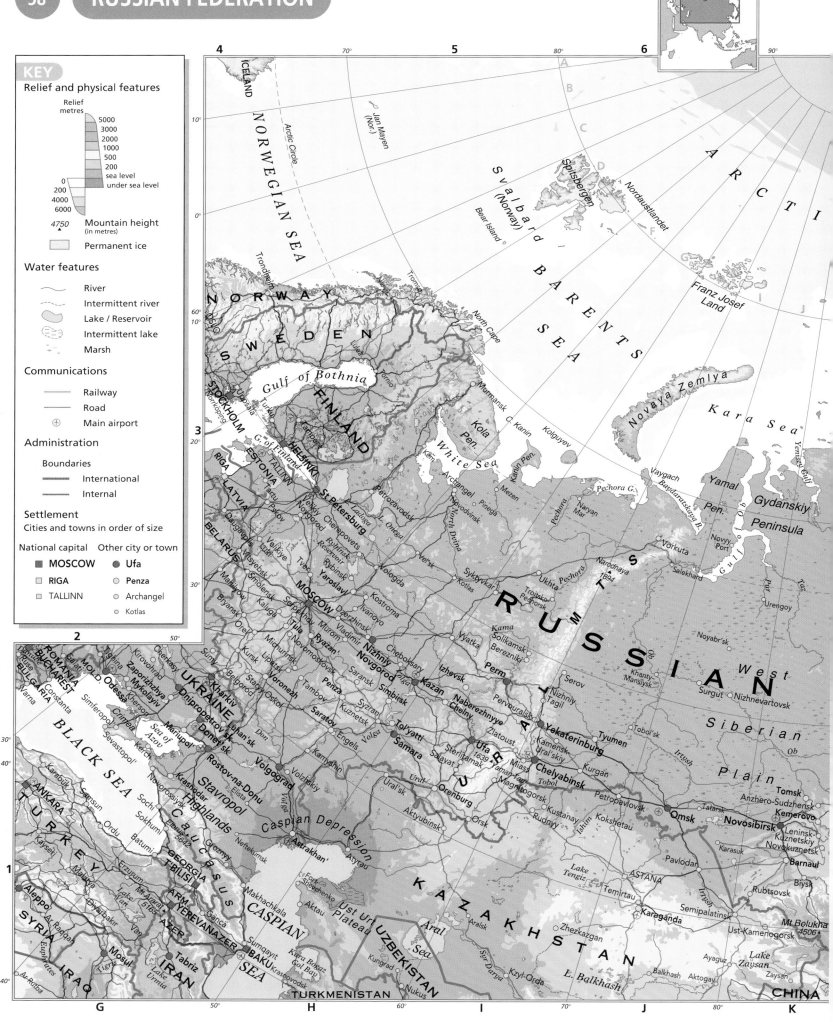

KEY

Relief and physical features

Relief
metres
5000
3000
2000
1000
500
200
sea level
under sea level
0
200
4000
6000

▲ 4750 Mountain height
(in metres)

Permanent ice

Water features

~~~~ River

Intermittent river

Lake / Reservoir

Intermittent lake

Marsh

Communications

Railway

Road

⊕ Main airport

Administration

Boundaries

International

Internal

Settlement

Cities and towns in order of size

National capital   Other city or town

■ MOSCOW   ● Ufa

□ RIGA   ○ Penza

□ TALLINN   ○ Archangel

○ Kotlas

ARCTIC OCEAN

Severnaya Zemlya
Komsomolets
October Revolution
Bolshevik
C. Chelyuskin

Taymyr Peninsula
Byrranga Mts
Lake Taymyr
Nordvik
Pyasina
Khatanga
Anabar
Olenek
Ust-Olenek
Olenek
Tiksi
Bulun

Dudinka
Noril'sk
Yenisey
Kotuy
Olenek

Laptev Sea

New Siberian Islands
Kotel'nyy
Bolshoi Lyakhovskiy
G. of Tona

East Siberian Sea

Wrangel I.
De Longa Str.
Chukchi Sea
Point Hope
Chukchi Pen.
Gulf of Anadyr
Anadyr

Novaya Sibir'
Ambarchik
Indigirka
Srednekolymsk
Kolyma
Omolon

Bering Strait
Seward Pen.
U.S.A.
St Lawrence I.
St Matthew I.
Nunivak

BERING SEA

Koryak Range
Anadyr
Kamennoye
Gizhiga
Penzhina
G. of Penzh.
Gizhiga Gulf
Palana

Kamchatka
Klyuchevskaya Sopka 4750
Kamchatka Peninsula
Ust-Kamchatsk
Petropavlovsk-Kamchatskiy

Yana
Verkhoyansk
Cherskogo Range
Mt Pobeda 3147
Yagodnoye
Magadan
Okhotsk

Verkhoyansk Range
Lena
Aldan

SEA OF OKHOTSK

Severo-Kuril'sk

Kuril Islands

CENTRAL
SIBERIAN
PLATEAU

FEDERATION

Tura
Lower Tunguska
Vilyuy
Vilyuysk
Nyurba
Olenek

Yakutsk
Lena
Olekminsk
Ust-Maya
Maya
Nelkan
Ayan

Dzhugdzhur Range

Shantar Islands

Amgun
Nikolayevsk-na-Amure
Okha
Aleksandrovsk-Sakhalinskiy

Sakhalin

Tatar Strait
Poronaysk
Uglegorsk
Yuzhno-Sakhalinsk
Kholmsk

Administered by Russia Claimed by Japan

Stony Tunguska
Lena
Vitim
Tynda

Stanovoy Range

Yenisey
Yeniseysk
Angara
Ust-Ilimsk
Ust-Kut

Svobodnyy
Skovorodino
Amur
Blagoveshchensk
Amur

Komsomolsk-na-Amure
Amur

Khabarovsk

Sikhote-Alin Range

Wakkanai
Asahikawa
Asahi-dake 2290
Hokkaido
Sapporo
Hakodate
Aomori

Achinsk
Kansk
Bratsk
Nizhneudinsk
Krasnoyarsk

Lake Baikal

Yablonovy Range

Da Hinggan Ling

Bei'an
Yichun
L. Khanka

Ussuriysk
Vladivostok
Nakhodka
Chongjin

Honshu
Hachinohe
Akita
Niigata

Abakan
Usol'ye-Sibirskoye
Irkutsk
Ulan-Ude

Chita
Karymskoye
Krasnokamensk

Hulun Nur

Qiqihar
Harbin
Mudanjiang
Jilin

Sea of Japan

Western Sayan
Eastern Sayan
Yenisey
Kyzyl
Uvs Nuur
Hövsgöl Nuur

ULAN BATOR
MONGOLIA
Bayanhongor

Changchun
Shenyang
Fushun
Anshan
NORTH KOREA

CHINA

JAPAN
Nagoya
Kobe

Altai Mts

Conic Equidistant projection

ARCTIC OCEAN

PACIFIC OCEAN

ATLANTIC OCEAN

Greenland

Baffin Island

Labrador

Hudson Bay

CANADIAN SHIELD

GREAT PLAINS

ROCKY MOUNTAINS

Great Basin

Great Plains

Gulf of Mexico

Bahamas

Cuba

Greater Antilles

Hispaniola

Puerto Rico

Lesser Antilles

Caribbean Sea

Jamaica

PACIFIC OCEAN

Relief

Relief metres
5000
3000
2000
1000
500
200
sea level
under sea level
0
200
4000
6000

Ice cap

B. BELIZE
C.R. COSTA RICA
D.R. DOMINICAN REPUBLIC
E.S. EL SALVADOR
G. GUATEMALA
H. HAITI
HO. HONDURAS
J. JAMAICA
N. NICARAGUA
P. PANAMA

GREENLAND

U.S.A.

CANADA

UNITED STATES OF AMERICA

MEXICO

THE BAHAMAS

CUBA

D.R.

J.

B. HO
G. N
E.S.
C.R. P.

SCALE 1 : 95 000 000

SCALE 1 : 40 000 000

Chamberlin Trimetric projection

1 TEMPERATURE AND PRESSURE : JANUARY

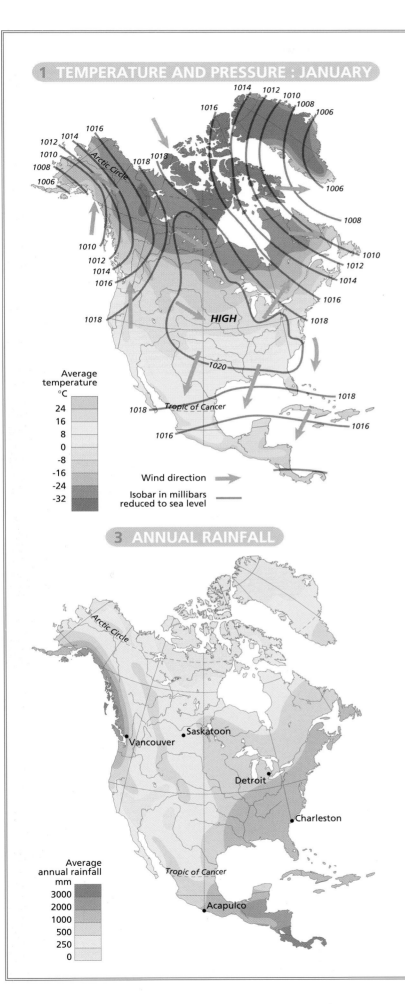

Average temperature °C

24
16
8
0
-8
-16
-24
-32

Wind direction →

Isobar in millibars reduced to sea level

2 TEMPERATURE AND PRESSURE : JULY

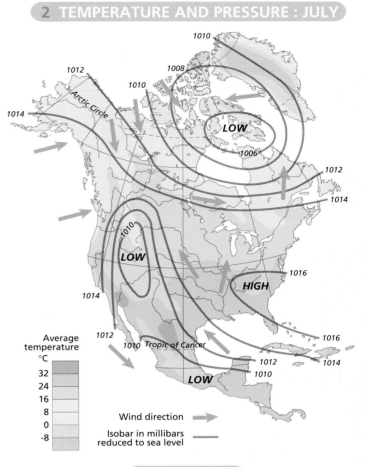

Average temperature °C

32
24
16
8
0
-8

Wind direction →

Isobar in millibars reduced to sea level

3 ANNUAL RAINFALL

Average annual rainfall mm

3000
2000
1000
500
250
0

4 STATISTICS

Saskatoon (515 metres)	Jan	Feb	Mar	Apr	May	Jun	Jul	Aug	Sep	Oct	Nov	Dec
Temperature - max. (°C)	-13	-11	-3	9	18	22	25	24	17	11	-1	-9
Temperature - min. (°C)	-24	-22	-14	-3	3	9	11	9	3	-3	-11	-19
Rainfall - (mm)	23	13	18	18	36	66	61	48	38	23	13	15

Vancouver (14 metres)	Jan	Feb	Mar	Apr	May	Jun	Jul	Aug	Sep	Oct	Nov	Dec
Temperature - max. (°C)	5	7	10	14	18	21	23	23	18	14	9	6
Temperature - min. (°C)	0	1	3	4	8	11	12	12	9	7	4	2
Rainfall - (mm)	218	147	127	84	71	64	31	43	91	147	211	224

Charleston (3 metres)	Jan	Feb	Mar	Apr	May	Jun	Jul	Aug	Sep	Oct	Nov	Dec
Temperature - max. (°C)	14	15	19	23	27	30	31	31	28	24	19	15
Temperature - min. (°C)	6	7	10	14	19	23	24	24	22	16	11	7
Rainfall - (mm)	74	84	86	71	81	119	185	168	130	81	58	71

Acapulco (3 metres)	Jan	Feb	Mar	Apr	May	Jun	Jul	Aug	Sep	Oct	Nov	Dec
Temperature - max. (°C)	31	31	31	32	32	33	32	33	32	32	32	31
Temperature - min. (°C)	22	22	22	23	25	25	25	25	24	24	23	22
Rainfall - (mm)	6	1	0	1	36	281	256	252	349	159	28	8

Detroit (189 metres)	Jan	Feb	Mar	Apr	May	Jun	Jul	Aug	Sep	Oct	Nov	Dec
Temperature - max. (°C)	-1	0	6	13	19	25	28	27	23	16	8	2
Temperature - min. (°C)	-7	-8	-3	3	9	14	17	17	13	7	1	-4
Rainfall - (mm)	53	53	64	64	84	91	84	69	71	61	61	58

SCALE 1 : 80 000 000

0 800 1600 2400 3200 km

Bonne projection

KEY

Relief and physical features

Relief metres
5000
3000
2000
1000
500
200
0 sea level
200 under sea level
4000
6000

6194 ▲ **Mountain height**
(in metres)

Permanent ice

Water features

~ River

Lake / Reservoir

Intermittent lake

Marsh

Communications

—— Railway

—— Road

⊕ Main airport

Administration

Boundaries

—— International

—— Internal

Settlement
Cities and towns in order of size

National capital   Other city or town

■ **OTTAWA**   ● **Montréal**

□ REYKJAVÍK   ○ **Winnipeg**

                  ○ Québec

                  ○ Churchill

SCALE 1 : 17 000 000

0   200   400   600   800 km

H 100° I 90° J 80° K 70° L 60° M 80° 50° N 40° O 5 30° P 20° 70° Q 4 10°

10°
60°

ue
Axel
Heiberg
Island
Amund
Ringnes I.
z
a
b
e
e
n
Ellesmere Island
Nares Strait
h
Cape
Parry
Qaanaaq
(Thule)
Cape York
Melville
Bay
Upernavik
GREENLAND
(Denmark)
Kong Christian IX Land
Gunnbjørn
Field
3700
Denmark Strait
Arctic Circle
Siglufjördhur
ICELAND
Akureyri
1763
Vatnajökull
Höfn
Isafjördhur
Faxaflói
REYKJAVÍK
Keflavík
20°

Bathurst I.
Islands
Jones Sound
Cornwallis I.
Devon Island
Lancaster Sound
Resolute
Bay
Somerset
Island
Prince
of Wales
Island
Brodeur Peninsula
Arctic Bay
Borden
Peninsula
Bylot
Island
Pond Inlet
(Mittimatalik)
Clyde River
Baffin
Bay
Disko I.
Saqqaq
Qasigiannguit
Sisimiut
Maniitsoq
Davis Strait
Kong Frederick VI Kyst
Tasiilaq
20°

Boothia
Peninsula
Gulf of Boothia
King
William
Island
Taloyoak
Melville
Peninsula
Hall
Beach
Prince
Charles I.
Baffin Island
Home Bay
Penny
Icecap
C. Dyer
Pangnirtung
NUUK
(Godthåb)
Paamiut
Ivittuut
Nanortalik
Cape Farewell
30°

A V U T
Back
Repulse
Bay
Foxe
Basin
Nettilling
Lake
Cumberland Sound
Amadjuak
Lake
Iqaluit
Frobisher Bay
Labrador
30°

Baker Lake
(Qamanittuaq)
Baker
Lake
Rankin Inlet
Arviat
Southampton
Island
Coral
Harbour
Foxe Channel
Fisher Str.
Coats I.
Mansel I.
Hudson Strait
Akpatok I.
Resolution
Island
C. Chidley
Sea
A T L A N T I C
50°

A D A
Cape Churchill
Churchill
TOBA
Nelson
Churchill
Hudson
Bay
Belcher
Islands
Cape
Henrietta Maria
Ottawa Is
Salluit
Puvurnituq
Inukjuak
Kangiqsujuaq
A
Ungava
Bay
Feuilles
Kangiqsualujjuaq
Kuujjuaq
Baleine
George
Nain
Hopedale
Cape Harrison
N
E
W
F
O
U
N
D
O C E A N
40°

Fort Severn
Severn
Winisk
Big Trout Lake
James
Bay
Akimiski
Island
Lac à
l'Eau Claire
Caniapiscau
Réservoir
Caniapiscau
Schefferville
Smallwood
Reservoir
Labrador
Churchill
Happy Valley-
Goose Bay
Port Hope
Simpson
St Anthony
L
A
N
D
Strait of Belle Isle
40°

Sandy Lake
Red Lake
Lac
St Joseph
Sioux
Lookout
ONTARIO
Ekwan
Fort
Albany
Albany
Moosonee
Rés. de
La Grande 2
Fort
George
(Chisasibi)
Rés. de
La Grande 3
Eastmain
Fort Rupert
(Waskaganish)
L. Evans
Rés. de
La Grande 4
Eastmain
Harricana
QUÉBEC
Gagnon
Labrador City
Wabush
St Anthony
Havre-St-Pierre
Île d'Anticosti
St Lawrence
Sept-Îles
Corner
Brook
Grand
Falls
Newfoundland
Channel-Port
aux Basques
Bonavista
Gander
St John's
2

Kenora
Lake of
the Woods
Fort Frances
Red Lakes
MINNESOTA
Bemidji
Duluth
Longlac
Nipigon
Kapuskasing
Groundhog
Lake
Superior
Thunder
Bay
Sault Ste
Marie
Sudbury
Timmins
Kirkland Lake
Chapleau
Missinaibi
Val d'Or
Amos
Réservoir
Gouin
Roberval
Chibougamau
L. Mistassini
Mistissini
(Baie-du-Posta)
Baie Comeau
Jonquière
Chicoutimi
Rimouski
Gaspé
Pen.
Gaspé
Gulf of
St Lawrence
Cabot Strait
St Pierre
& Miquelon
(Fr.)
Glace Bay
Sydney
Cape Breton
Island
Sable I.
40°

Albert Lea
WISCONSIN
Green Bay
Cadillac
Bay
City
Flint
GAN
Georgian B.
Owen
Sound
North Bay
Ottawa
Peterborough
Oshawa
Rivière-du-Loup
Montmagny
Québec
Trois Rivières
Sherbrooke
MAINE
Edmundston
Bathurst
NEW
BRUNSWICK
Moncton
Fredericton
St John
PRINCE EDWARD
ISLAND
Charlottetown
Truro
NOVA
SCOTIA
Halifax
Cape Sable
50°

Minneapolis-
St Paul
Marquette
Escanaba
Lake Michigan
Lake Huron
Sudbury
North Bay
OTTAWA
Montréal
VER. 1917
Mt Washington
N.H.
Manchester
Bangor
Augusta
Portland
Yarmouth
Bay of Fundy
St John
40°

IOWA
ICA
Milwaukee
Rockford
Cedar
Rapids
Green Bay
Grand
Rapids
Detroit
Toledo
Chicago
MI
Cadillac
Bay
City
Flint
L. Huron
Sault Ste
Marie
London
Toronto
L. Ontario
Rochester
Buffalo
Erie
Cleveland
PENNSYLVANIA
Williamsport
Scranton
N.J.
New York
NEW
YORK
Syracuse
Albany
Binghamton
MASS.
Worcester
Hartford
New Haven
CONN.
Providence
Boston
Lowell
Long Island
Cape Cod
50°

TATES
ONTARIO
Nipigon
Thunder
Bay
MINNESOTA
Duluth
Ashland
WISCONSIN

Chamberlin Trimetric projection

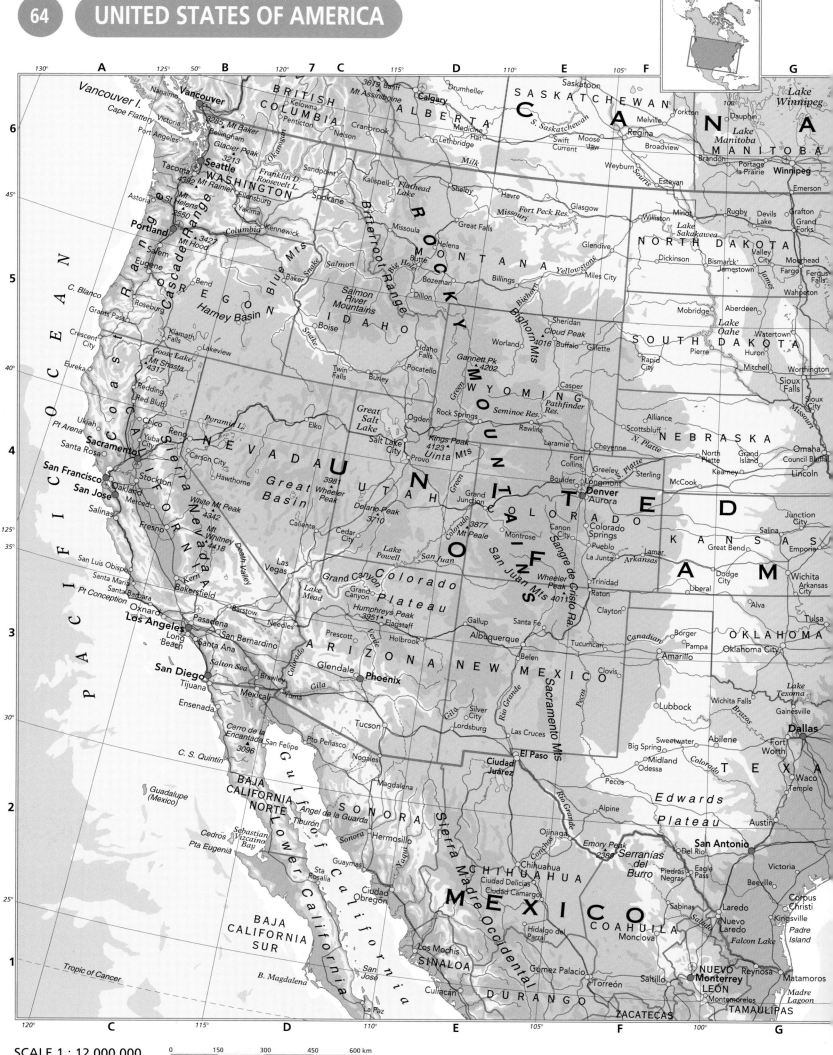

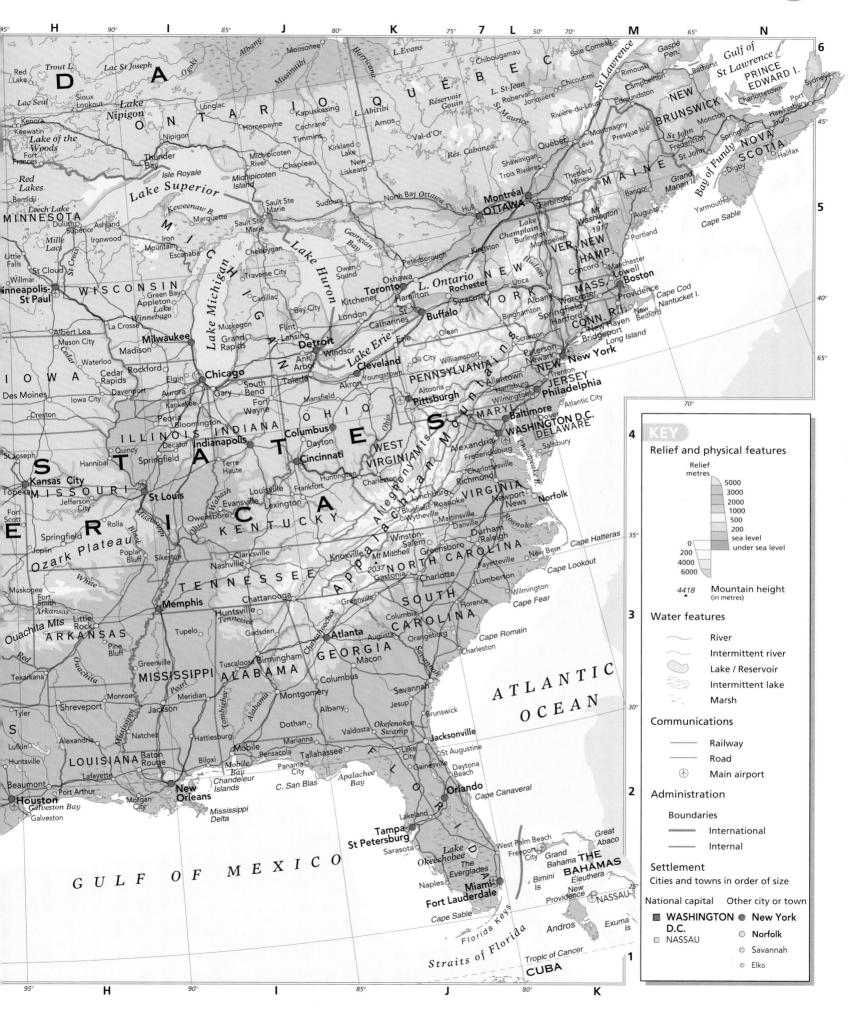

KEY

Relief and physical features

Relief
metres
5000
3000
2000
1000
500
200
sea level
under sea level
200
4000
6000

▲ 4418 Mountain height
(in metres)

Water features

~~~~~ River

~~~~~ Intermittent river

Lake / Reservoir

Intermittent lake

Marsh

Communications

—— Railway

—— Road

✈ Main airport

Administration

Boundaries

—— International

—— Internal

Settlement

Cities and towns in order of size

National capital   Other city or town

■ **WASHINGTON** ● New York
   D.C.
□ NASSAU   ◉ **Norfolk**
                  ○ Savannah
                  ○ Elko

Lambert Conformal Conic projection

1 POPULATION DENSITY

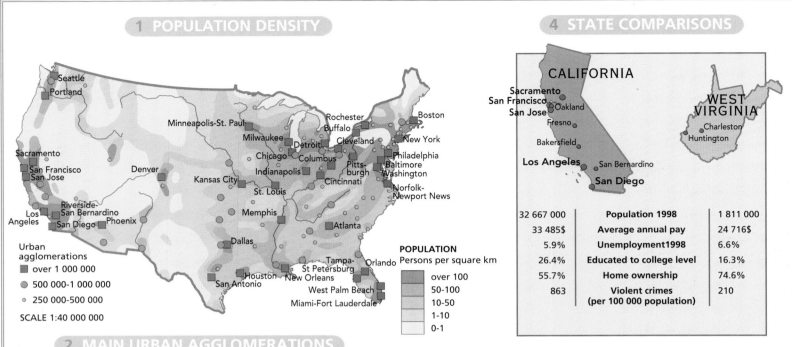

Urban agglomerations
- ■ over 1 000 000
- ● 500 000-1 000 000
- • 250 000-500 000

SCALE 1:40 000 000

POPULATION
Persons per square km
- over 100
- 50-100
- 10-50
- 1-10
- 0-1

4 STATE COMPARISONS

32 667 000	**Population 1998**	1 811 000
33 485$	**Average annual pay**	24 716$
5.9%	**Unemployment 1998**	6.6%
26.4%	**Educated to college level**	16.3%
55.7%	**Home ownership**	74.6%
863	**Violent crimes (per 100 000 population)**	210

2 MAIN URBAN AGGLOMERATIONS

Urban agglomeration	1980	1998	% change
New York	15 600 000	16 626 000	6.6
Los Angeles	9 500 000	13 129 000	38.2
Chicago	6 780 000	6 945 000	2.4
Philadelphia	4 116 000	4 398 000	6.9
San Francisco	3 201 000	4 051 000	26.6
Washington	2 777 000	3 927 000	41.4
Dallas	2 468 000	3 912 000	58.5
Detroit	3 806 000	3 785 000	-0.6
Houston	2 424 000	3 365 000	38.8
San Diego	1 718 000	2 983 000	73.6
Boston	2 681 000	2 915 000	8.7

5 POPULATION GROWTH

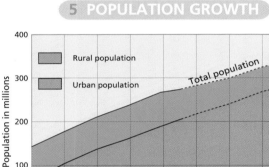

Rural population
Urban population
Total population

Population in millions

3 POPULATION CHANGE

POPULATION CHANGE
1985 - 1998
%
- over 50
- 31-50
- 16-30
- 0-15
- -1 - -10

SCALE 1:40 000 000

CO.	CONNECTICUT	PENN.	PENNSYLVANIA
DEL.	DELAWARE	R.I.	RHODE ISLAND
MASS.	MASSACHUSETTS	S.C.	SOUTH CAROLINA
N.H.	NEW HAMPSHIRE	V.	VERMONT
N.J.	NEW JERSEY	W.V.	WEST VIRGINIA

6 IMMIGRATION

IMMIGRATION INTO U.S.A BY COUNTRY 1997
Total 798 400

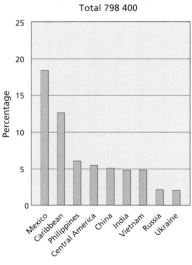

Percentage

Mexico, Caribbean, Philippines, Central America, China, India, Vietnam, Russia, Ukraine

7 ECONOMIC ACTIVITY

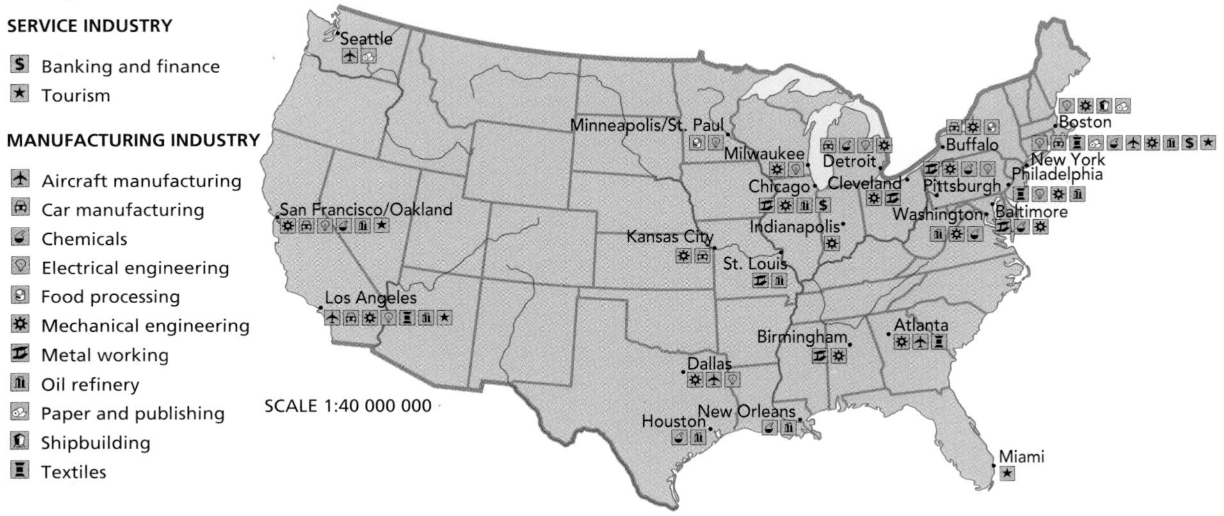
• Major industrial centre
SERVICE INDUSTRY
Banking and finance
Tourism
MANUFACTURING INDUSTRY
Aircraft manufacturing
Car manufacturing
Chemicals
Electrical engineering
Food processing
Mechanical engineering
Metal working
Oil refinery
Paper and publishing
Shipbuilding
Textiles
Seattle
Minneapolis/St. Paul
Milwaukee
Detroit
Chicago
Cleveland
Buffalo
Boston
New York
Philadelphia
Pittsburgh
Baltimore
Washington
San Francisco/Oakland
Kansas City
Indianapolis
St. Louis
Los Angeles
Birmingham
Atlanta
Dallas
Houston
New Orleans
Miami
SCALE 1:40 000 000

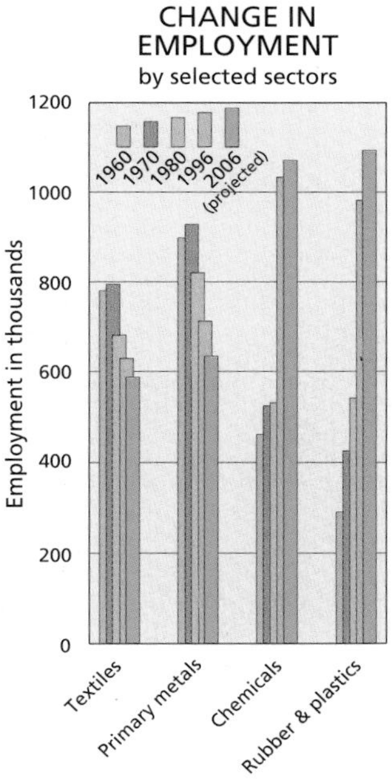
CHANGE IN EMPLOYMENT
by selected sectors
1960
1970
1980
1996
2006 (projected)
Employment in thousands
1200
1000
800
600
400
200
0
Textiles
Primary metals
Chemicals
Rubber & plastics

8 TRADE

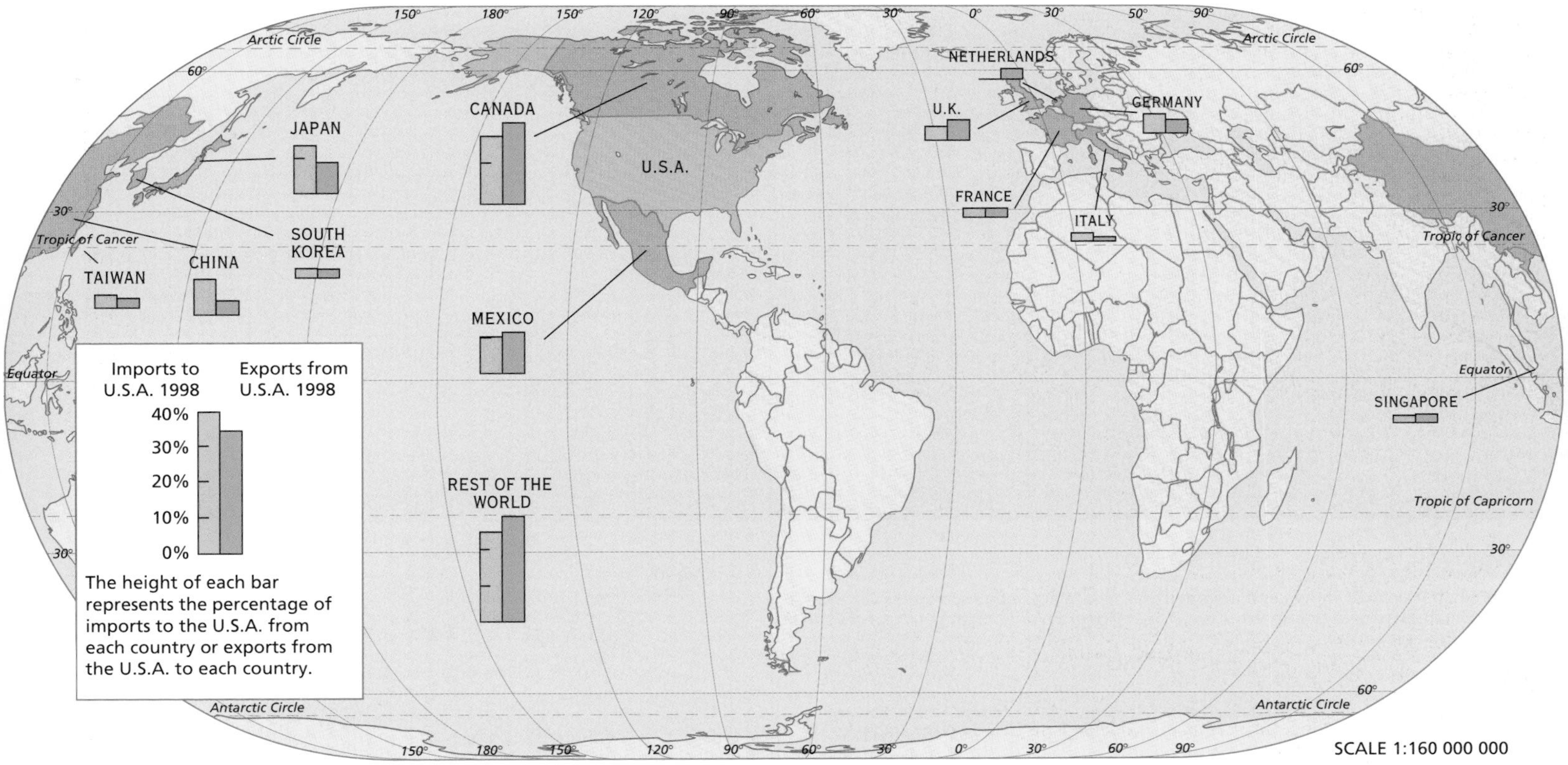
Arctic Circle
Tropic of Cancer
Equator
Tropic of Capricorn
Antarctic Circle
CANADA
U.S.A.
MEXICO
REST OF THE WORLD
JAPAN
SOUTH KOREA
TAIWAN
CHINA
NETHERLANDS
U.K.
GERMANY
FRANCE
ITALY
SINGAPORE
Imports to U.S.A. 1998
Exports from U.S.A. 1998
40%
30%
20%
10%
0%
The height of each bar represents the percentage of imports to the U.S.A. from each country or exports from the U.S.A. to each country.
SCALE 1:160 000 000

Exports 1960
Total: US$ 20 717 million

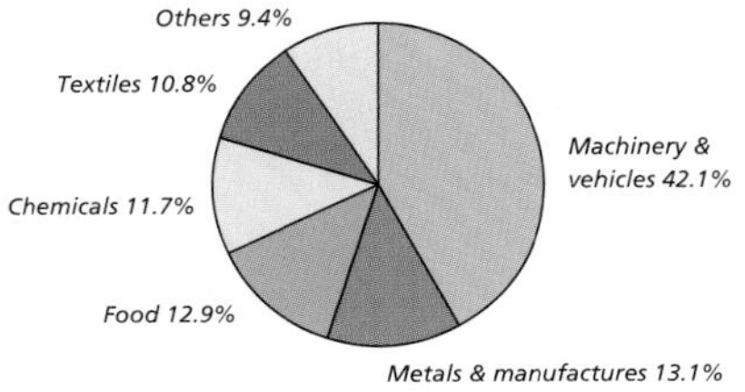
Others 9.4%
Textiles 10.8%
Chemicals 11.7%
Food 12.9%
Metals & manufactures 13.1%
Machinery & vehicles 42.1%

Exports 1998
Total: US$ 682 977 million

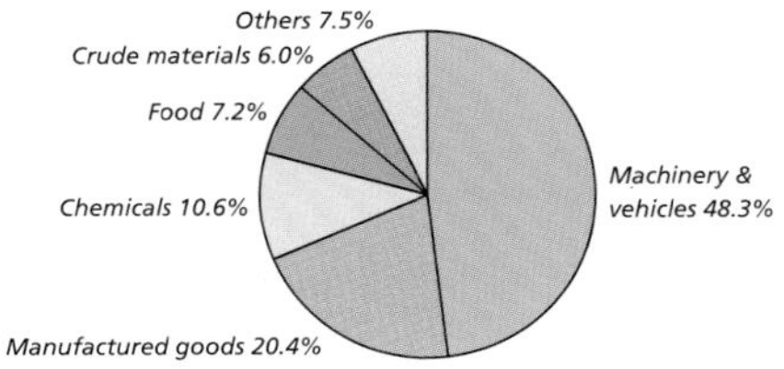
Others 7.5%
Crude materials 6.0%
Food 7.2%
Chemicals 10.6%
Manufactured goods 20.4%
Machinery & vehicles 48.3%

Imports 1998
Total: US$ 913 828 million

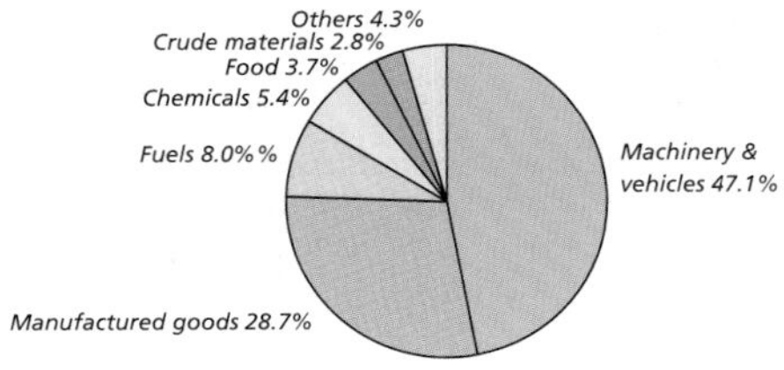
Others 4.3%
Crude materials 2.8%
Food 3.7%
Chemicals 5.4%
Fuels 8.0%%
Manufactured goods 28.7%
Machinery & vehicles 47.1%

Built-up area

The built up area shown as blue/green on the satellite image surrounds San Francisco Bay and extends south to San Jose. Three bridges link the main built up areas across San Francisco Bay.

Woodland

Areas of dense woodland cover much of the Santa Cruz Mountains to the west of the San Andreas Fault Zone. Other areas of woodland are found on the ridges to the east of San Francisco Bay.

Marsh / Salt Marsh

Areas of dark green on the satellite image represent marshland in the Coyote Creek area and salt marshes between the San Mateo and Dumbarton Bridges.

Reservoir / lake

Lakes and reservoirs stand out from the surrounding land. Good examples are the Upper San Leandro Reservoir east of Piedmont and the San Andreas Lake which lies along the fault line.

Airport

A grey blue colour shows San Francisco International Airport as a flat rectangular strip of land jutting out into the bay.

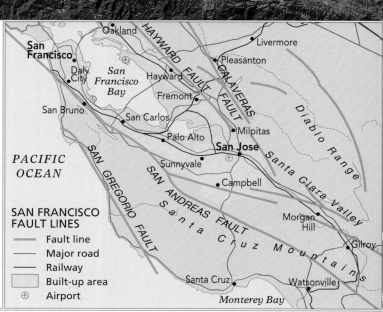

SAN FRANCISCO
FAULT LINES

— Fault line
— Major road
— Railway
☐ Built-up area
⊕ Airport

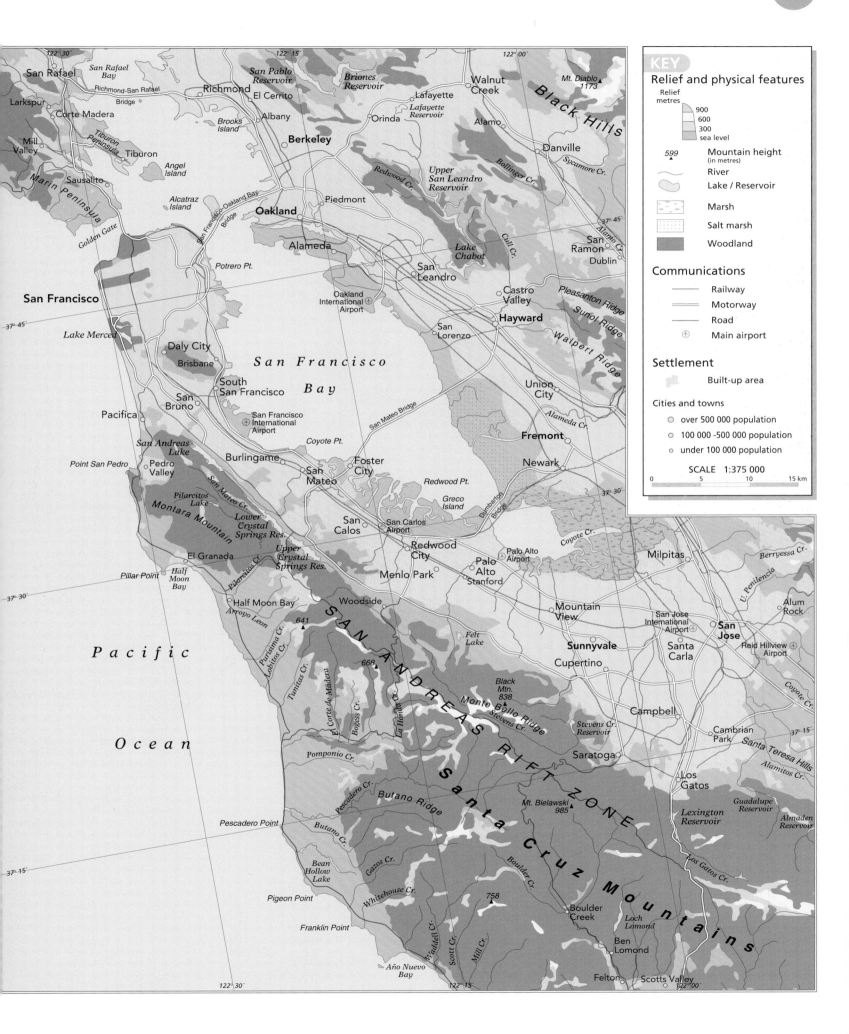

KEY

Relief and physical features

Relief
metres
900
600
300
sea level

599 ▲ Mountain height (in metres)

River

Lake / Reservoir

Marsh

Salt marsh

Woodland

Communications

Railway

Motorway

Road

⊕ Main airport

Settlement

Built-up area

Cities and towns

◉ over 500 000 population

◎ 100 000 -500 000 population

○ under 100 000 population

SCALE 1:375 000

0 5 10 15 km

San Rafael
San Rafael Bay
Larkspur
Richmond-San Rafael Bridge
Corte Madera
Mill Valley
Tiburon Peninsula
Tiburon
Marin Peninsula
Sausalito
Angel Island
Golden Gate
San Francisco
Lake Mercea
Daly City
Brisbane
South San Francisco
San Bruno
Pacifica
San Andreas Lake
Point San Pedro
Pedro Valley
Montara Mountain
Pilarcitos Lake
Lower Crystal Springs Res.
El Granada
Pillar Point
Half Moon Bay
Half Moon Bay
Arroyo Leon
641 ▲
Purisima Cr.
Lobitos Cr.
Tunitas Cr.
668 ▲
El Corte de Madera
Bogess Cr.
La Honda Cr.
Pomponio Cr.
Pescadero Point
Butano Ridge
Butano Cr.
Pescadero Cr.
Bean Hollow Lake
Gazos Cr.
Whitehouse Cr.
Pigeon Point
Franklin Point
Waddell Cr.
Scott Cr.
Mill Cr.
Año Nuevo Bay

Richmond
El Cerrito
Albany
Brooks Island
Berkeley
Oakland
Alameda
San Francisco-Oakland Bay Bridge
Alcatraz Island
Potrero Pt.
Piedmont
Redwood Cr.
San Leandro
San Lorenzo
Oakland International Airport
San Francisco Bay
Burlingame
San Mateo
Coyote Pt.
San Francisco International Airport
Foster City
San Mateo Cr.
San Mateo Bridge
Redwood Pt.
Greco Island
Coyote Pt.
San Calos
San Carlos Airport
Dumbarton Bridge
Redwood City
Menlo Park
Woodside
Felt Lake
Stevens Cr.
Black Mtn. 838
Monte Bello Ridge
Mt. Bielawski 985
Boulder Cr.
758 ▲
Boulder Creek
Loch Lomond
Ben Lomond
Felton
Scotts Valley

San Pablo Reservoir
Briones Reservoir
Lafayette
Lafayette Reservoir
Orinda
Upper San Leandro Reservoir
Lake Chabot
Castro Valley
Hayward
Union City
Alameda Cr.
Fremont
Newark
Palo Alto Airport
Palo Alto
Stanford
Mountain View
Coyote Cr.
Milpitas
Berryessa Cr.
San Jose International Airport
Sunnyvale
San Jose
Cupertino
Santa Carla
Reid Hillview Airport
Campbell
Stevens Cr. Reservoir
Saratoga
Cambrian Park
Los Gatos
Santa Teresa Hills
Alamitos Cr.
Lexington Reservoir
Guadalupe Reservoir
Almaden Reservoir
Los Gatos Cr.
U. Penilencia
Alum Rock

Walnut Creek
Mt. Diablo ▲ 1173
Black Hills
Alamo
Danville
Sycamore Cr.
Bollinger Cr.
San Ramon
Dublin
Pleasanton Ridge
Sunol Ridge
Walpert Ridge
Cull Cr.
Alamo Cr.

SAN ANDREAS RIFT ZONE
Santa Cruz Mountains

Pacific Ocean

37° 45'
37° 30'
37° 15'
122° 30'
122° 15'
122° 00'
37° 45'
37° 30'
37° 15'

TENNESSEE

OKLAHOMA

UNITED STATES OF AMERICA

ARKANSAS

MISSISSIPPI

ALABAMA

TEXAS

LOUISIANA

GULF OF MEXICO

UNITED STATES OF AMERICA

San Diego
Tijuana
Ensenada
Cerro. de la Encantada 3096
San Felipe

Glendale
Phoenix
Mexicali
Tucson
Pto Peñasco
Nogales

Gila
Lordsburg
Las Cruces
El Paso
Ciudad Juárez

Clovis
Lubbock
Big Spring
Sweetwater
Midland
Odessa
Pecos
Alpine

Oklahoma City
Wichita Falls
Abilene
Fort Worth
Dallas
Waco
Temple
Austin

Fort Smith
Little Rock
Pine Bluff
Texarkana
Shreveport
Monroe
Alexandria
Lufkin
Huntsville
Beaumont

Memphis
Greenville
Jackson
Meridian
Hattiesburg
Baton Rouge
Lafayette
Port Arthur

Huntsville
Tupelo
Birmingham
Tuscaloosa
Montgomery
Mobile
Biloxi
Pensacola

BAJA CALIFORNIA NORTE
Sonora
Hermosillo
Ciudad Obregón
Guaymas
Sta Rosalía

ARIZONA
NEW MEXICO
Silver City
Magdalena
Angel de la Guarda
Tiburón

Chihuahua
Ciudad Delicias
Ciudad Camargo
Hidalgo del Parral

CHIHUAHUA
COAHUILA
Monclova
Saltillo
Gómez Palacio
Torreón

Emory Peak 2389
Serranías del Burro
Piedras Negras
Sabinas
Laredo
Nuevo Laredo
Falcon Lake

San Antonio
Del Rio
Eagle Pass
Victoria
Beeville
Kingsville
Corpus Christi

Houston
Galveston
Galveston Bay
Port Arthur

New Orleans
Chandeleur Islands
Morgan City
Mobile Bay
Mississippi Delta

BAJA CALIFORNIA SUR
B. Magdalena
Sebastián Vizcaíno B.
Sta Vizcaíno
San José
Espíritu Santo
La Paz
Cerralvo
C. San Lucas

Gulf of California
Lower California
Los Mochis
Culiacán
Durango
Mazatlán

SINALOA
DURANGO
ZACATECAS
Zacatecas
Aguascalientes

Reynosa
Matamoros
NUEVO LEÓN
Monterrey
Montemorelos
Madre Lagoon
C. Peña Nevada 3644
Ciudad Victoria

GULF OF MEXICO

Tropic of Cancer

MEXICO

Sierra Madre Occidental
Sierra Madre Oriental

NAYARIT
Tepic
I. Marías

SAN LUIS POTOSÍ
San Luis Potosí
Ciudad Madero
Tampico
Tamiahua Lagoon

León
GUANAJUATO
Guanajuato
Irapuato
Celaya
QUERÉTARO
Querétaro
HIDALGO
Pachuca
Poza Rica

Banderas Bay
C. Corrientes
Guadalajara
JALISCO
L. de Chapala

Colima 3839
COLIMA
Uruapán
MICHOACÁN
Morelia
Toluca
MEXICO CITY
MEXICO
Cuernavaca
MORELOS
Popocatépetl 5452
Puebla
PUEBLA
Tlaxcala
Jalapa
Córdoba
Orizaba C.
Veracruz

Campeche Bay

Mérida
YUCATÁN
Campeche
Cancún
C. Catoche
Cozumel I.

PACIFIC OCEAN

Revillagigedo Is (Mexico)
I. San Benedicto
I. Socorro

L. Infiernillo
Balsas
GUERRERO
Chilpancingo
Acapulco
Sierra Madre del Sur
OAXACA
Oaxaca

R Coatzacoalcos
Minatitlán
TABASCO
Villahermosa
Coatzacoalcos

Yucatán
QUINTANA ROO
Banco Chinchorro

CAMPECHE
Hondo
Ambergris Cay
Belize Cay
BELMOPAN
BELIZE
Turneffe Is

Gulf of Tehuantepec
Cd Ixtepec
Juchitán
CHIAPAS
Tuxtla Gutiérrez
Tapachula

Maya Mts
L. Izabal 4210
GUATEMALA
Quezaltenango
GUATEMALA CITY
Sipacate
Santa Ana
San Salvador
EL SALVADOR
San Miguel

Gulf of Honduras
Bay Is.
San Pedro Sula
HOND
TEGUCIGALPA

G. of Fonseca
L. Managua
MANAGUA
C. Sta Elena

Mexican States numbered on map
1. AGUASCALIENTES
2. DISTRITO FEDERAL
3. TLAXCALA

KEY

Relief and physical features

Relief metres
5000
3000
2000
1000
500
200
sea level
under sea level
200
4000
6000

▲ 5775   Mountain height
(in metres)

Water features

〰 River
Intermittent river
Lake / Reservoir
Intermittent lake
Marsh

Communications

Railway
Road
⊕ Main airport

Administration

Boundaries
International
Internal

Settlement
Cities and towns in order of size

National capital   Other city or town
■ HAVANA   ● Puebla
□ BELMOPAN   ◎ El Paso
□ CASTRIES   ○ Acapulco
                        ○ Guanajuato

SCALE 1 : 13 000 000

0   200   400   600   800 km

ATLANTIC

OCEAN

Bermuda
(UK) • Hamilton

CARIBBEAN SEA

Leeward Islands

Lesser Antilles

Lesser Antilles

Windwards Is

HAVANA
Pinar del Rio
Santa Clara
CUBA
Matanzas
Arch. de Sabana
Arch. de Camagüey
G. of Batabanó
Cienfuegos
Arch. de los Canarreos
Ciego de Ávila
Camagüey
Victoria de las Tunas
Holguín
C. San Antonio
Isle of Pines
Jardines de la Riena
Bayamo
Guantánamo
Sa Maestra
2005 Turquino Santiago de Cuba

THE BAHAMAS
Grand Bahama
Freeport City
Great Abaco
Bimini Is
New Providence
NASSAU
Andros
Exuma Is
Exuma Sd
Eleuthera
Cat I.
San Salvador
Rum Cay
Long I.
Great Exuma
Crooked I. Pass.
Crooked I.
Mayaguana
Acklins I.
Caicos Is
Little Inagua
Great Inagua

Turks and Caicos Islands (UK)
Cockburn Town
Turks Is

Hispaniola
Port-de-Paix
Cap Haïtien
Santiago
Gonaïves
HAITI
3175 Pico Duarte
SANTO DOMINGO
DOMINICAN REPUBLIC
PORT-AU-PRINCE
2680
Jérémie
La Selle
La Romana
Les Cayes
Jacmel
C. Beata
Beata I.

Windward Passage

SAN JUAN
Virgin Is (UK)
Anegada
Anguilla (UK)
Saint Martin (Fr.)
St Barthélémy (Fr.)
Virgin Is (USA)
Sint Maarten (Neth.)
Mayagüez
Mona Passage
Mona
Ponce
PUERTO RICO (USA)
Barbuda
ANTIGUA & BARBUDA
ST JOHN'S
Antigua
ST KITTS-NEVIS
Montserrat (UK)
Guadeloupe (Fr.)
Pointe-à-Pitre
Marie Galante (Fr.)

DOMINICA
ROSEAU
Martinique (Fr.)
Fort-de-France
CASTRIES
ST LUCIA
BARBADOS
Kingstown
BRIDGETOWN
ST VINCENT & THE GRENADINES
The Grenadines
GRENADA
ST GEORGE'S

Little Cayman
Cayman Brac
Cayman Is (UK)
Grand Cayman
Montego Bay
JAMAICA
KINGSTON

Swan Is (Honduras)

C A R I B B E A N

G R E A T E R   A N T I L L E S

Straits of Florida
Florida Keys
West Palm Beach

NORTH CAROLINA
Knoxville
▲Mt Mitchell 2037
Raleigh
Chattanooga
Charlotte
Fayetteville
New Bern
Cape Hatteras
Greenville
SOUTH CAROLINA
Columbia
Lumberton
Cape Lookout
Gadsden
Wilmington
Cape Fear
Atlanta
Augusta
Orangeburg
GEORGIA
Macon
Charleston
Cape Romain
Columbus
Savannah
Albany
Jesup
Dothan
Brunswick
Marianna
Valdosta
Okefenokee Swamp
Tallahassee
Lake City
Jacksonville
St Augustine
C. San Blas
Apalachee Bay
Gainesville
Daytona Beach
Panama City
FLORIDA
Orlando
Lakeland
Cape Canaveral
Tampa-St Petersburg
Sarasota
Lake Okeechobee
The Everglades
Naples
Miami-Fort Lauderdale
Cape Sable

Channel
Cayos Miskito
I. de Providencia (Colombia)
I. de San Andrés (Colombia)
Is del Maíz (Nic.)
Perlas Pt

URAS
NICARAGUA
L. Nicaragua
Cord. Isabelia
Río Grande
Coco
Caratasca Lagoon
Mosquitos Coast

COSTA RICA
SAN JOSÉ
Chirripó
3820
Baru
▲3475
David
Gulf of Nicoya
Coronada Bay
Osa Pen.
Gulf of Chiriquí
Azuero Pen.
I. Coiba
Mariato Pt

P A N A M A
Panama Canal
Colón
PANAMA CITY
Gulf of Mosquitos
Gulf of Darién
Gulf of Panamá
Mala Pt

Gallinas Pt
Guajira Pen.
Los Taques
Aruba (Neth.)
Netherlands Antilles
Curaçao
Bonaire
Orchila (Ven.)
Blanquilla (Ven.)
Margarita
Los Testigos
Tobago
TRINIDAD & TOBAGO

Ríohacha
Maicao
Santa Marta
Barranquilla
Cristóbal Colón
5775
Cartagena
Valledupar
Sincelejo
Montería
Gulf of Morrosquillo
Cauca
Sa. de Perijá

G. of Venezuela
Maracaibo
Cabimas
L. Maracaibo
Valera
Coro
Los Teques
CARACAS
Valencia
Maracay
Barquisimeto
Acarigua
Guanare
Barinas
Cord. de Mérida
Mérida
San Fernando de Apure

Paria Pen.
Cumaná
Güiria
PORT OF SPAIN
Barcelona
Maturín
Trinidad
La Tortuga
Los Roques (Ven.)
Maiquetía
C. Cadera
Guanipa
Tigre
Zaraza
Orinoco Delta
Ciudad Bolívar
Ciudad Guayana
Orinoco

COLOMBIA
Bucaramanga
Sierra Nevada del Cocuy
5493
Cúcuta
San Cristóbal
Cord. Oriental
Medellín
Bello
Quibdó
Tunja
Meta
Meta

V E N E Z U E L A
Cerro Yavi
2283
La Gran Sabana
2810
Mt Roraima

Gulf of Cupica

SCALE 1 : 80 000 000

SCALE 1 : 35 000 000

Lambert Azimuthal Equal Area projection

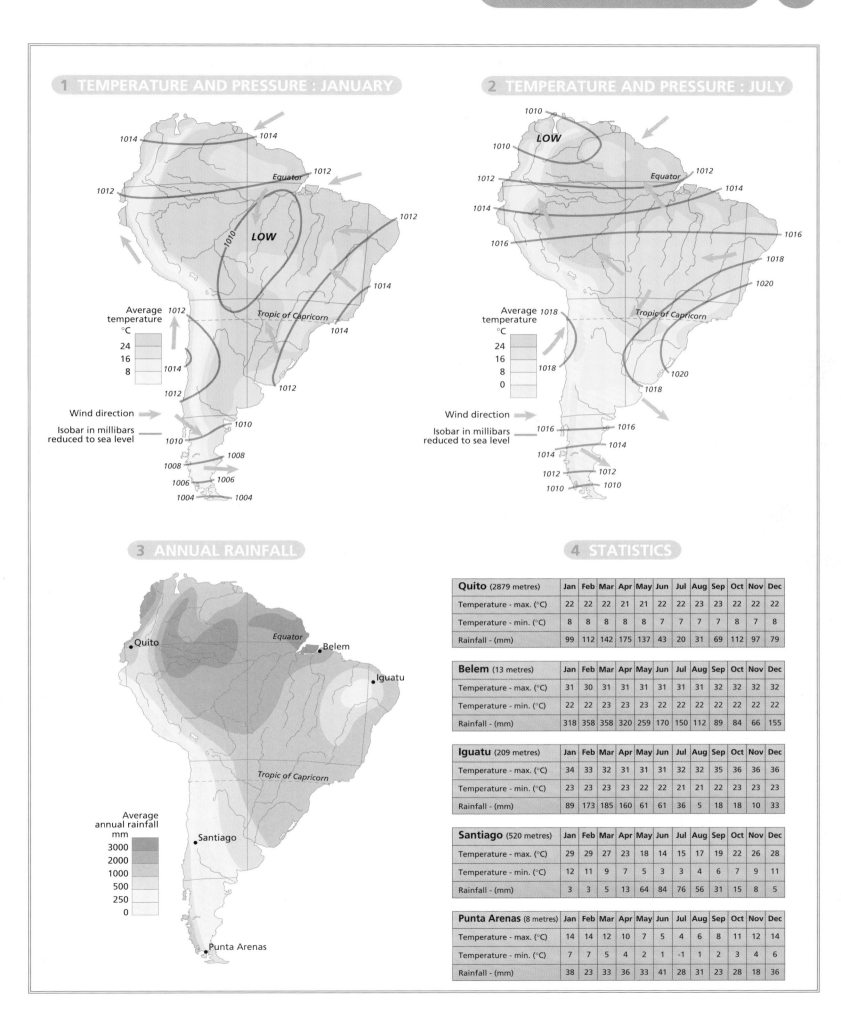

1 TEMPERATURE AND PRESSURE : JANUARY

1014
1014
1012
Equator
1012
1012
1010
LOW
1012
1014
Tropic of Capricorn
1014

Average temperature °C
24
16
8

1012
1014
1012
1014
1012

Wind direction
Isobar in millibars reduced to sea level

1010
1010
1008
1008
1006
1006
1004
1004

2 TEMPERATURE AND PRESSURE : JULY

1010
LOW
1010
1012
Equator
1012
1014
1014
1016
1016
1018
1020

Average temperature °C
24
16
8
0

1018
1018
1020
1018

Wind direction
Isobar in millibars reduced to sea level

1016
1016
1014
1014
1012
1012
1010
1010

3 ANNUAL RAINFALL

Equator
Quito
Belem
Iguatu
Tropic of Capricorn

Average annual rainfall mm
3000
2000
1000
500
250
0

Santiago
Punta Arenas

4 STATISTICS

Quito (2879 metres)	Jan	Feb	Mar	Apr	May	Jun	Jul	Aug	Sep	Oct	Nov	Dec
Temperature - max. (°C)	22	22	22	21	21	22	22	23	23	22	22	22
Temperature - min. (°C)	8	8	8	8	8	7	7	7	7	8	7	8
Rainfall - (mm)	99	112	142	175	137	43	20	31	69	112	97	79

Belem (13 metres)	Jan	Feb	Mar	Apr	May	Jun	Jul	Aug	Sep	Oct	Nov	Dec
Temperature - max. (°C)	31	30	31	31	31	31	31	31	32	32	32	32
Temperature - min. (°C)	22	22	23	23	23	22	22	22	22	22	22	22
Rainfall - (mm)	318	358	358	320	259	170	150	112	89	84	66	155

Iguatu (209 metres)	Jan	Feb	Mar	Apr	May	Jun	Jul	Aug	Sep	Oct	Nov	Dec
Temperature - max. (°C)	34	33	32	31	31	31	32	32	35	36	36	36
Temperature - min. (°C)	23	23	23	23	22	22	21	21	22	23	23	23
Rainfall - (mm)	89	173	185	160	61	61	36	5	18	18	10	33

Santiago (520 metres)	Jan	Feb	Mar	Apr	May	Jun	Jul	Aug	Sep	Oct	Nov	Dec
Temperature - max. (°C)	29	29	27	23	18	14	15	17	19	22	26	28
Temperature - min. (°C)	12	11	9	7	5	3	3	4	6	7	9	11
Rainfall - (mm)	3	3	5	13	64	84	76	56	31	15	8	5

Punta Arenas (8 metres)	Jan	Feb	Mar	Apr	May	Jun	Jul	Aug	Sep	Oct	Nov	Dec
Temperature - max. (°C)	14	14	12	10	7	5	4	6	8	11	12	14
Temperature - min. (°C)	7	7	5	4	2	1	-1	1	2	3	4	6
Rainfall - (mm)	38	23	33	36	33	41	28	31	23	28	18	36

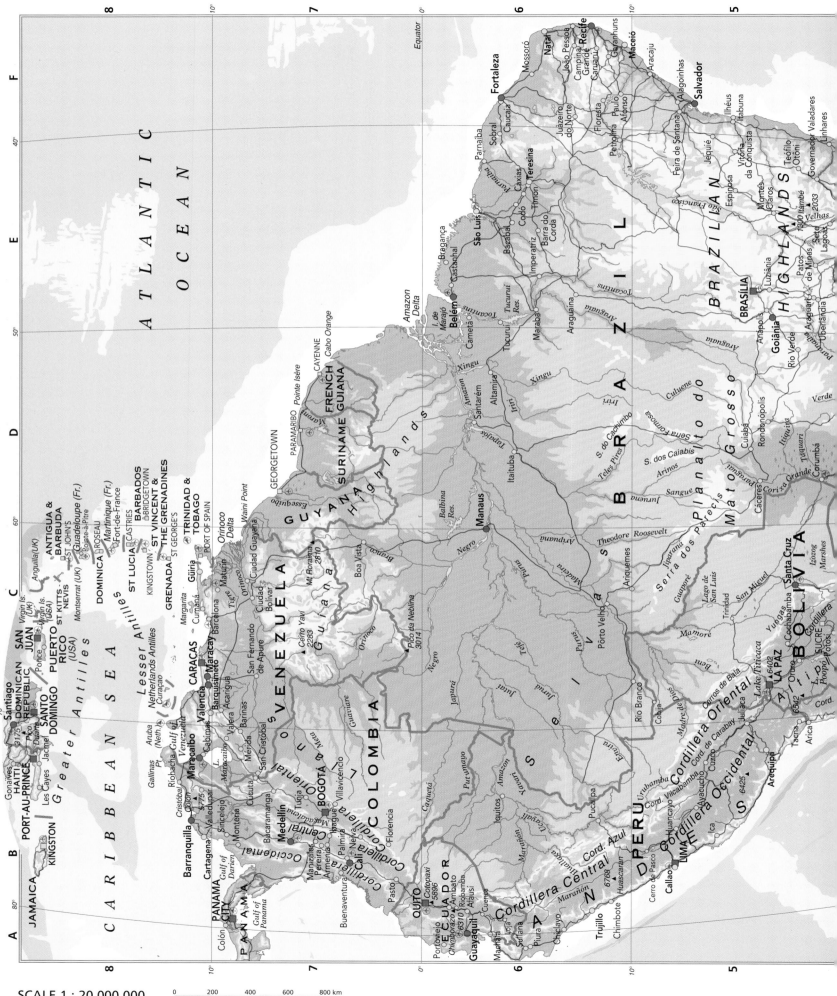

SCALE 1 : 20 000 000

0   200   400   600   800 km

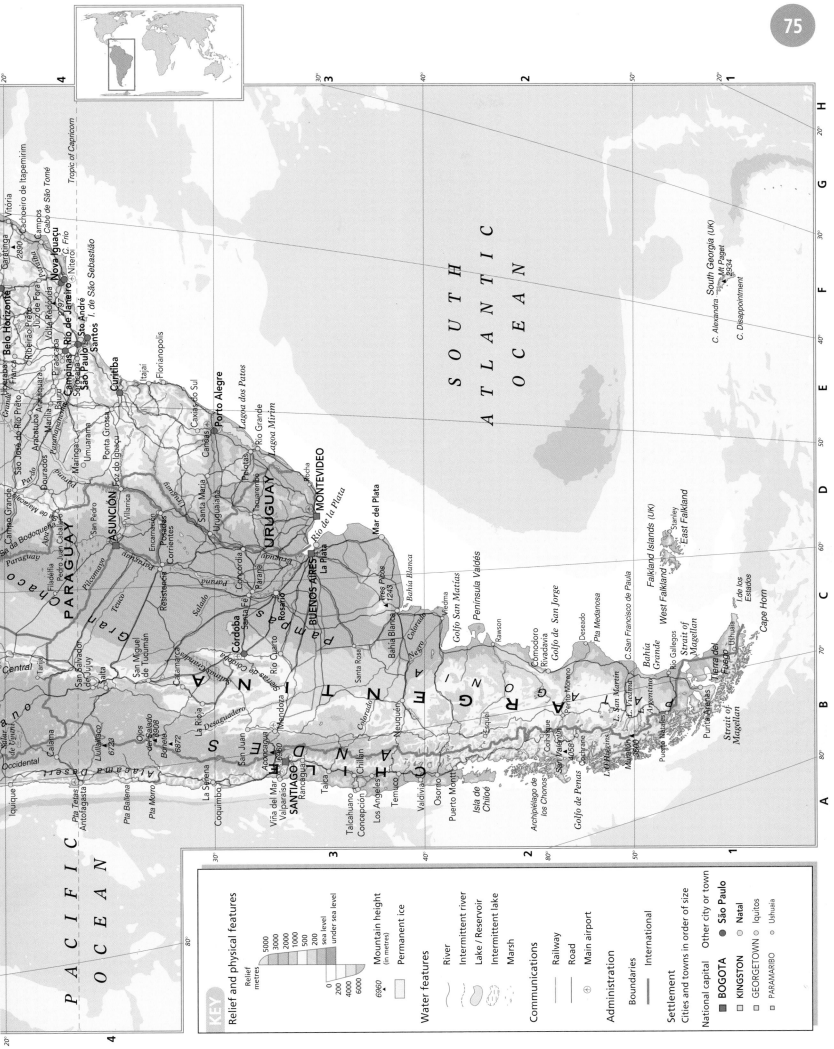

Sinusoidal projection

PACIFIC OCEAN

SOUTH ATLANTIC OCEAN

KEY

Relief and physical features

Relief
metres
5000
3000
2000
1000
500
200
sea level
under sea level

0
200
4000
6000

6960 ▲ Mountain height
(in metres)

Permanent ice

Water features

River

Intermittent river

Lake / Reservoir

Intermittent lake

Marsh

Communications

Railway

Road

⊕ Main airport

Administration

Boundaries

International

Settlement

Cities and towns in order of size

National capital   Other city or town

■ **BOGOTA**   ● São Paulo

□ KINGSTON   ○ Natal

□ GEORGETOWN   ○ Iquitos

□ PARAMARIBO   ○ Ushuaia

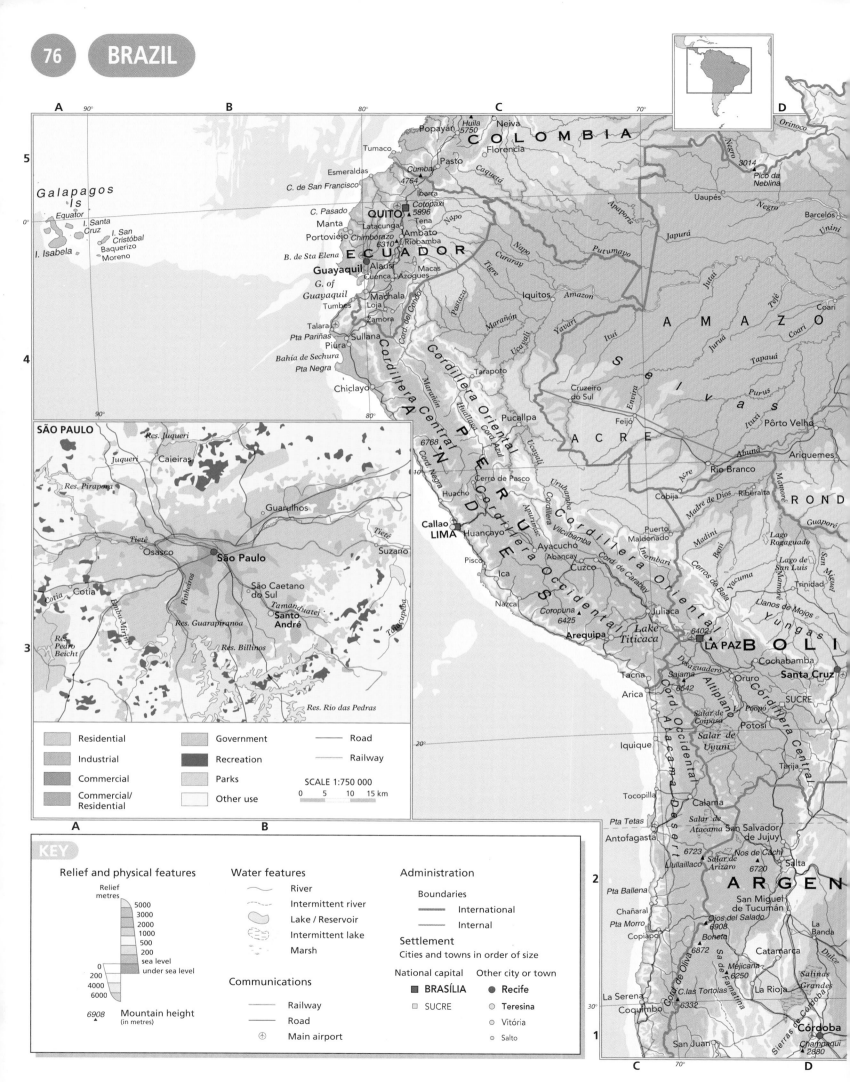

KEY

Relief and physical features

Relief
metres
5000
3000
2000
1000
500
200
0 sea level
     under sea level
200
4000
6000

6908 ▲ Mountain height
(in metres)

Water features

~~~ River
- - - Intermittent river
Lake / Reservoir
Intermittent lake
Marsh

Communications

Railway
Road
⊕ Main airport

Administration

Boundaries
International
Internal

Settlement
Cities and towns in order of size

National capital
■ BRASÍLIA
□ SUCRE

Other city or town
● Recife
◉ Teresina
◎ Vitória
○ Salto

SÃO PAULO inset key

Residential	Government	Road
Industrial	Recreation	Railway
Commercial	Parks	
Commercial/ Residential	Other use	

SCALE 1:750 000
0   5   10   15 km

SCALE 1 : 15 000 000
0   150   300   450   600 km

Lambert Azimuthal Equal Area projection

1 POPULATION DENSITY

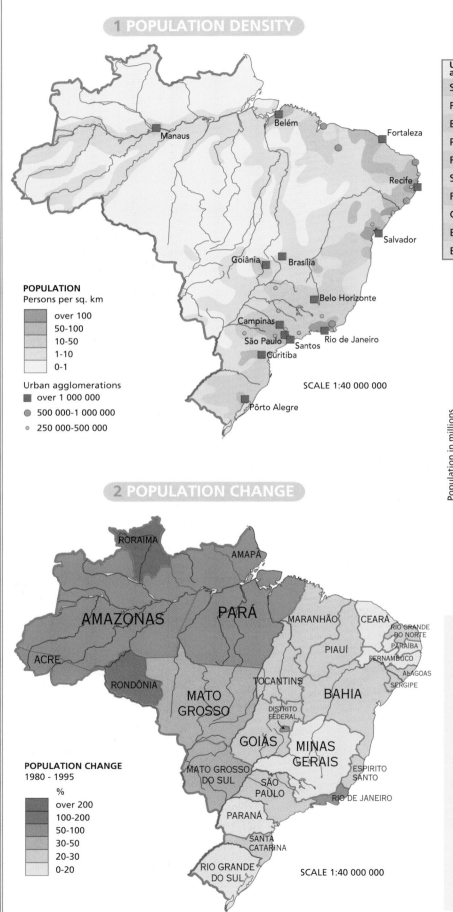

POPULATION
Persons per sq. km

- over 100
- 50-100
- 10-50
- 1-10
- 0-1

Urban agglomerations

- ■ over 1 000 000
- ● 500 000-1 000 000
- ○ 250 000-500 000

SCALE 1:40 000 000

2 POPULATION CHANGE

POPULATION CHANGE
1980 - 1995

%

- over 200
- 100-200
- 50-100
- 30-50
- 20-30
- 0-20

SCALE 1:40 000 000

3 MAIN URBAN AGGLOMERATIONS

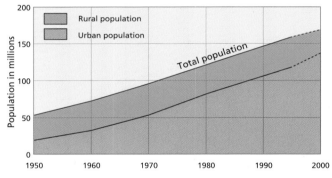

Urban agglomeration	1980	1995	% change
São Paulo	12 497 000	16 417 000	31.4
Rio de Janeiro	8 741 000	9 888 000	13.1
Belo Horizonte	2 588 000	3 899 000	50.7
Pôrto Alegre	2 273 000	3 349 000	47.3
Recife	2 337 000	3 168 000	35.6
Salvador	1 754 000	2 819 000	60.7
Fortaleza	1 569 000	2 660 000	69.5
Curitiba	1 427 000	2 270 000	59.1
Brasília	1 162 000	1 778 000	53.0
Belém	992 000	1 574 000	58.7

4 POPULATION GROWTH

- Rural population
- Urban population

Total population

Population in millions

5 MIGRATION

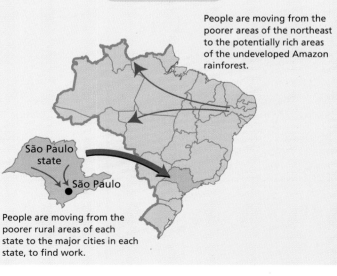

People are moving from the poorer areas of the northeast to the potentially rich areas of the undeveloped Amazon rainforest.

São Paulo state

São Paulo

People are moving from the poorer rural areas of each state to the major cities in each state, to find work.

6 REGIONAL COMPARISONS

São Luís
Fortaleza
Teresina
Natal
NORTH EAST
Recife
Maceio
Salvador
SOUTH EAST
Belo Horizonte
Campinas
São Paulo
Rio de Janeiro
Santos

North East		South East
44 768 201	Population 1996	67 003 069
60.4	Infant mortality rate (per 1000 population)	25.8
64.5	Life expectancy (years)	68.8
86.4%	School enrolment	94.1%
28.7%	Illiteracy rate (population aged 15 and over)	8.7%
56.2%	Access to safe water	86.5%
81.7%	Access to electricity	97.8%

7 ECONOMIC ACTIVITY

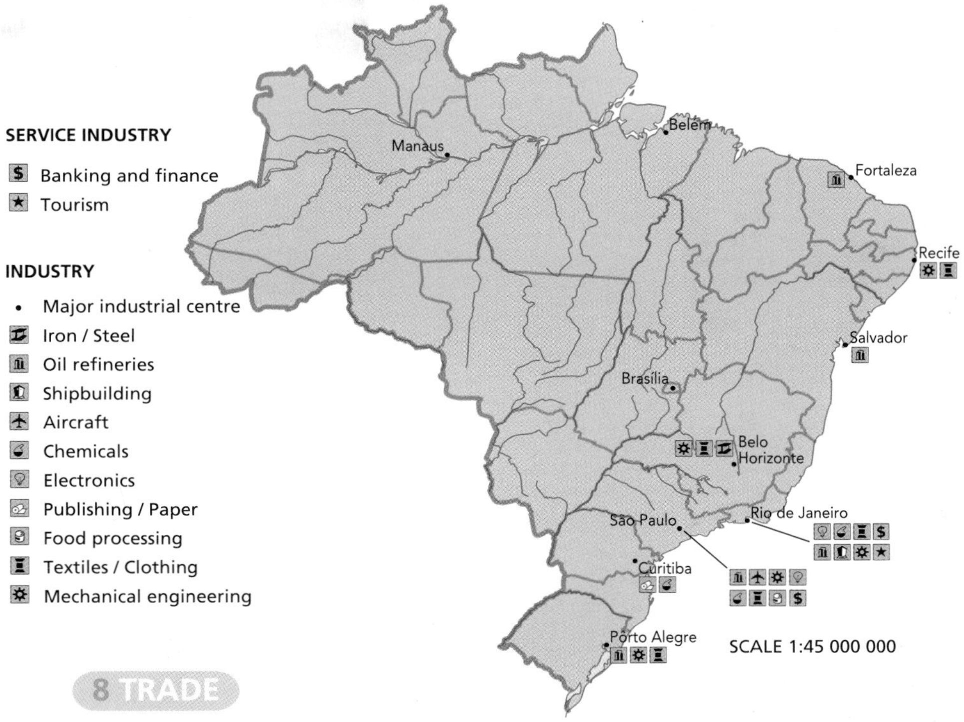
SERVICE INDUSTRY
Banking and finance
Tourism
INDUSTRY
Major industrial centre
Iron / Steel
Oil refineries
Shipbuilding
Aircraft
Chemicals
Electronics
Publishing / Paper
Food processing
Textiles / Clothing
Mechanical engineering
Manaus
Belém
Fortaleza
Recife
Salvador
Brasília
Belo Horizonte
Rio de Janeiro
São Paulo
Curitiba
Porto Alegre
SCALE 1:45 000 000

8 TRADE

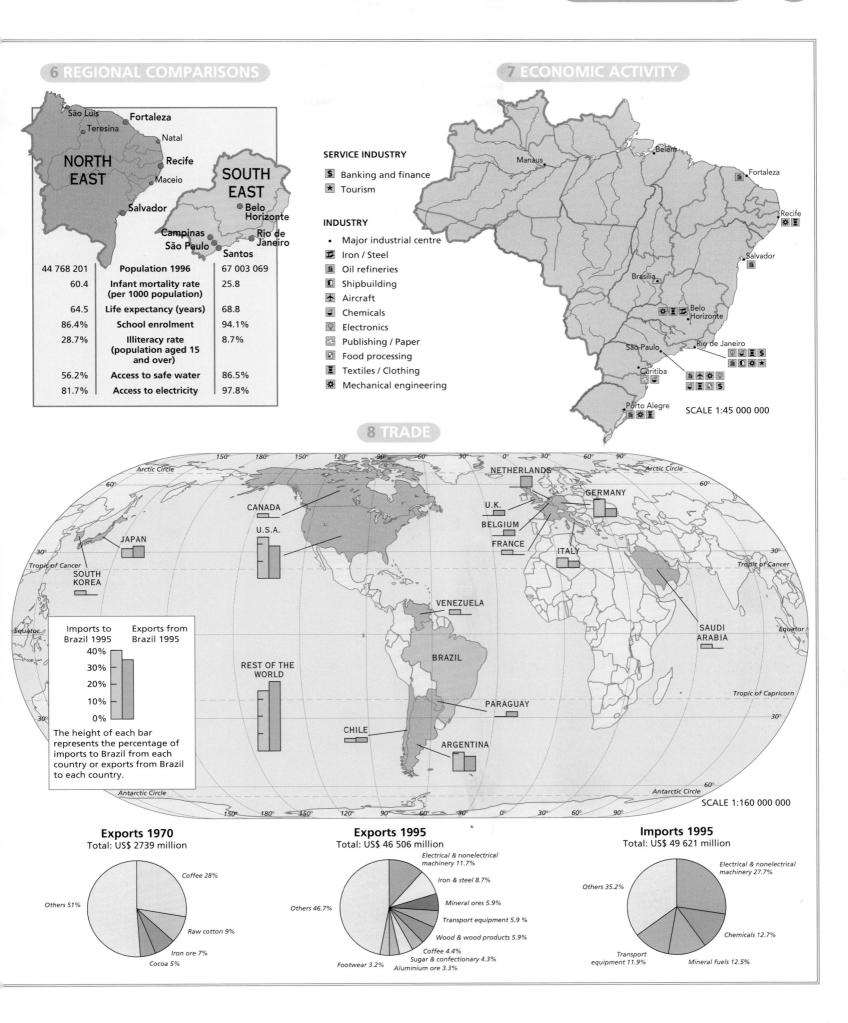

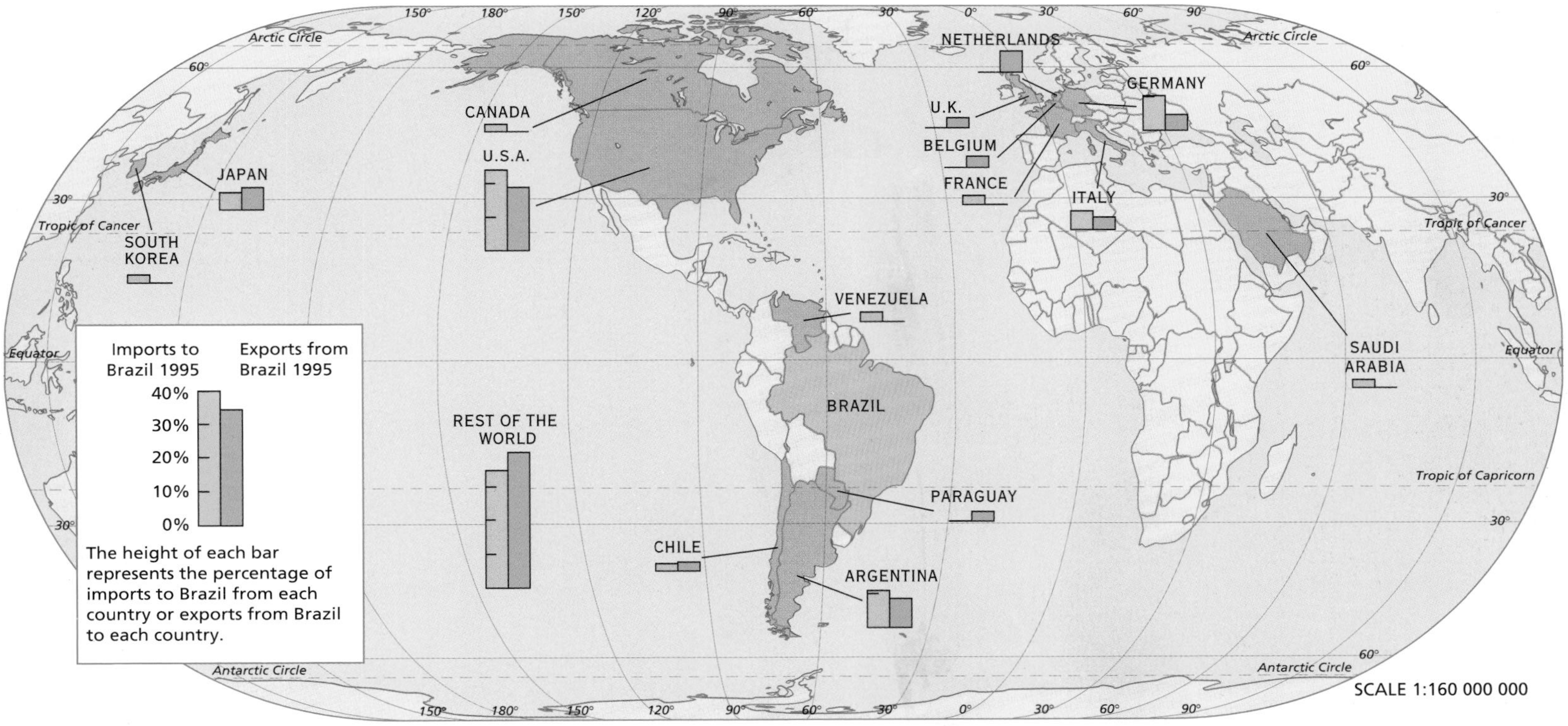
Arctic Circle
Tropic of Cancer
Equator
Tropic of Capricorn
Antarctic Circle
NETHERLANDS
GERMANY
U.K.
BELGIUM
FRANCE
ITALY
CANADA
U.S.A.
JAPAN
SOUTH KOREA
VENEZUELA
SAUDI ARABIA
BRAZIL
REST OF THE WORLD
PARAGUAY
CHILE
ARGENTINA
Imports to Brazil 1995
Exports from Brazil 1995
40%
30%
20%
10%
0%
The height of each bar represents the percentage of imports to Brazil from each country or exports from Brazil to each country.
SCALE 1:160 000 000

Exports 1970
Total: US$ 2739 million

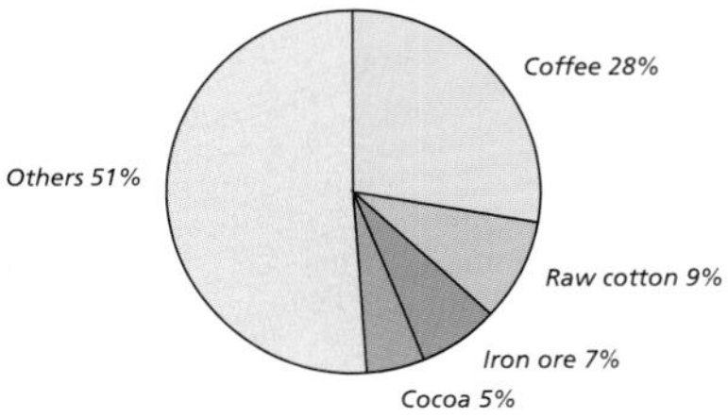
Coffee 28%
Raw cotton 9%
Iron ore 7%
Cocoa 5%
Others 51%

Exports 1995
Total: US$ 46 506 million

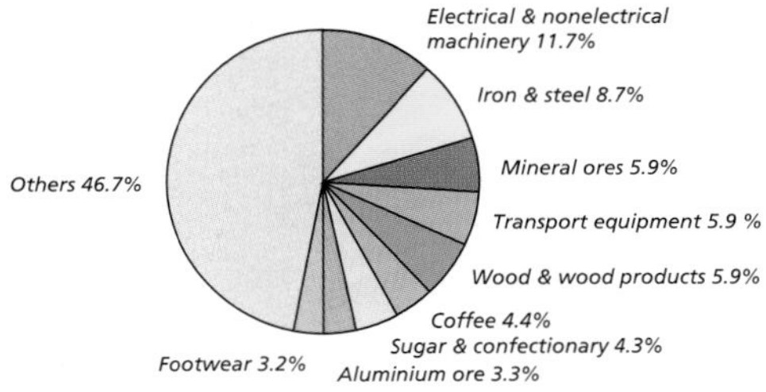
Electrical & nonelectrical machinery 11.7%
Iron & steel 8.7%
Mineral ores 5.9%
Transport equipment 5.9 %
Wood & wood products 5.9%
Coffee 4.4%
Sugar & confectionary 4.3%
Aluminium ore 3.3%
Footwear 3.2%
Others 46.7%

Imports 1995
Total: US$ 49 621 million

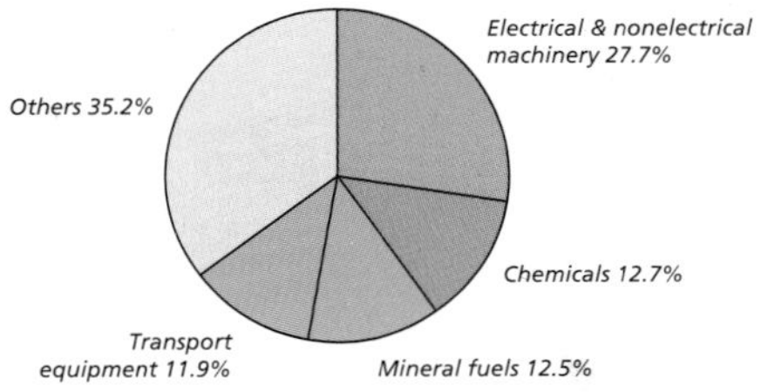
Electrical & nonelectrical machinery 27.7%
Chemicals 12.7%
Mineral fuels 12.5%
Transport equipment 11.9%
Others 35.2%

Forest
Dense forest covers much of this area and the courses of the many tributaries of the river Guaporé can be followed cutting through the forest areas.

Marshy Savanna
An area of marshy savanna lies between the forest and the river Guaporé. Similar areas can also be seen south of the river around Laguna Bella Vista.

Deforested Areas
Large rectangular areas of pale blue on the satellite image are areas of deforestation, probably from commercial logging. In the bottom right of the image the pale blue line patterns are systematic deforestation due to the practice of slash and burn farming.

Highland
The highland of the Serra dos Parecis can be seen at the top right of the image.

Lakes
Several small dark blue/black outlines of lakes can be seen along the course of the river Guaporé. Laguna Bella Vista stands out clearly as a much larger feature.

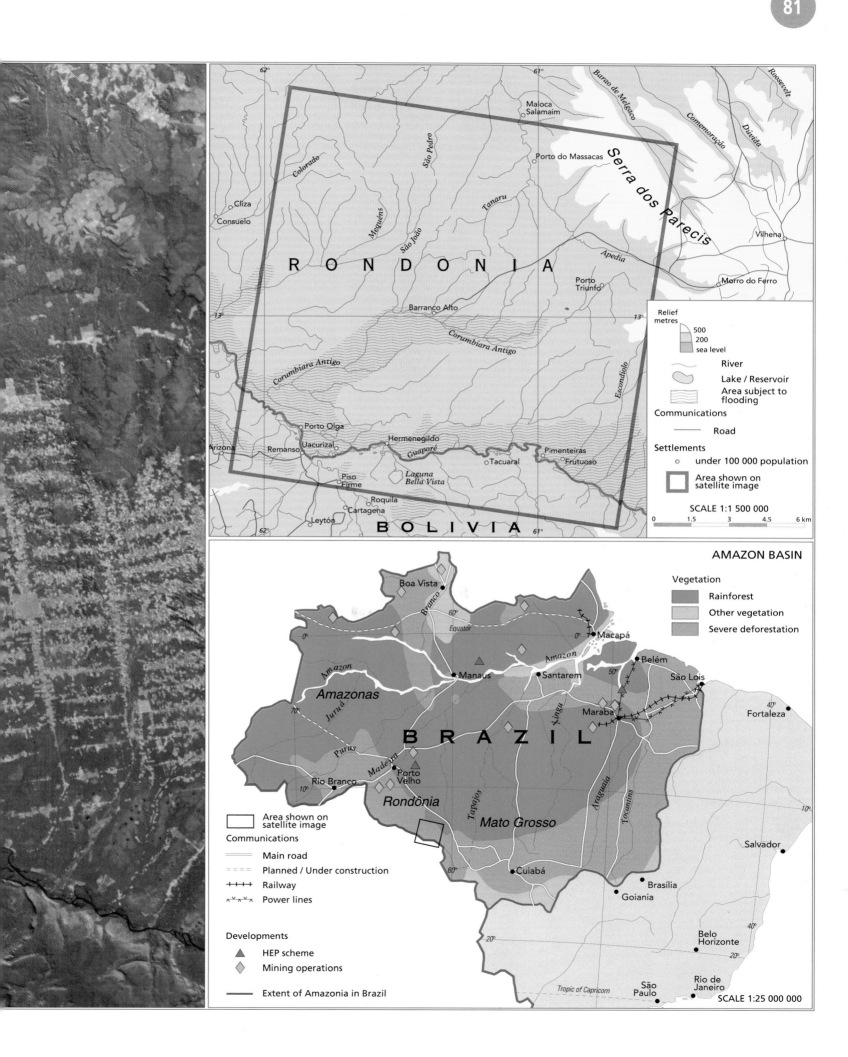

SCALE 1:1 500 000

Relief
metres

500
200
sea level

River

Lake / Reservoir

Area subject to
flooding

Communications

Road

Settlements

○ under 100 000 population

□ Area shown on
satellite image

0   1.5   3   4.5   6 km

AMAZON BASIN

Vegetation

Rainforest

Other vegetation

Severe deforestation

□ Area shown on
satellite image

Communications

Main road

Planned / Under construction

Railway

Power lines

Developments

▲ HEP scheme

◆ Mining operations

— Extent of Amazonia in Brazil

SCALE 1:25 000 000

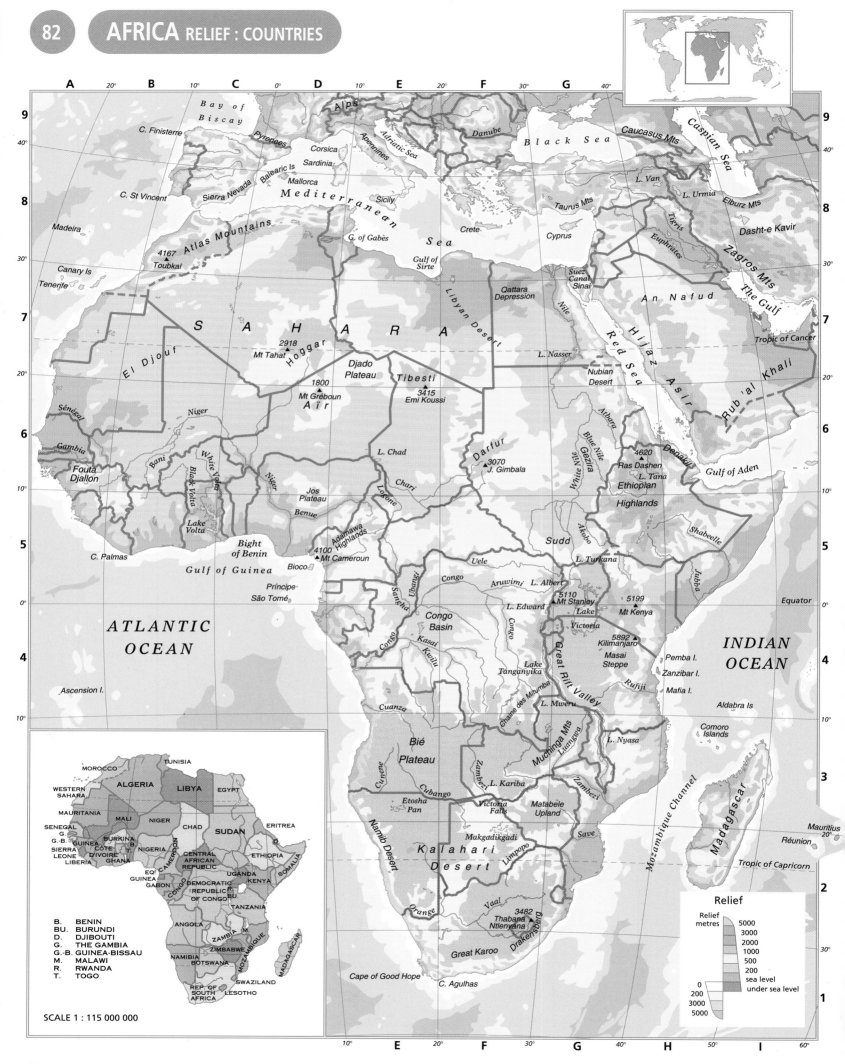

A   20°   B   10°   C   0°   D   10°   E   20°   F   30°   G   40°

Bay of Biscay

Alps

C. Finisterre

Pyrenees

Corsica

Apennines

Adriatic Sea

Sardinia

Danube

Black Sea

Caucasus Mts

Caspian Sea

C. St Vincent

Sierra Nevada

Balearic Is

Mallorca

Sicily

M e d i t e r r a n e a n

Crete

Taurus Mts

L. Van

L. Urmia

Elburz Mts

Madeira

Atlas Mountains

4167 ▲ Toubkal

G. of Gabès

S e a

Cyprus

Tigris

Euphrates

Dasht-e Kavir

Zagros Mts

Canary Is

Tenerife

Gulf of Sirte

Libyan Desert

Qattara Depression

Suez Canal
Sinai

An Nafud

The Gulf

Tropic of Cancer

S   A   H   A   R   A

El Djouf

2918
Mt Tahat ▲   Hoggar

Djado Plateau

L. Nasser

Nubian Desert

Red Sea

Hijaz Asir

Rub 'al Khali

Sénégal

1800
▲ Mt Gréboun
Aïr

Tibesti

3415
▲ Emi Koussi

Niger

Darfur

3070
▲ J. Gimbala

Athara

Blue Nile

Gezira

4620
▲ Ras Dashen
L. Tana
Ethiopian
Highlands

Denakil

Gambia

Fouta Djallon

Bani

White Volta

Black Volta

Niger

L. Chad

Chari

Jos Plateau

Benue

Logone

White Nile

Akobo

Shabeelle

Gulf of Aden

C. Palmas

Lake Volta

Bight of Benin

Gulf of Guinea

Adamawa Highlands

4100
Mt Cameroun ▲

Bioco

Sudd

Uele

L. Turkana

Jubba

Príncipe

São Tomé

Sangha

Ubangi

Congo

Aruwimi   L. Albert

5110
▲ Mt Stanley
L. Edward

5199
▲ Mt Kenya

Equator

ATLANTIC OCEAN

Congo Basin

Kasai

Kwilu

Congo

Lake
Victoria

5892
▲ Kilimanjaro
Masai Steppe

Pemba I.

Zanzibar I.

Mafia I.

INDIAN OCEAN

Ascension I.

Cuanza

Lake Tanganyika

Great Rift Valley

Rufiji

Aldabra Is

Chaîne des Mitumba

L. Mweru

Comoro Islands

Bié Plateau

Muchinga Mts

Luangwa

L. Nyasa

Cunene

Cubango

Zambezi

L. Kariba

Zambezi

Madagascar

Mozambique Channel

Mauritius

Réunion

Etosha Pan

Victoria Falls

Matabèle Upland

Save

Tropic of Capricorn

Namib Desert

Makgadikgadi

K a l a h a r i

D e s e r t

Limpopo

Orange

Vaal

3482
Thabana ▲
Ntlenyana   Drakensberg

Great Karoo

Cape of Good Hope

C. Agulhas

10°   E   20°   F   30°   G   40°   H   50°   I   60°

Inset map – Africa Countries:

MOROCCO

WESTERN SAHARA

TUNISIA

ALGERIA   LIBYA   EGYPT

MAURITANIA

MALI   NIGER   CHAD   SUDAN   ERITREA

SENEGAL
G.
G.-B.   GUINEA
SIERRA LEONE
LIBERIA   CÔTE D'IVOIRE   GHANA   BURKINA
B.   NIGERIA   CENTRAL AFRICAN REPUBLIC   ETHIOPIA

EQ. GUINEA
CONGO   CAMEROON
GABON   DEMOCRATIC REPUBLIC OF CONGO   UGANDA   KENYA   SOMALIA

R.
BU.   TANZANIA

ANGOLA   ZAMBIA   M.
ZIMBABWE   MOZAMBIQUE   MADAGASCAR
NAMIBIA   BOTSWANA
SWAZILAND
REP. OF SOUTH AFRICA   LESOTHO

B.   BENIN
BU.   BURUNDI
D.   DJIBOUTI
G.   THE GAMBIA
G.-B.   GUINEA-BISSAU
M.   MALAWI
R.   RWANDA
T.   TOGO

SCALE 1 : 115 000 000

Lambert Azimuthal Equal Area projection

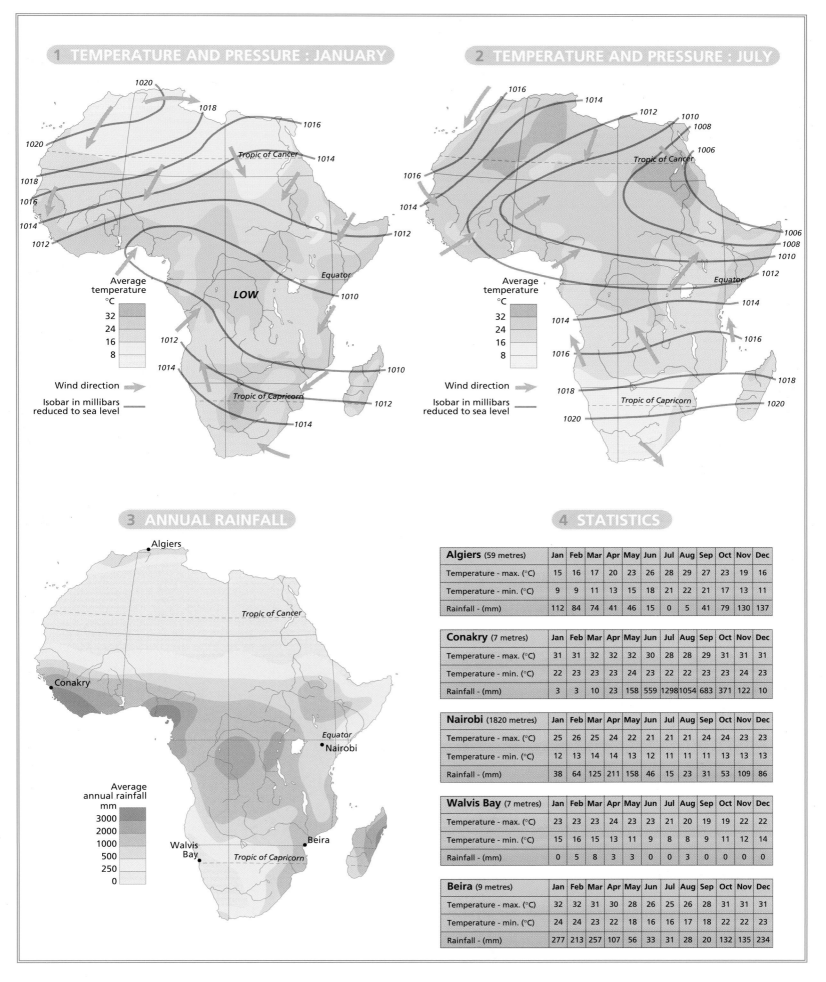

1 TEMPERATURE AND PRESSURE : JANUARY

1020
1018
1016
1020
Tropic of Cancer   1014
1018
1016
1014
1012

Average
temperature
°C
32
24
16
8

LOW

Equator   1010

1012
1014
1010
Tropic of Capricorn   1012
1014

Wind direction
Isobar in millibars
reduced to sea level

2 TEMPERATURE AND PRESSURE : JULY

1016
1014
1012   1010
1008
1006
Tropic of Cancer
1016
1014
1006
1008
1010
1012
Equator
1014
1014
1016
1016
1018
Tropic of Capricorn   1018
1020
1020

Average
temperature
°C
32
24
16
8

Wind direction
Isobar in millibars
reduced to sea level

3 ANNUAL RAINFALL

Algiers
Tropic of Cancer
Conakry
Equator
Nairobi
Average
annual rainfall
mm
3000
2000
1000
500
250
0
Walvis
Bay
Beira
Tropic of Capricorn

4 STATISTICS

Algiers (59 metres)	Jan	Feb	Mar	Apr	May	Jun	Jul	Aug	Sep	Oct	Nov	Dec
Temperature - max. (°C)	15	16	17	20	23	26	28	29	27	23	19	16
Temperature - min. (°C)	9	9	11	13	15	18	21	22	21	17	13	11
Rainfall - (mm)	112	84	74	41	46	15	0	5	41	79	130	137

Conakry (7 metres)	Jan	Feb	Mar	Apr	May	Jun	Jul	Aug	Sep	Oct	Nov	Dec
Temperature - max. (°C)	31	31	32	32	32	30	28	28	29	31	31	31
Temperature - min. (°C)	22	23	23	23	24	23	22	22	23	23	24	23
Rainfall - (mm)	3	3	10	23	158	559	1298	1054	683	371	122	10

Nairobi (1820 metres)	Jan	Feb	Mar	Apr	May	Jun	Jul	Aug	Sep	Oct	Nov	Dec
Temperature - max. (°C)	25	26	25	24	22	21	21	21	24	24	23	23
Temperature - min. (°C)	12	13	14	14	13	12	11	11	11	13	13	13
Rainfall - (mm)	38	64	125	211	158	46	15	23	31	53	109	86

Walvis Bay (7 metres)	Jan	Feb	Mar	Apr	May	Jun	Jul	Aug	Sep	Oct	Nov	Dec
Temperature - max. (°C)	23	23	23	24	23	23	21	20	19	19	22	22
Temperature - min. (°C)	15	16	15	13	11	9	8	8	9	11	12	14
Rainfall - (mm)	0	5	8	3	3	0	0	3	0	0	0	0

Beira (9 metres)	Jan	Feb	Mar	Apr	May	Jun	Jul	Aug	Sep	Oct	Nov	Dec
Temperature - max. (°C)	32	32	31	30	28	26	25	26	28	31	31	31
Temperature - min. (°C)	24	24	23	22	18	16	16	17	18	22	22	23
Rainfall - (mm)	277	213	257	107	56	33	31	28	20	132	135	234

SCALE 1 : 77 000 000

0   1000   2000   3000 km

Lambert Azimuthal Equal Area projection

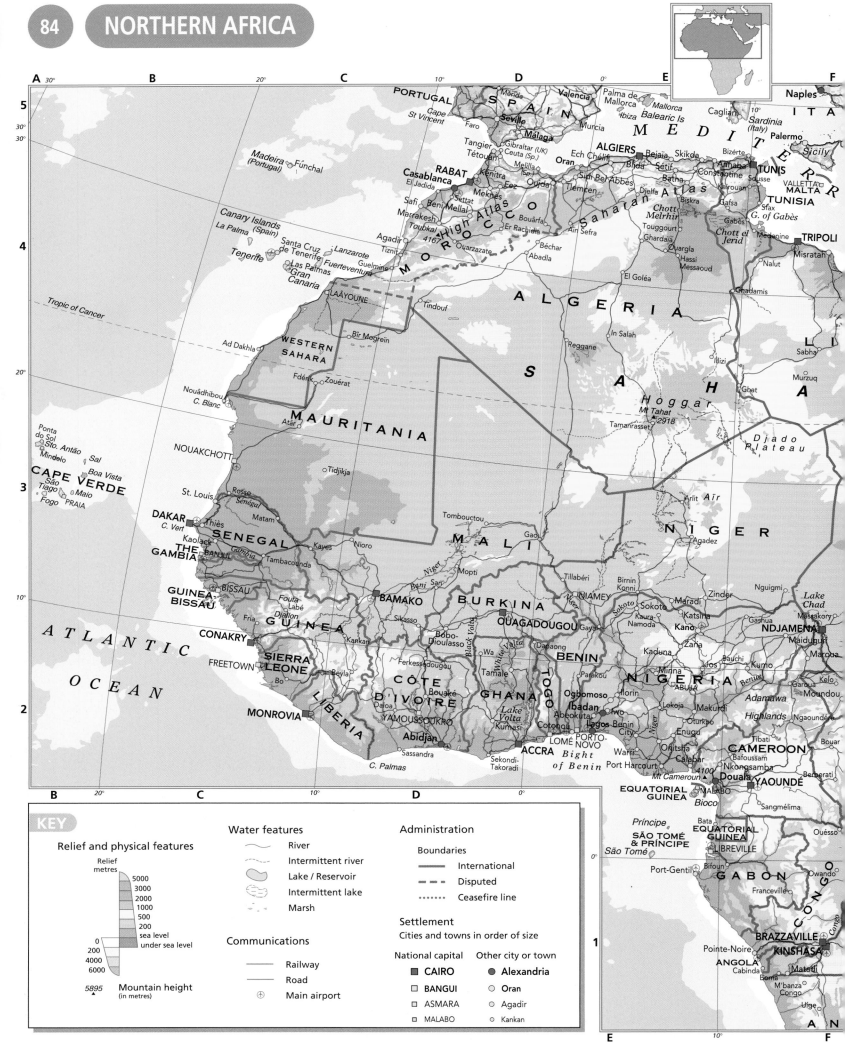

KEY

Relief and physical features

Relief
metres

5000
3000
2000
1000
500
200
sea level
0
under sea level
200
4000
6000

5895 ▲ Mountain height
(in metres)

Water features

～ River
～ Intermittent river
Lake / Reservoir
Intermittent lake
Marsh

Communications

—— Railway
—— Road
⊕ Main airport

Administration

Boundaries

—— International
- - - Disputed
····· Ceasefire line

Settlement
Cities and towns in order of size

National capital   Other city or town

■ **CAIRO**   ● Alexandria
□ **BANGUI**   ○ Oran
□ ASMARA   ○ Agadir
□ MALABO   ○ Kankan

SCALE 1 : 20 000 000

0   200   400   600   800 km

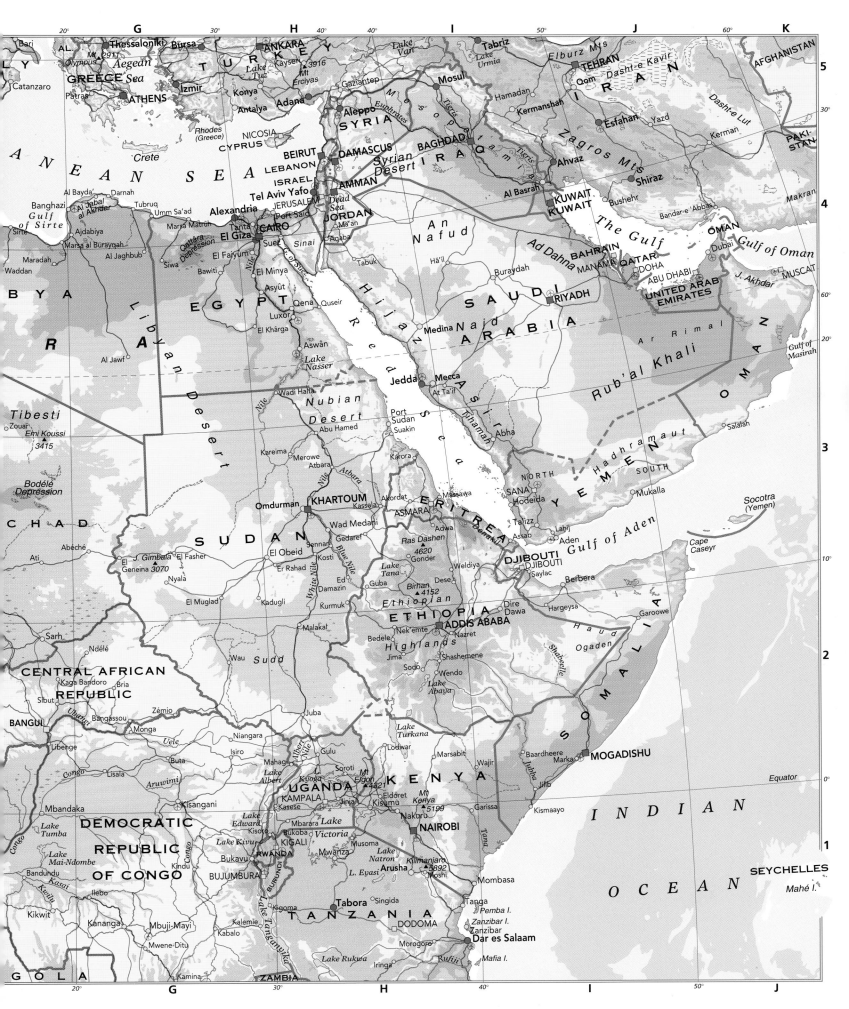

Millers Stereographic projection

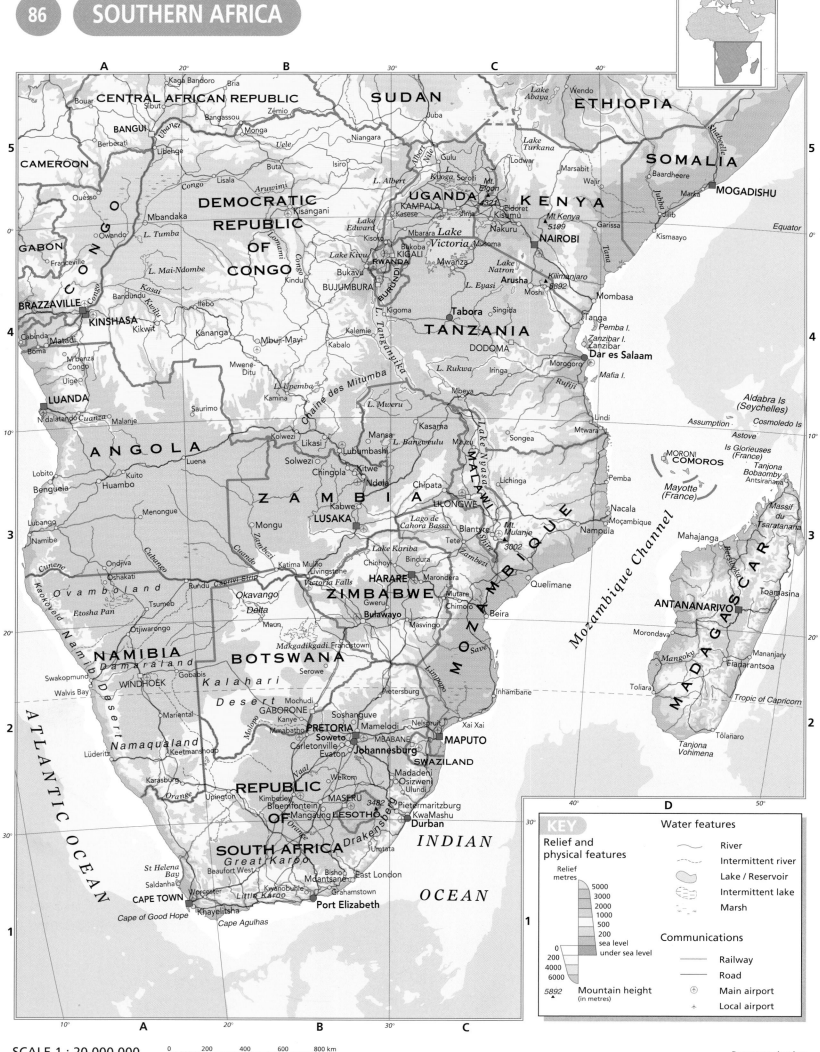

CENTRAL AFRICAN REPUBLIC
SUDAN
ETHIOPIA
BANGUI
CAMEROON
SOMALIA
MOGADISHU
GABON
DEMOCRATIC
REPUBLIC
OF
CONGO
UGANDA
KAMPALA
KENYA
KIGALI
RWANDA
NAIROBI
BRAZZAVILLE
KINSHASA
BUJUMBURA
BURUNDI
Arusha
Tabora
TANZANIA
DODOMA
Dar es Salaam
LUANDA
ANGOLA
ZAMBIA
MALAWI
LUSAKA
LILONGWE
Blantyre
MOZAMBIQUE
COMOROS
MORONI
Mayotte
(France)
NAMIBIA
BOTSWANA
HARARE
ZIMBABWE
Bulawayo
Beira
MADAGASCAR
ANTANANARIVO
Kalahari
Desert
GABORONE
WINDHOEK
PRETORIA
Soweto
Johannesburg
MAPUTO
SWAZILAND
MBABANE
REPUBLIC
OF
SOUTH AFRICA
MASERU
LESOTHO
Durban
ATLANTIC OCEAN
INDIAN
OCEAN
CAPE TOWN
Port Elizabeth

KEY
Relief and
physical features
Relief
metres
5000
3000
2000
1000
500
200
sea level
0
under sea level
200
4000
6000
5892 Mountain height
(in metres)

Water features
River
Intermittent river
Lake / Reservoir
Intermittent lake
Marsh

Communications
Railway
Road
Main airport
Local airport

0   200   400   600   800 km

Bonne projection

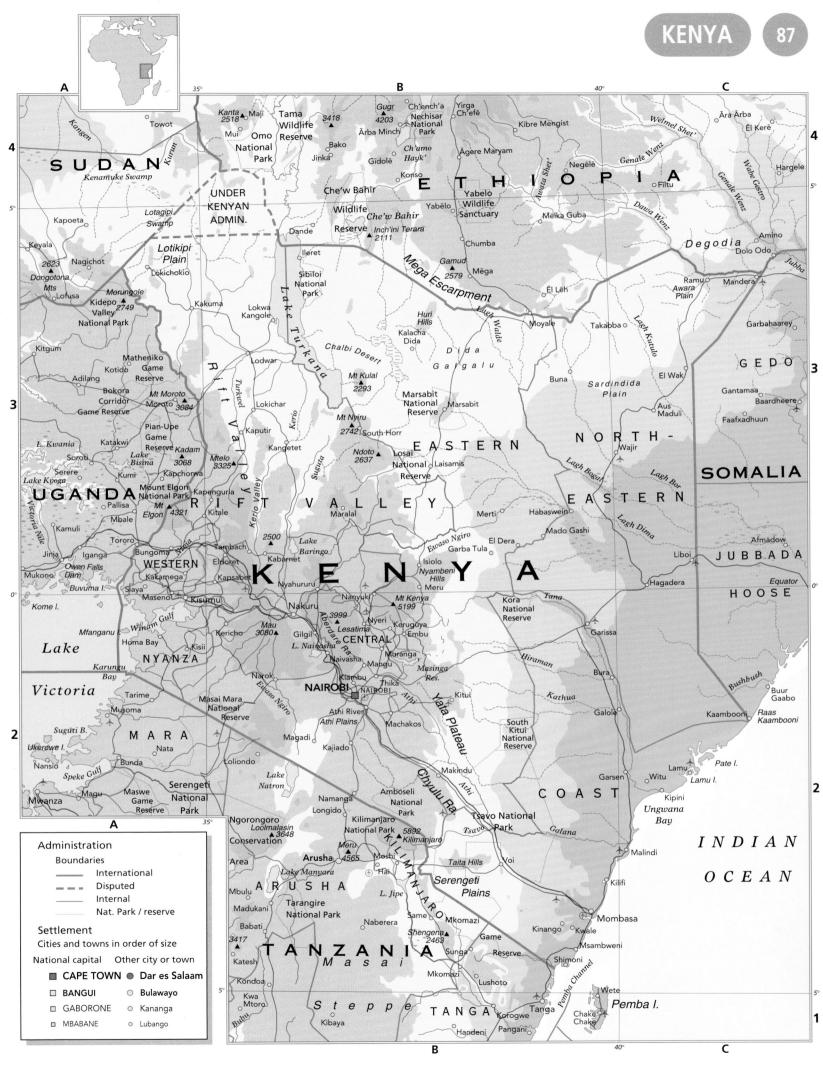

SUDAN

ETHIOPIA

SOMALIA

UGANDA

KENYA

Lake Victoria

TANZANIA

INDIAN OCEAN

UNDER KENYAN ADMIN.

Towot
Kanta 2518
Majī
Tama Wildlife Reserve
3418
Gugr 4203
Ch'ench'a
Nechisar National Park
Yirga Ch'efē
Kibre Mengist
Welmel Shet'
Āra Ārba
Ēl Kerē
Kangen
Mui
Omo National Park
Ārba Minch
Āgere Maryam
Negēlē
Genale Wenz
Hargele
Kurum
Kenamuke Swamp
Bako
Jinka
Gīdolē
Ch'amo Hayk'
Konso
Yabelo
Yabēlo Wildlife Sanctuary
Melka Guba
Fīltu
Wabē Gestro
Kapoeta
Lotagipi Swamp
Dande
Che'w Bahīr
Wildlife Reserve
Che'w Bahīr
Inch'ini Terara 2111
Chumba
Melka Guba
Degodia
Amino
Dolo Odo
Keyala
Lotikipi Plain
Ileret
Sibīloi National Park
Gamud 2579
Mēga
Ēl Lēh
Moyale
Ramu
Awara Plain
Mandera
Jubba
2623 Dongotona Mts
Nagichot
Lokichokio
Lake Turkana
Chalbi Desert
Kalacha Dida
Dida Galgalu
Buna
Sardindida Plain
El Wak
GEDO
Garbahaarey
Lofusa
Kidepo Valley National Park
Morungole 2749
Kakuma
Lokwa Kangole
Huri Hills
Takabba
Lagh Kutulo
Gantamaa
Baardheere
Kitgum
Matheniko Game Reserve
Kotido
Adilang
Bokora Corridor Game Reserve
Mt Moroto 3084
Moroto
Lodwar
Turkwel
Mt Kulal 2293
Marsabit National Reserve
Marsabit
Wajir
Aus Maduli
Faafxadhuun
L. Kwania
Katakwi
Pian-Upe Game Reserve
Kadam 3068
Lokichar
Kerio
Mt Nyiru 2742
South Horr
EASTERN
NORTH-
SOMALIA
Soroti
Serere
Lake Bisina
Kapchorwa
Kaputir
Kangetet
Ndoto 2637
Losai National Reserve
Laisamis
Lagh Bogal
Lagh Bor
Baardheere
Lake Kyoga
Kumi
Mount Elgon National Park
Mtelo 3325
Sugutá
EASTERN
Mado Gashi
Afmadow
UGANDA
Pallisa
Mt Elgon 4321
Kapenguria
Maralal
Merti
Habaswein
Lagh Dima
JUBBADA
Mbale
Kitale
RIFT VALLEY
2500
Lake Baringo
Ewaso Ngiro
El Dera
Garba Tula
Isiolo
Liboi
Kamuli
Tororo
Tambach
Kabarnet
Nyambeni Hills
Meru
Hagadera
Equator
HOOSE
Jinja
Iganga
Bungoma
Eldoret
Kapsabet
Nyahururu
KENYA
Tana
Kora National Reserve
Garissa
Mukono
Owen Falls Dam
Buvuma I.
Kakamega
WESTERN
Nakuru
Nanyuki
Mt Kenya 5199
Kome I.
Siaya
Maseno
Kisumu
Mau 3080
Gilgil
3999
Aberdare Ra.
Lesatima
Nyeri
Keruguya
Embu
Bura
Kathua
Galole
Kaambooni
Lake Victoria
Winam Gulf
Kericho
Kisii
NYANZA
L. Naivasha
Naivasha
Mangu
Murang'a
Masinga Res.
Hiraman
Buur Gaabo
Mfanganu I.
Homa Bay
Karungu Bay
Narok
Ewaso Narok
Kiambu
Thika
NAIROBI
Kitui
Galana
Raas Kaambooni
Suguti B.
Tarime
Musoma
Ewaso Ngiro
Athi River
Athi Plains
Machakos
South Kitui National Reserve
Bushbush
Ukerewe I.
Nansio
MARA
Nata
Magadi
Kajiado
Makindu
Athi
Garsen
Lamu
Pate I.
Lamu I.
Mwanza
Magu
Maswe Game Reserve
Serengeti National Park
Loliondo
Lake Natron
Namanga
Longido
Amboseli National Park
Chyulu Ra.
Tsavo National Park
COAST
Witu
Kipini
Ungwana Bay
Speke Gulf
Ngorongoro Conservation Area
Loolmalasin 3648
Kilimanjaro National Park
5892 Kilimanjaro
Tsavo
Voi
Malindi
Meru 4565
Arusha
Moshi
Hai
Taita Hills
Kilifi
Serengeti Plains
ARUSHA
Lake Manyara
KILIMANJARO
Shengena 2468
Game Reserve
Kinango
Kwale
Mombasa
Mbulu
Madukani
L. Jipe
Same
Mkomazi
Msambweni
Tarangire National Park
Naberera
Shimoni
Babati
3417
TANZANIA
Sunga
Lushoto
Chake Chake
Wete
Katesh
Masai
Mkomazi
Kwa Mtoro
Kondoa
Steppe
TANGA
Korogwe
Tanga
Pemba I.
Kibaya
Bubu
Kwa Mtoro
Handeni
Pangani
Pemba Channel

Administration
Boundaries
International
Disputed
Internal
Nat. Park / reserve
Settlement
Cities and towns in order of size
National capital   Other city or town
CAPE TOWN   ● Dar es Salaam
□ BANGUI   ○ Bulawayo
□ GABORONE   ○ Kananga
□ MBABANE   ○ Lubango

SCALE 1 : 5 000 000
0   50   100   150   200 km

Oblated Stereographic projection

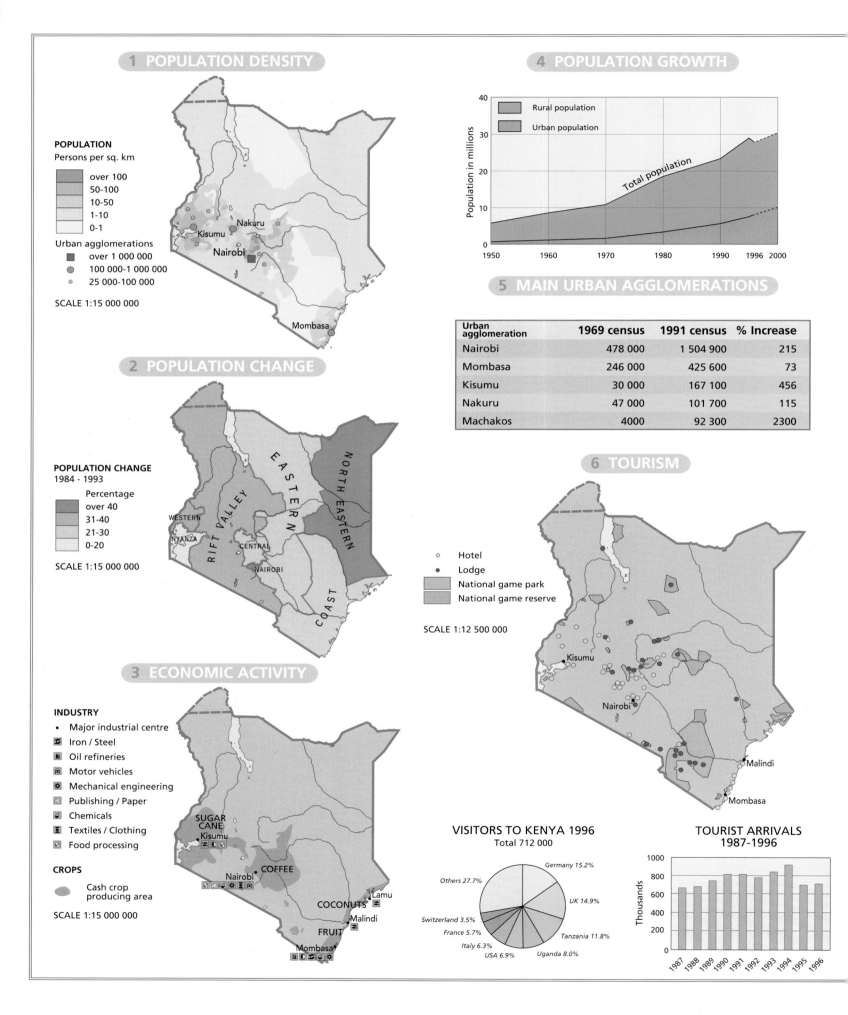

1 POPULATION DENSITY

POPULATION
Persons per sq. km

- over 100
- 50-100
- 10-50
- 1-10
- 0-1

Urban agglomerations
- over 1 000 000
- 100 000-1 000 000
- 25 000-100 000

SCALE 1:15 000 000

Nakuru
Kisumu
Nairobi
Mombasa

2 POPULATION CHANGE

POPULATION CHANGE
1984 - 1993

Percentage
- over 40
- 31-40
- 21-30
- 0-20

SCALE 1:15 000 000

EASTERN
NORTH EASTERN
WESTERN
RIFT VALLEY
NYANZA
CENTRAL
NAIROBI
COAST

3 ECONOMIC ACTIVITY

INDUSTRY
- • Major industrial centre
- Iron / Steel
- Oil refineries
- Motor vehicles
- Mechanical engineering
- Publishing / Paper
- Chemicals
- Textiles / Clothing
- Food processing

CROPS
- Cash crop producing area

SCALE 1:15 000 000

SUGAR CANE
Kisumu
Nairobi
COFFEE
COCONUTS
Lamu
Malindi
FRUIT
Mombasa

4 POPULATION GROWTH

Rural population
Urban population
Total population

Population in millions

1950 1960 1970 1980 1990 1996 2000

5 MAIN URBAN AGGLOMERATIONS

Urban agglomeration	1969 census	1991 census	% Increase
Nairobi	478 000	1 504 900	215
Mombasa	246 000	425 600	73
Kisumu	30 000	167 100	456
Nakuru	47 000	101 700	115
Machakos	4000	92 300	2300

6 TOURISM

- ○ Hotel
- ● Lodge
- National game park
- National game reserve

SCALE 1:12 500 000

Kisumu
Nairobi
Malindi
Mombasa

VISITORS TO KENYA 1996
Total 712 000

Germany 15.2%
UK 14.9%
Tanzania 11.8%
Uganda 8.0%
USA 6.9%
Italy 6.3%
France 5.7%
Switzerland 3.5%
Others 27.7%

TOURIST ARRIVALS 1987-1996

Thousands

1987 1988 1989 1990 1991 1992 1993 1994 1995 1996

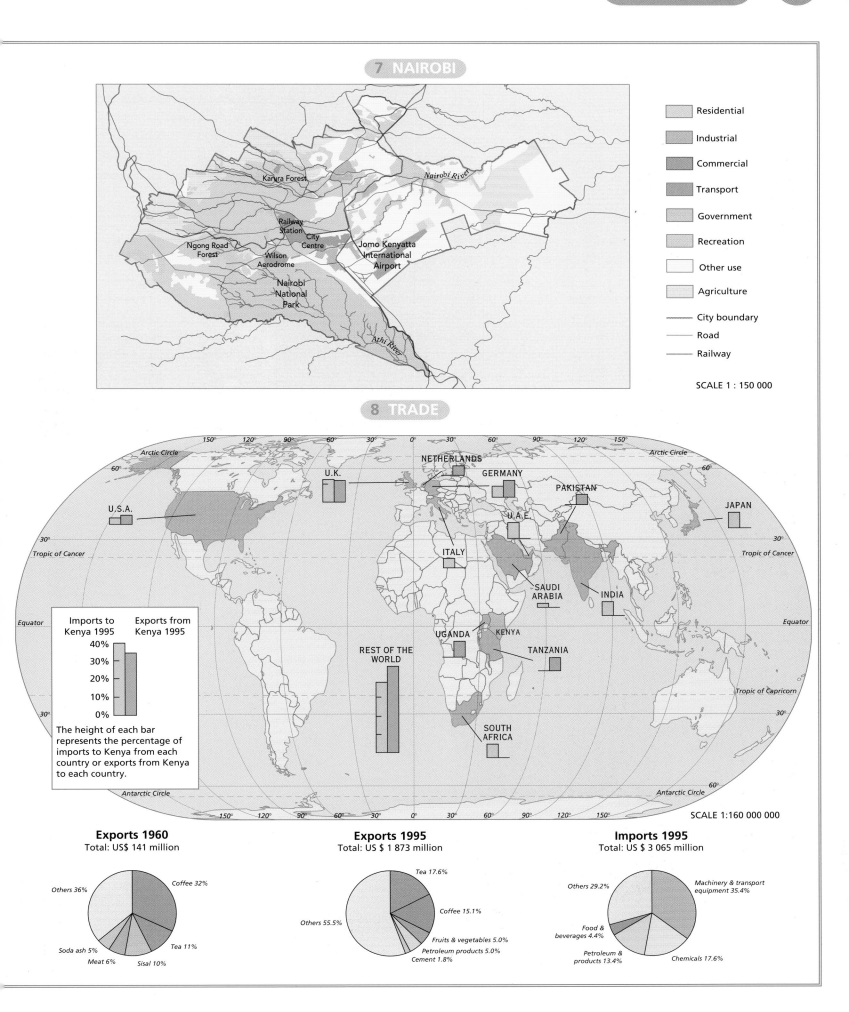

7 NAIROBI

Karura Forest

Nairobi River

Railway Station

City Centre

Ngong Road Forest

Wilson Aerodrome

Jomo Kenyatta International Airport

Nairobi National Park

Athi River

Residential
Industrial
Commercial
Transport
Government
Recreation
Other use
Agriculture
City boundary
Road
Railway

SCALE 1 : 150 000

8 TRADE

NETHERLANDS
U.K.
GERMANY
PAKISTAN
U.S.A.
JAPAN
U.A.E.
ITALY
SAUDI ARABIA
INDIA
UGANDA
KENYA
TANZANIA
REST OF THE WORLD
SOUTH AFRICA

Imports to Kenya 1995   Exports from Kenya 1995

40%
30%
20%
10%
0%

The height of each bar represents the percentage of imports to Kenya from each country or exports from Kenya to each country.

SCALE 1:160 000 000

Exports 1960
Total: US$ 141 million

Coffee 32%
Others 36%
Tea 11%
Sisal 10%
Meat 6%
Soda ash 5%

Exports 1995
Total: US $ 1 873 million

Tea 17.6%
Coffee 15.1%
Others 55.5%
Fruits & vegetables 5.0%
Petroleum products 5.0%
Cement 1.8%

Imports 1995
Total: US $ 3 065 million

Machinery & transport equipment 35.4%
Others 29.2%
Food & beverages 4.4%
Petroleum & products 13.4%
Chemicals 17.6%

SCALE 1 : 40 000 000

0   400   800   1200   1600 km

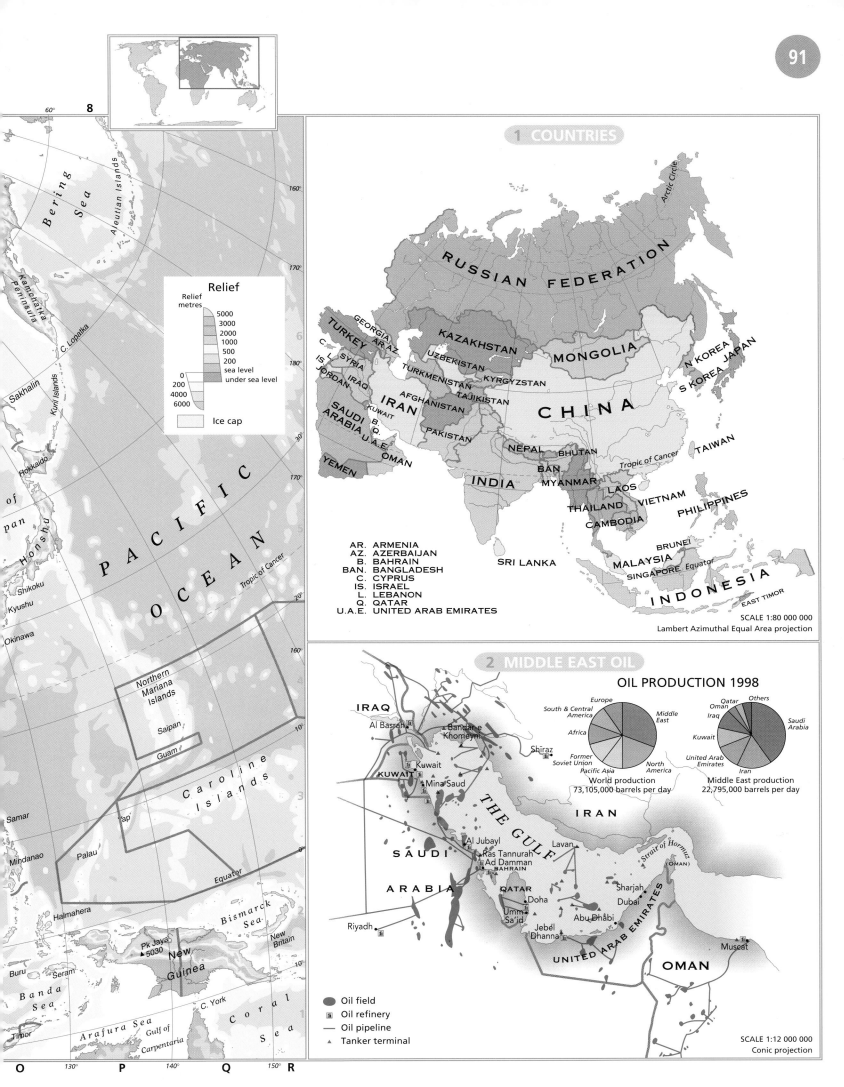

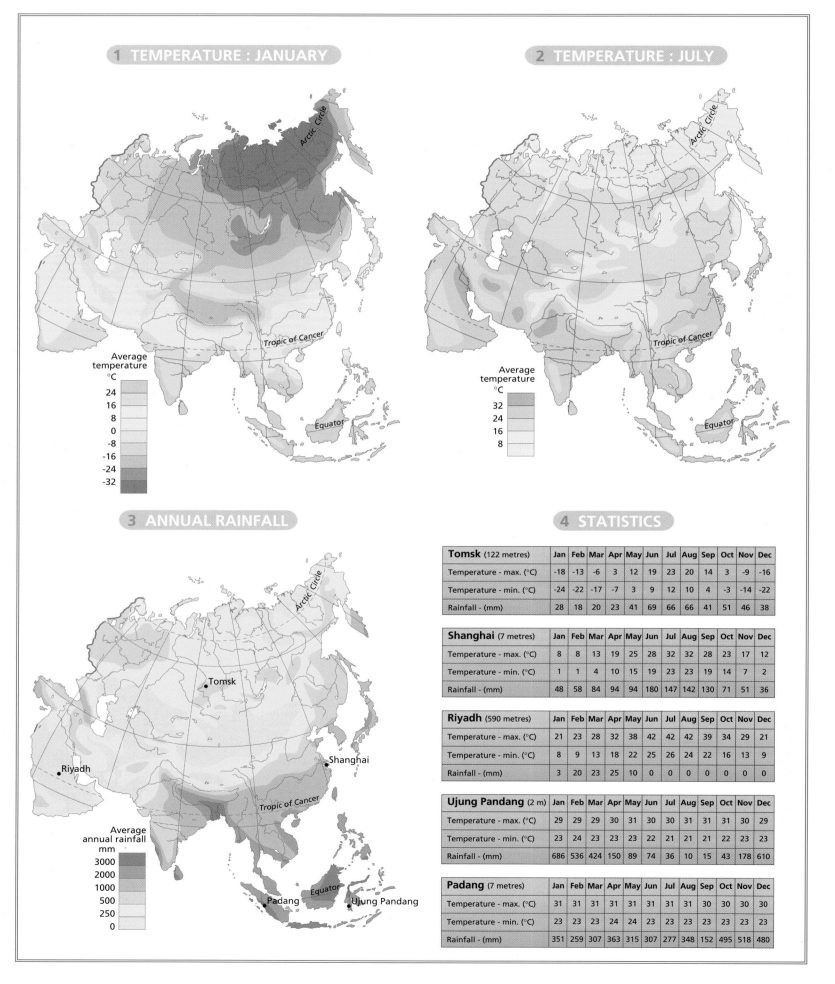

1 TEMPERATURE : JANUARY

Average
temperature
°C

24
16
8
0
-8
-16
-24
-32

Tropic of Cancer

Equator

2 TEMPERATURE : JULY

Average
temperature
°C

32
24
16
8

Tropic of Cancer

Equator

3 ANNUAL RAINFALL

Tomsk

Riyadh

Shanghai

Tropic of Cancer

Average
annual rainfall
mm

3000
2000
1000
500
250
0

Padang   Equator   Ujung Pandang

4 STATISTICS

Tomsk (122 metres)	Jan	Feb	Mar	Apr	May	Jun	Jul	Aug	Sep	Oct	Nov	Dec
Temperature - max. (°C)	-18	-13	-6	3	12	19	23	20	14	3	-9	-16
Temperature - min. (°C)	-24	-22	-17	-7	3	9	12	10	4	-3	-14	-22
Rainfall - (mm)	28	18	20	23	41	69	66	66	41	51	46	38

Shanghai (7 metres)	Jan	Feb	Mar	Apr	May	Jun	Jul	Aug	Sep	Oct	Nov	Dec
Temperature - max. (°C)	8	8	13	19	25	28	32	32	28	23	17	12
Temperature - min. (°C)	1	1	4	10	15	19	23	23	19	14	7	2
Rainfall - (mm)	48	58	84	94	94	180	147	142	130	71	51	36

Riyadh (590 metres)	Jan	Feb	Mar	Apr	May	Jun	Jul	Aug	Sep	Oct	Nov	Dec
Temperature - max. (°C)	21	23	28	32	38	42	42	42	39	34	29	21
Temperature - min. (°C)	8	9	13	18	22	25	26	24	22	16	13	9
Rainfall - (mm)	3	20	23	25	10	0	0	0	0	0	0	0

Ujung Pandang (2 m)	Jan	Feb	Mar	Apr	May	Jun	Jul	Aug	Sep	Oct	Nov	Dec
Temperature - max. (°C)	29	29	29	30	31	30	30	31	31	31	30	29
Temperature - min. (°C)	23	24	23	23	23	22	21	21	22	23	23	23
Rainfall - (mm)	686	536	424	150	89	74	36	10	15	43	178	610

Padang (7 metres)	Jan	Feb	Mar	Apr	May	Jun	Jul	Aug	Sep	Oct	Nov	Dec
Temperature - max. (°C)	31	31	31	31	31	31	31	31	30	30	30	30
Temperature - min. (°C)	23	23	23	24	24	23	23	23	23	23	23	23
Rainfall - (mm)	351	259	307	363	315	307	277	348	152	495	518	480

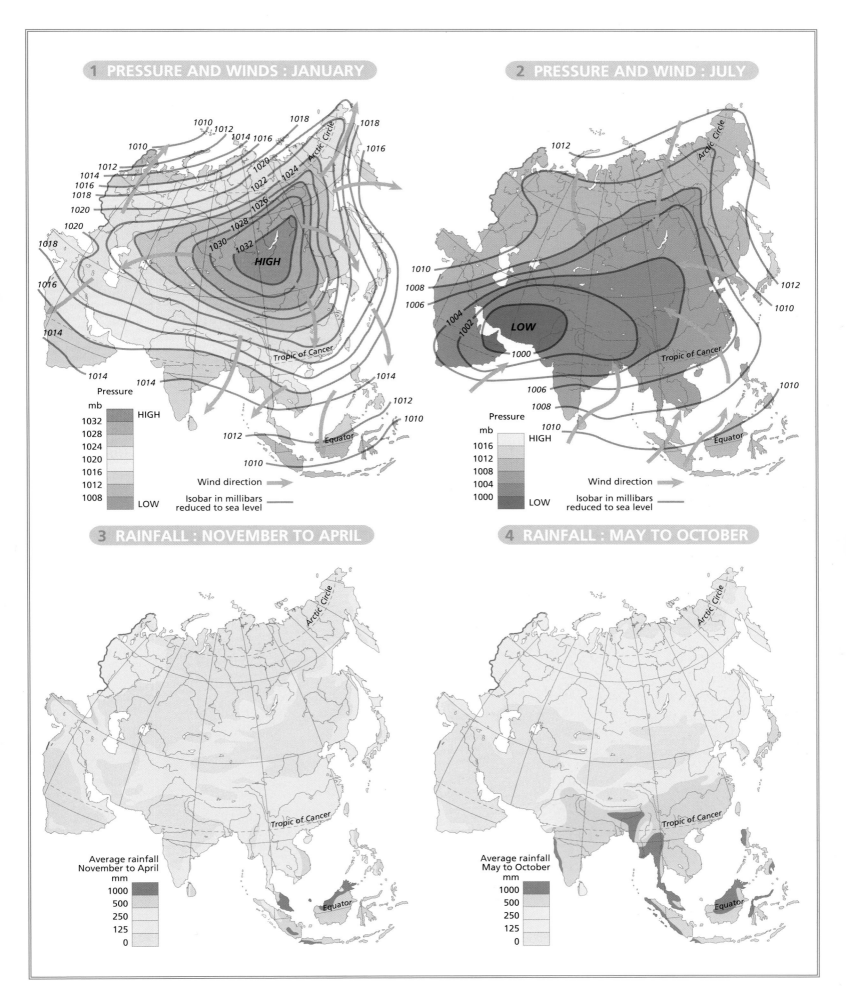

1 PRESSURE AND WINDS : JANUARY

1010 1012 1014 1016 1018 1018 1016
1010
1012
1020
1022
1024
1020
1014
1016
1018
1020
1020
1026
1018
1030 1028
1032
1016
HIGH
1014
Arctic Circle
1014 1014
Tropic of Cancer
1014
1012
1010
1012
Equator
1010

Pressure
mb
1032	HIGH
1028	
1024	
1020	
1016	
1012	
1008	LOW

Wind direction →
Isobar in millibars
reduced to sea level

2 PRESSURE AND WIND : JULY

1012
Arctic Circle
1010
1008
1006
1004 1002
1000
LOW
1012
1010
Tropic of Cancer
1006
1008
1010
1010
Equator

Pressure
mb
1016	HIGH
1012	
1008	
1004	
1000	LOW

Wind direction →
Isobar in millibars
reduced to sea level

3 RAINFALL : NOVEMBER TO APRIL

Arctic Circle

Tropic of Cancer

Average rainfall
November to April
mm
1000	
500	
250	
125	
0	

Equator

4 RAINFALL : MAY TO OCTOBER

Arctic Circle

Tropic of Cancer

Average rainfall
May to October
mm
1000	
500	
250	
125	
0	

Equator

SCALE 1 : 100 000 000

0 1000 2000 3000 4000 km

Lambert Azimuthal Equal Area projection

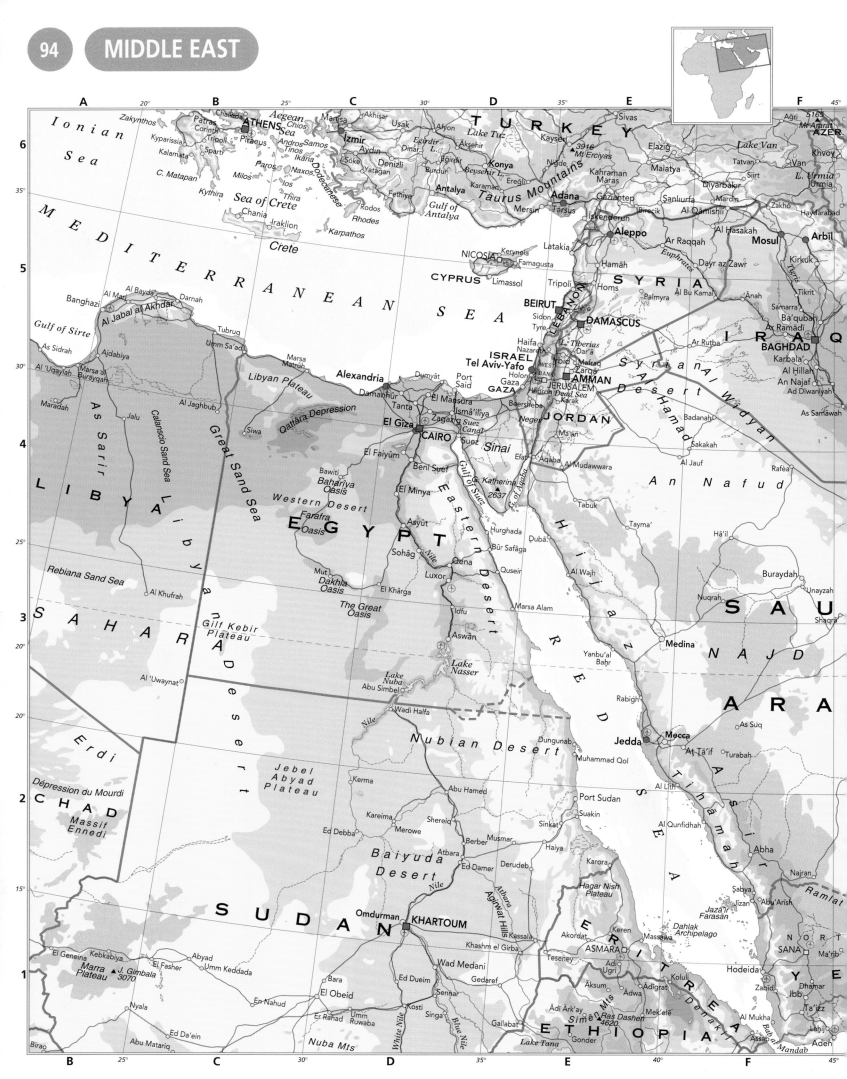

SCALE 1 : 12 000 000

0   100   200   300   400 km

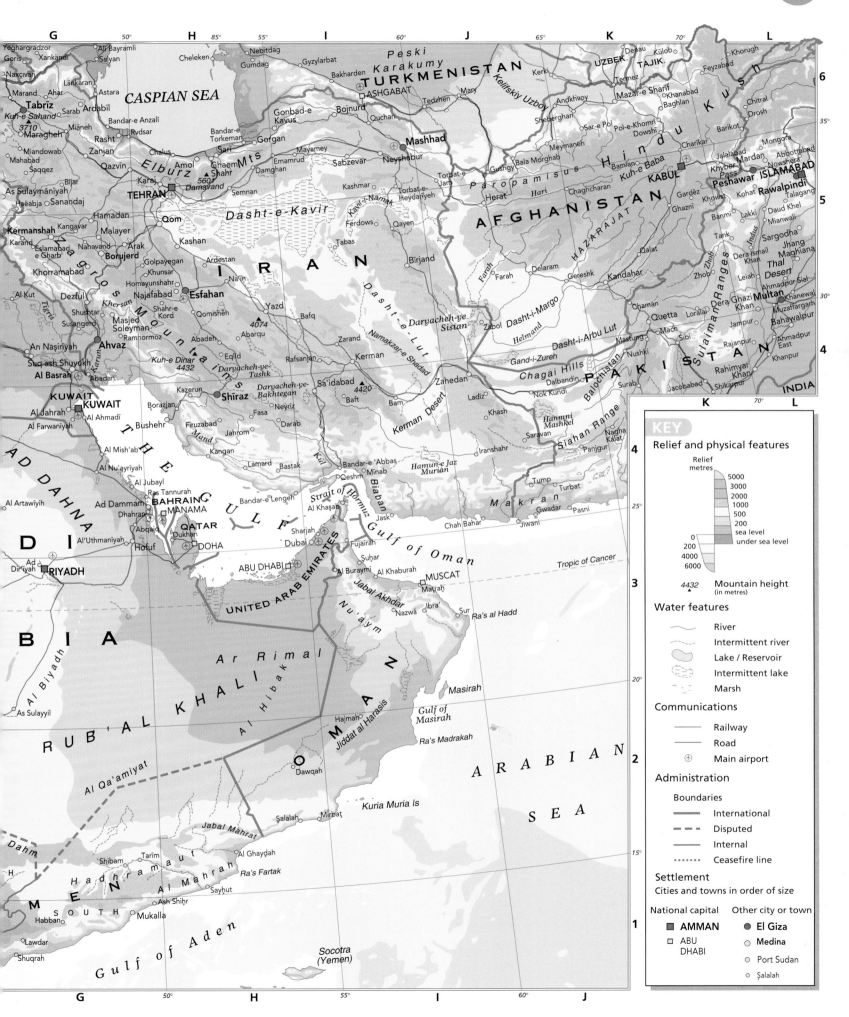

G 50° **H** 85° 55° **I** 60° **J** 65° **K** 70° **L**

Yeghargradzor
Goris
Naxçivan
Marand · Ahar
Ali Bayramli
Salyan
Xankändi
Lankaran
Astara
Cheleken
Nebitdag
Gumdag
Gyzylarbat
Bakharden
Peski
Karakumy
TURKMENISTAN
Kerki
Denau · Külob
Khorugh
UZBEK. **TAJIK.**
Termez
Feyzabad
Khorugh
6

35°

CASPIAN SEA
Tabrīz
Kuh-e Sahand
3710
Maragheh
Sarab · Ardabil
Bandar-e Anzali
ASHGABAT
Gonbad-e Kavus
Bojnurd
Quchan
Tedzhen
Mary
Andkhvoy
Sheberghan
Mazar-e Sharif
Khanabad
Baghlan
Chitral
Drosh
35°

Miandowab
Mahabad
Saqqez
Zanjan
Miāneh
Rasht
Qazvin
Rvdsar
Chalus
Amol
Sari
Ghaem Shahr
Bandar-e Torkeman
Gorgan
Mayamey
Emamrud
Sabzevar
Neyshabur
Mashhad
Torbat-e Jam
Gushgy
Bala Morghab
Meymaneh
Sar-e Pol
Pol-e-Khomri
Dowshi
Bamian
Charikar
Jalalabad
Mardan
Abbottabad
Nowshera

As Sulaymānīyah
Haēabja · Sanandaj
Bijar
Hamadan
Karaj
TEHRAN
5601
Damavand
Semnan
Damghan
Kashmar
Torbat-e Heydariyeh
Ferdows
Elburz Mts
Dasht-e Kavir
Kavil-i-Namak
Herat
Paropamisus
Chaghcharan
Kuh-e Baba
KABUL
Gardēz
Hindu Kush
Khyber Pass
Peshawar **ISLAMABAD**
Kohat **Rawalpindi**
Talagang
5

Kermanshah
Karand
Nahavand
Malayer
Arak
Qom
AFGHANISTAN
HAZARAJAT
Ghazni
Khowst
Banmi
Daud Khel
Mianwali

Eslamabad
e Gharb
Borujerd
Khorramabad
Golpayegan
Kashan
Qayen
Tabas
Bīrjand
Farah
Qalat
Lakki
Tank
Sargodha
Jhang
Maghiana

Al Kut
Dezful
Homayunshahr
Najafabad
Ardestan
Na'in
Esfahan
Yazd
4074
Farah
Delaram
Gereshk
Kandahar
Chaman
Zhob
Leiah
Dera Ismail Khan
Thal Desert
Ahmadpur Sial

Shushtar
Masjed Soleyman
Ramhormoz
Shahr-e Kord
Qomisheh
Abadeh
Abarqu
Dasht-e-Lut
Daryacheh-ye Sistan
Zabol
Dasht-i-Margo
Helmand
Dalbandin
Mastung
Quetta
Mach
Sibi
Loralai
Dera Ghazi Khan
Multan
Khanewal
Muzaffargarh

Susangerd
Ahvaz
Eqlid
Kuh-e Dinar
4432
Daryacheh-ye Tashk
Rafsanjan
Zarand
Namakzar-e Shadad
Gand-i-Zureh
Chagai Hills
Dasht-i-Arbu Lut
Nok Kundi
Nushki
Surab
Nagha Kalat
Jacobabad
Rahimyar Khan
Jampur
Rajanpur
Bahawalpur

Suq ash Shuyukh
Al Basrah
Abadan
Kazerun
Daryacheh-ye Bakhtegan
Sa'idabad
4420
Kerman
Kerman Desert
Zahedan
Khash
Ladiz
Saravan
Siahan Range
Panjgur
Dalbandin
Mastung
Ahmadpur East
Khanpur
Shikarpur
INDIA

KUWAIT
Al Jahrah
KUWAIT
Al Ahmadi
Borazjan
Shīrāz
Fasa
Neyriz
Baft
Bam
4

Al Farwaniyah
Bushehr
Firuzabad
Jahrom
Darab
Iranshahr
Hammi Mashkel
Saravan
Siahan Range
Panjgur

Al Mish'ab
Al Nu'ayriyah
THE
Kangan
Bastak
Hamun-e Jaz Murian
Tump
Turbat
4

Al Jubayl
Lamard
Kil
Bandar-e 'Abbas
Minab
Jask
Makran
Gwadar
Pasni

Al Artawiyih
Ras Tannurah
GULF
Bandar-e Lengeh
Qeshm
Biaban
Chah Bahar
Jiwani
25°

Ad Dammam
BAHRAIN
Al Khasab
Strait of Hormuz
Gulf of Oman

AD DAHNA
Dhahran
MANAMA
Sharjah
UNITED ARAB EMIRATES
Dubai
Fujairah
Suhar

Al Uthmaniyah
Abqaiq
Hofuf
QATAR
Dukhan
DOHA
ABU DHABI
Al Buraymi
Al Khaburah
MUSCAT
Matrah
Tropic of Cancer

Ad Dir'iyah
RIYADH
D
I
B
I
A
Al Biyadh
As Sulayyil
Ar Rimal
RUB' AL KHALI
Al Hibak
Nu'aym
Jabal Akhdar
Nazwa
Ibra'
Sur
Ra's al Hadd
3

Al Qa'amiyat
OMAN
Jiddat al Harasis
Gulf of Masirah
Masirah
20°

Dawqah
Ra's Madrakah
ARABIAN
2

Dahm
YEMEN
Al Qa'amiyat
Salalah
Mirbat
Kuria Muria Is
SEA
15°

Shibam
Tarim
Hadhramaut
Al Mahrah
Al Ghaydah
Ra's Fartak
Jabal Mahrat
Sayhut
Ash Shihr
1

Habban
SOUTH
Mukalla
M
E
N
Socotra
(Yemen)

Lawdar
Shuqrah
Gulf of Aden

G 50° **H** 55° **I** 60° **J**

Relief and physical features

Relief metres

5000
3000
2000
1000
500
200
0 sea level
under sea level
200
4000
6000

4432 ▲ Mountain height (in metres)

Water features

~ River
Intermittent river
Lake / Reservoir
Intermittent lake
Marsh

Communications

Railway
Road
⊕ Main airport

Administration

Boundaries

International
Disputed
Internal
Ceasefire line

Settlement

Cities and towns in order of size

National capital
■ **AMMAN**
□ **ABU DHABI**

Other city or town
● **El Giza**
○ **Medina**
○ Port Sudan
○ Salalah

Albers Conic Equal Area projection

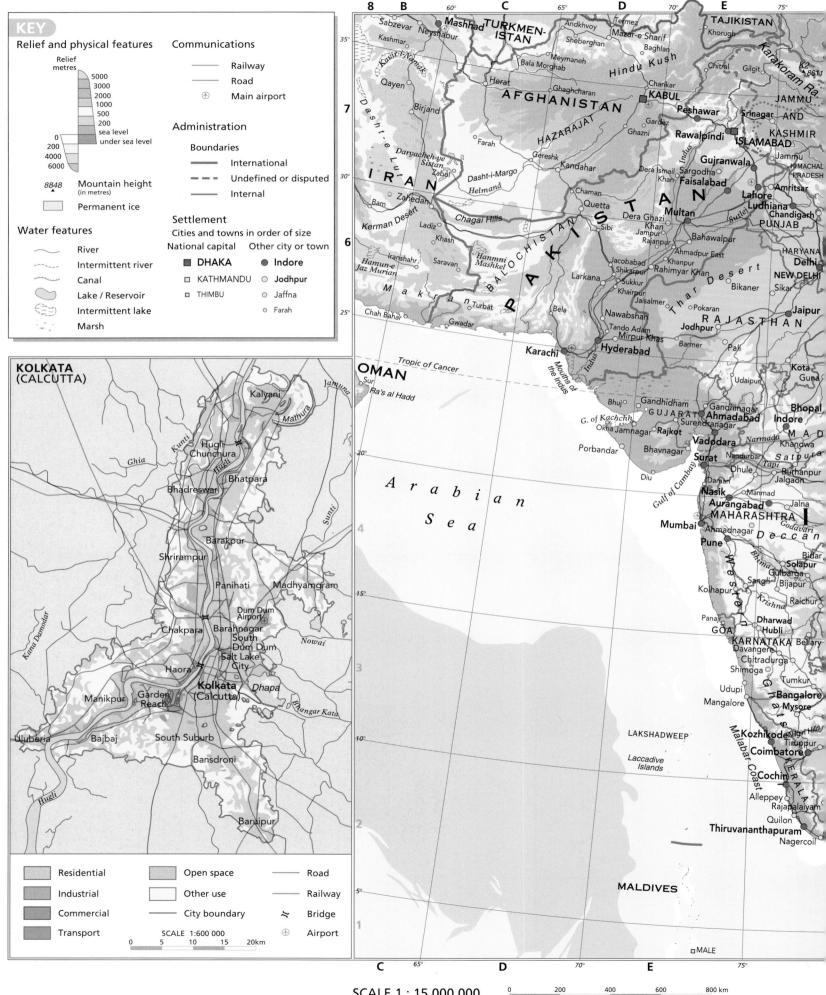

KEY

Relief and physical features

Relief
metres
5000
3000
2000
1000
500
200
sea level
0
under sea level
200
4000
6000

8848 ▲ Mountain height
(in metres)

Permanent ice

Water features

River

Intermittent river

Canal

Lake / Reservoir

Intermittent lake

Marsh

Communications

Railway

Road

⊕ Main airport

Administration

Boundaries

International

Undefined or disputed

Internal

Settlement

Cities and towns in order of size

National capital	Other city or town
■ DHAKA	● Indore
□ KATHMANDU	● Jodhpur
□ THIMBU	○ Jaffna
	○ Farah

KOLKATA (CALCUTTA)

Kalyani
Jamuna
Hugli
Chunchura
Bhatpara
Bhadreswar
Kunti
Mathura
Ghia
Barakpur
Shrirampur
Panihati
Madhyamgram
Dum Dum Airport
Barahnagar
Chakpara
South Dum Dum
Salt Lake City
Nowai
Haora
Manikpur
Garden Reach
Kolkata (Calcutta)
Dhapa
Bhangar Kata
Kana Damodar
Uluberia
Bajbaj
South Suburb
Bansdroni
Hugli
Baruipur

Residential	Open space	Road
Industrial	Other use	Railway
Commercial	City boundary	⊼ Bridge
Transport		⊕ Airport

SCALE 1:600 000
0   5   10   15   20km

Sabzevar • Mashhad TURKMEN-ISTAN
Kashmar Neyshabur
Andkhvoy Termez Mazar-e Sharif TAJIKISTAN Khorugh
Meymaneh Sheberghan Baghlan
Kunt-i-Namak Bala Morghab Hindu Kush Chitral Gilgit K2 8611
Qayen Herat Ghaghcharan Charikar KABUL JAMMU
Birjand AFGHANISTAN Gardez KABUL Peshawar Srinagar AND
HAZARAJAT Ghazni Rawalpindi ISLAMABAD KASHMIR
Daryacheh-ye Sistan Farah Gereshk Dera Ismail Sargodha Jammu HIMACHAL
Zabol Dasht-i-Margo Helmand Khan Faisalabad PRADESH
Zahedan Chaman Quetta Multan Lahore Amritsar
Bam Kandahar Dera Ghazi Khan Bahawalpur Ludhiana Chandigarh
Kerman Desert Chagai Hills Sibi Jampur Rajanpur Ahmadpur East PUNJAB
Ladiz Khash Jacobabad Khanpur Rahimyar Khan HARYANA
Iranshahr Hanmni Mashkel Larkana Shikarpur Delhi
Hamun-e Jaz Murian Saravan Sukkur Khairpur Bikaner NEW DELHI
Turbat Nawabshah Jaisalmer Sikar
Chah Bahar Gwadar Bela Tando Adam Jodhpur Jaipur
Karachi Mirpur Khas Barmer Pali RAJASTHAN
OMAN Hyderabad Kota
Tropic of Cancer Mouths of the Indus Udaipur Guna
Sur Ra's al Hadd Bhuj Gandhidham Gandhinagar Bhopal
G. of Kachchh Surendranagar Ahmadabad Indore M A D
Okha Jamnagar Rajkot Narmada Khandwa
Porbandar Bhavnagar Vadodara Satpura
Diu Surat Nandurbar Burhanpur
Daman Dhule Tapi Jalgaon
Arabian Nasik Manmad
Sea Aurangabad Jalna
Mumbai Ahmadnagar MAHARASHTRA Godavari
Pune Deccan
Bhima Bidar
Kolhapur Sangli Solapur
Krishna Gulbarga
Panaji Bijapur Raichur
GOA Dharwad
Hubli KARNATAKA Bellary
Davangere
Chitradurga
Shimoga
Udupi Tumkur
Mangalore Bangalore
Mysore
Nilgiri Hills
LAKSHADWEEP Kozhikode Tiruppur
Laccadive Coimbatore
Islands KERALA Cochin
Malabar Coast Alleppey Rajapalaiyam
Quilon
Thiruvananthapuram
Nagercoil
MALDIVES
□ MALE

Conic projection

1 POPULATION DENSITY

POPULATION
Persons per sq. km
- over 100
- 50-100
- 10-50
- 1-10
- 0-1

Urban agglomerations
- ■ over 5 000 000
- ● 1 000 000-5 000 000
- ● 500 000-1 000 000

Delhi

Kolkata (Calcutta)

Mumbai (Bombay)

Hyderabad

Bangalore

Chennai (Madras)

SCALE 1:24 000 000

3 POPULATION GROWTH

Rural population

Urban population

Total population

Population in millions

1 000
800
600
400
200
0

1950   1960   1970   1980   1990   1996 2000

2 MAIN URBAN AGGLOMERATIONS

Urban agglomeration	1985 census	1995 census	% Increase
Mumbai (Bombay)	10 137 000	15 093 000	49
Kolkata (Calcutta)	10 462 000	11 673 000	12
Delhi	6 993 000	9 882 000	41
Chennai (Madras)	4 983 000	5 906 000	19
Hyderabad	3 022 000	5 343 000	77
Bangalore	3 685 000	4 749 000	29
Ahmadabad	3 037 000	3 688 000	21
Kanpur	1 481 789	2 356 000	59
Nagpur	1 219 461	1 847 000	52
Pune	1 203 351	1 566 651	30

2 REGIONAL COMPARISON

Calicut

KERALA

Cochin
Alleppey
Quilon
Trivandrum

Muzaffarpur   ●Darbhanga
Patna ●Munger
●Bhagalpur
●Gaya
BIHAR ●Dhanbad
Ranchi

30 555 000	**Population**	93 080 000
94%	**Literacy rate Male**	52%
86%	**Female**	23%
Free up to 14 years	**Education**	Free up to 11 years
Hindu 58%	**Religions**	Hindu 83%
Christian 21%		Moslem 14%
Moslem 21%		
Malayalam	**Languages**	Hindi
Tamil		Urdu
Kannada		Bengali

5 POPULATION CHANGE

JAMMU AND KASHMIR

HIMACHAL PRADESH

PUNJAB

HARYANA

ARUNACHAL PRADESH

RAJASTHAN

UTTAR PRADESH

ASSAM
MEGHALAYA

N.
MA.

BIHAR

T.   MI.

GUJARAT

MADHYA PRADESH

WEST BENGAL

MAHARASHTRA

ORISSA

ANDHRA PRADESH

GOA

KARNATAKA

TAMIL NADU

KERALA

C. CHANDIGARGH
D. DELHI
MA. MANIPUR
MI. MIZORAM
N. NAGALAND
S. SIKKIM
T. TRIPURA

POPULATION CHANGE
1981 - 1994

Percentage
- over 50
- 41-50
- 31-40
- 0-30

SCALE 1:30 000 000

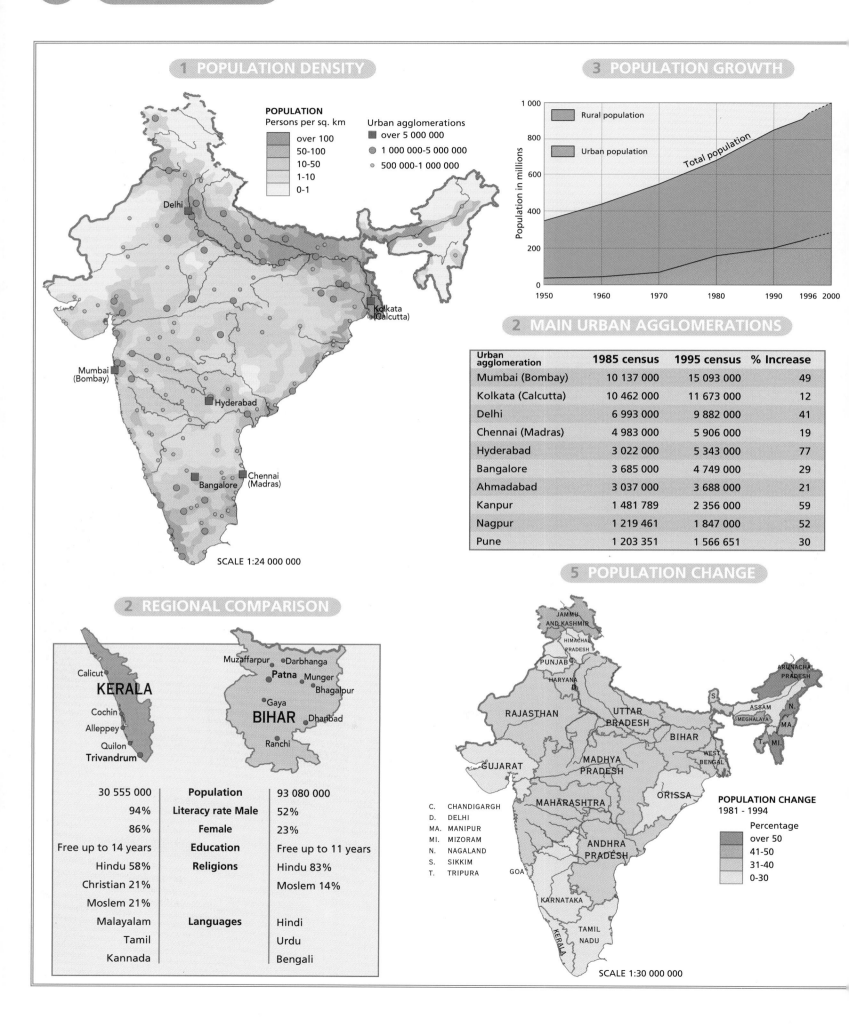

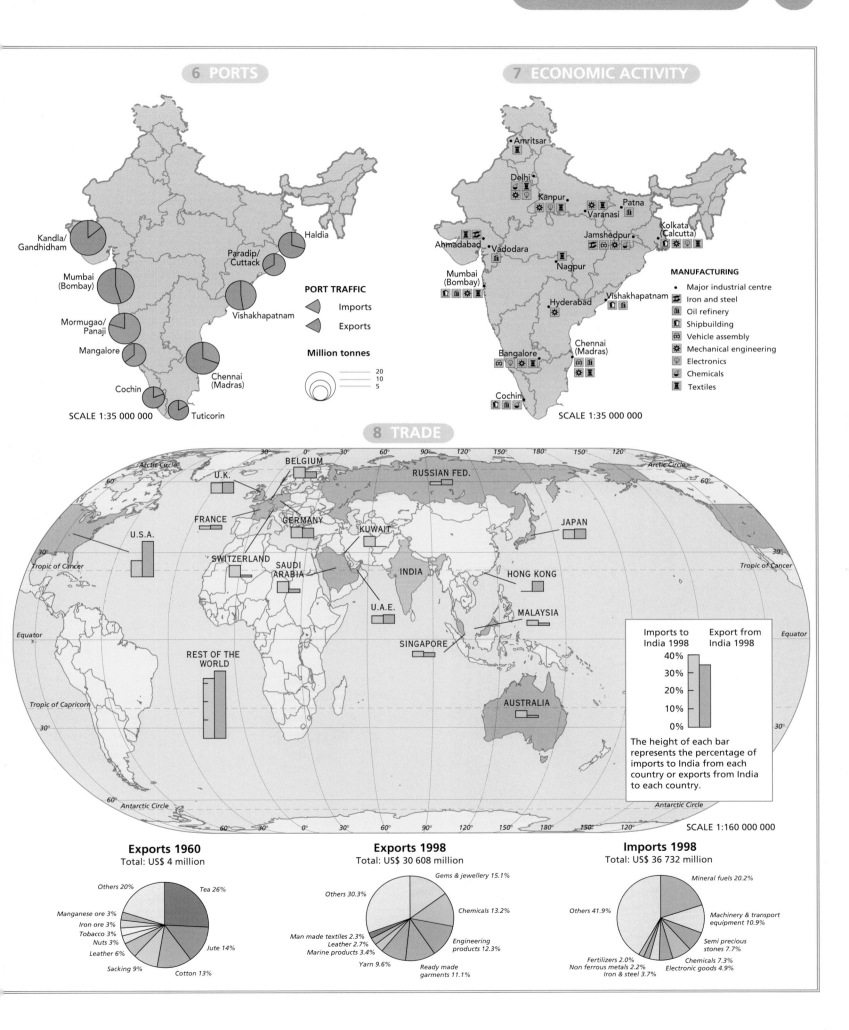

6 PORTS

Kandla/Gandhidham
Mumbai (Bombay)
Mormugao/Panaji
Mangalore
Cochin
Tuticorin
Chennai (Madras)
Vishakhapatnam
Paradip/Cuttack
Haldia

PORT TRAFFIC

Imports

Exports

Million tonnes

20
10
5

SCALE 1:35 000 000

7 ECONOMIC ACTIVITY

Amritsar
Delhi
Kanpur
Patna
Varanasi
Kolkata (Calcutta)
Jamshedpur
Ahmadabad
Vadodara
Nagpur
Mumbai (Bombay)
Hyderabad
Vishakhapatnam
Bangalore
Chennai (Madras)
Cochin

MANUFACTURING

- Major industrial centre
- Iron and steel
- Oil refinery
- Shipbuilding
- Vehicle assembly
- Mechanical engineering
- Electronics
- Chemicals
- Textiles

SCALE 1:35 000 000

8 TRADE

Arctic Circle
BELGIUM
U.K.
RUSSIAN FED.
Arctic Circle
FRANCE
GERMANY
JAPAN
U.S.A.
KUWAIT
SWITZERLAND
SAUDI ARABIA
INDIA
HONG KONG
Tropic of Cancer
Tropic of Cancer
U.A.E.
MALAYSIA
Equator
REST OF THE WORLD
SINGAPORE
Equator
Tropic of Capricorn
AUSTRALIA
Antarctic Circle
Antarctic Circle

Imports to India 1998 | Export from India 1998

40%
30%
20%
10%
0%

The height of each bar represents the percentage of imports to India from each country or exports from India to each country.

SCALE 1:160 000 000

Exports 1960
Total: US$ 4 million

Others 20%
Tea 26%
Manganese ore 3%
Iron ore 3%
Tobacco 3%
Nuts 3%
Leather 6%
Sacking 9%
Cotton 13%
Jute 14%

Exports 1998
Total: US$ 30 608 million

Gems & jewellery 15.1%
Others 30.3%
Chemicals 13.2%
Man made textiles 2.3%
Leather 2.7%
Marine products 3.4%
Engineering products 12.3%
Yarn 9.6%
Ready made garments 11.1%

Imports 1998
Total: US$ 36 732 million

Mineral fuels 20.2%
Others 41.9%
Machinery & transport equipment 10.9%
Semi precious stones 7.7%
Fertilizers 2.0%
Non ferrous metals 2.2%
Iron & steel 3.7%
Chemicals 7.3%
Electronic goods 4.9%

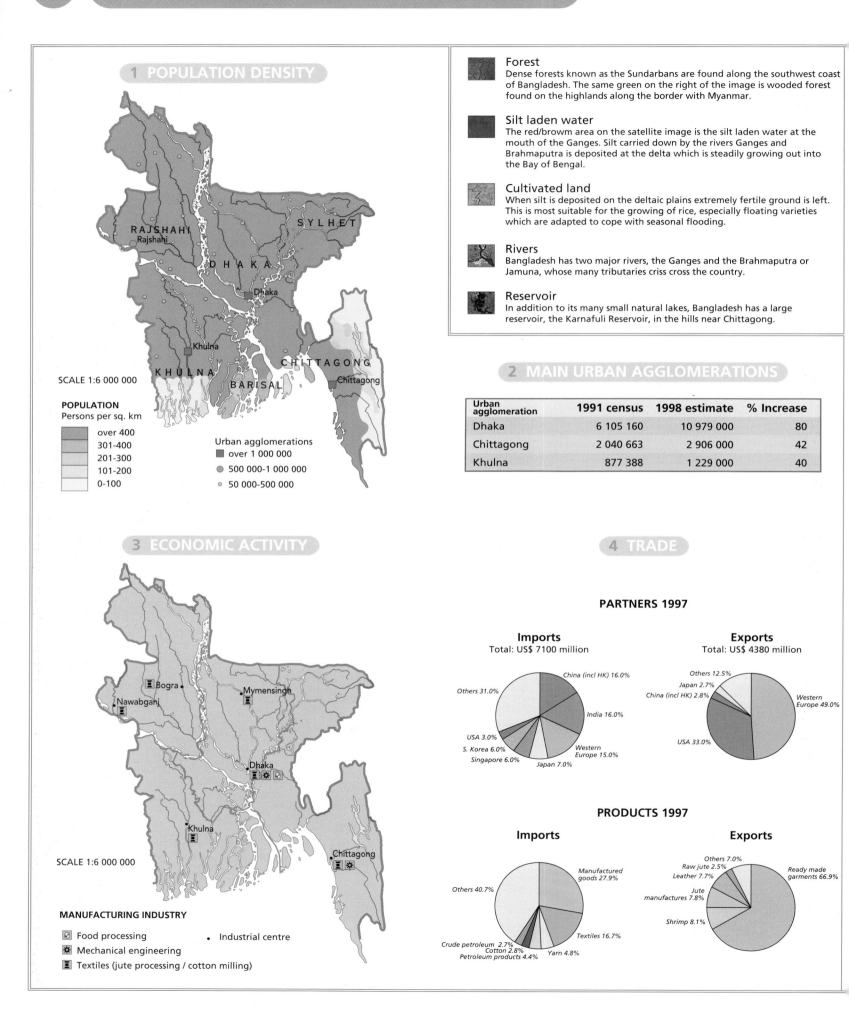

1 POPULATION DENSITY

RAJSHAHI
Rajshahi

SYLHET

D H A K A

Dhaka

Khulna

K H U L N A

C H I T T A G O N G

B A R I S A L

Chittagong

SCALE 1:6 000 000

POPULATION
Persons per sq. km

over 400
301-400
201-300
101-200
0-100

Urban agglomerations
■ over 1 000 000
● 500 000-1 000 000
∘ 50 000-500 000

Forest
Dense forests known as the Sundarbans are found along the southwest coast of Bangladesh. The same green on the right of the image is wooded forest found on the highlands along the border with Myanmar.

Silt laden water
The red/brown area on the satellite image is the silt laden water at the mouth of the Ganges. Silt carried down by the rivers Ganges and Brahmaputra is deposited at the delta which is steadily growing out into the Bay of Bengal.

Cultivated land
When silt is deposited on the deltaic plains extremely fertile ground is left. This is most suitable for the growing of rice, especially floating varieties which are adapted to cope with seasonal flooding.

Rivers
Bangladesh has two major rivers, the Ganges and the Brahmaputra or Jamuna, whose many tributaries criss cross the country.

Reservoir
In addition to its many small natural lakes, Bangladesh has a large reservoir, the Karnafuli Reservoir, in the hills near Chittagong.

2 MAIN URBAN AGGLOMERATIONS

Urban agglomeration	1991 census	1998 estimate	% Increase
Dhaka	6 105 160	10 979 000	80
Chittagong	2 040 663	2 906 000	42
Khulna	877 388	1 229 000	40

3 ECONOMIC ACTIVITY

Bogra

Mymensingh

Nawabganj

Dhaka

Khulna

Chittagong

SCALE 1:6 000 000

MANUFACTURING INDUSTRY

▣ Food processing
❁ Mechanical engineering
▤ Textiles (jute processing / cotton milling)

• Industrial centre

4 TRADE

PARTNERS 1997

Imports
Total: US$ 7100 million

China (incl HK) 16.0%
Others 31.0%
India 16.0%
USA 3.0%
S. Korea 6.0%
Singapore 6.0%
Japan 7.0%
Western Europe 15.0%

Exports
Total: US$ 4380 million

Others 12.5%
Japan 2.7%
China (incl HK) 2.8%
Western Europe 49.0%
USA 33.0%

PRODUCTS 1997

Imports

Manufactured goods 27.9%
Others 40.7%
Textiles 16.7%
Yarn 4.8%
Petroleum products 4.4%
Cotton 2.8%
Crude petroleum 2.7%

Exports

Others 7.0%
Raw jute 2.5%
Leather 7.7%
Ready made garments 66.9%
Jute manufactures 7.8%
Shrimp 8.1%

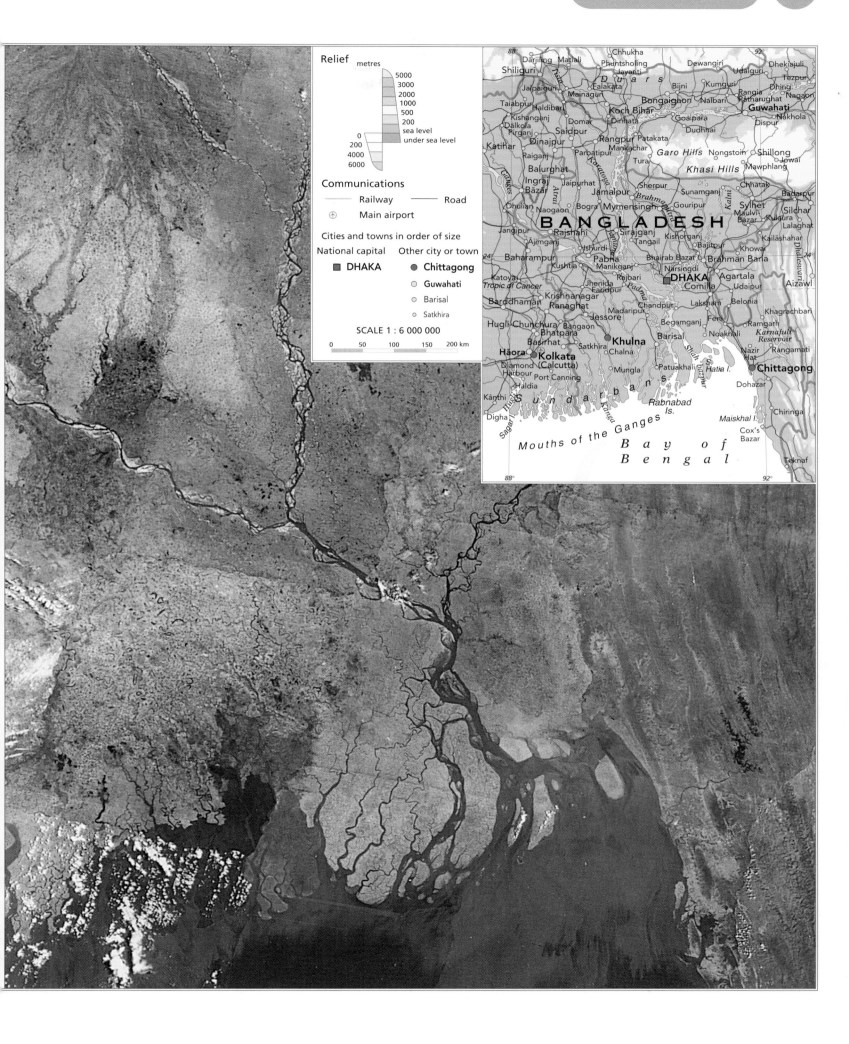

Relief
metres
5000
3000
2000
1000
500
200
sea level
0
under sea level
200
4000
6000

Communications
———— Railway ———— Road
✈ Main airport

Cities and towns in order of size
National capital   Other city or town
■ DHAKA   ● Chittagong
                     ○ Guwahati
                     ○ Barisal
                     ○ Satkhira

SCALE 1 : 6 000 000
0   50   100   150   200 km

Chhukha
Darjiling Matiali   Dewangiri
Shiliguri   Phuntsholing   Dhekiajuli
        Jayanti   Udalguri
Jalpaiguri   Bijni Kumguri Rangia   Tezpur
Taiabpur Mainagur   Bongaigaon Nalbari Patharughat Dhing Nagaon
    Kishanganj Koch Bihar   Goalpara   Guwahati Nakhola
Dālkola Haldibari   Dinhata   Dudhnai   Dispur
Pirganj Domar   Garo Hills Nongstoin Shillong
Katihar Dinajpur Rangpur Patakata Mankachar   Jowai
    Raiganj   Parbatipur   Tura   Khasi Hills Mawphlang
    Balurghat   Chhatak
Ingraj Jaipurhat Jamalpur   Sherpur Sunamganj   Badarpur
Bāzār Naogaon Bogra Mymensingh   Gouripur Sylhet Silchar
Dhulian   Brahmaputra   Maulvi Kulaura
Jangipur   Rajshahi Sirajganj   Kishorganj Bazar Lalaghat
    Ajimganj   Tangail Bajitpur Khowai
Baharampur Ishurdi Pabna   Bhairab Bazar Brahman Barla Kailashahar
Katoya   Kushtia Manikganj   Laksham
Tropic of Cancer   Rajbari   Narsingdi DHAKA Agartala Aizawl
Barddhaman Krishnanagar Jhenida Faridpur Chandpur Comilla Udaipur
Hugli-Chunchura Ranaghat Madaripur   Lakshan Belonia
    Bhatpara Bangaon Jessore Begamganj Feni Khagrachari
Hāora Basirhat Satkhira Khulna Barisal Noakhali Ramgarh Karnafuli Rangamati
    Kolkata   Chalna   Nazir Reservoir
Diamond (Calcutta) Port Canning Mungla   Hat Chittagong
Harbour   Sundarbans   Dohazar
Kānthi   Rabnabad   Maiskhal I. Chiringa
Digha Sagar I.   Is.   Cox's
        Mouths of the Ganges   Bazar
        Bay of   Teknaf
        Bengal

BANGLADESH

Duars
Padma

Bay of Bengal

KEY

Relief and physical features

Relief
metres
5000
3000
2000
1000
500
200
sea level
0
under sea level
200
4000
6000

8848 ▲ Mountain height
(in metres)

Permanent ice

Water features

~~~ River

- - - Intermittent river

~~~ Canal

Lake / Reservoir

Intermittent lake

Marsh

Communications

Railway

Road

⊕ Main airport

Administration

Boundaries

International

- - - Disputed

Internal

······ Ceasefire line

Settlement

Cities and towns in order of size

National capital   Other city or town

■ BEIJING   ● Dalian
□ BISHKEK   ○ Yantai
▫ KATHMANDU   ○ Guilin
▫ THIMBU   ○ Anxi

SCALE 1 : 15 000 000

0   200   400   600   800 km

Conic projection

SCALE 1 : 15 000 000

0   200   400   600   800 km

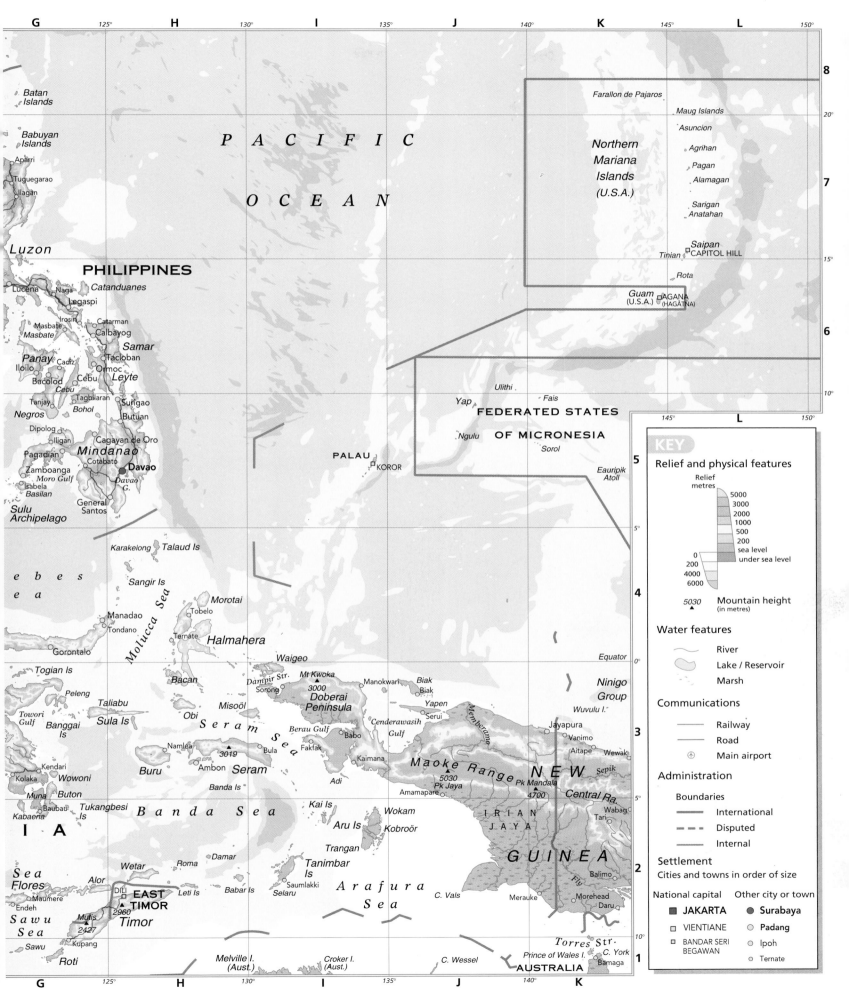

G 125° H 130° I 135° J 140° K 145° L 150°

Batan Islands

PACIFIC

Babuyan Islands
Aparri

OCEAN

Northern
Mariana
Islands
(U.S.A.)

Farallon de Pajaros

Maug Islands

Asuncion

Agrihan

Tuguegarao
Ilagan

Pagan

Alamagan

Luzon

Sarigan
Anatahan

PHILIPPINES

Saipan
Tinian □ CAPITOL HILL

Lucena Naga *Catanduanes*
Legaspi
Irosin   Catarman
Masbate   Calbayog
Masbate   *Samar*
Panay   Cadiz Tacloban
Iloilo   Ormoc
Bacolod   Cebu   *Leyte*
Cebu
Tanjay   Tagbilaran Surigao
Negros Bohol Butuan
Dipolog   Cagayan de Oro
Iligan
Pagadian   *Mindanao* Cotabato
Zamboanga   ●Davao
Moro Gulf   *Davao G.*
Isabela *Basilan*
General
Sulu   Santos
Archipelago

Rota

Guam □ AGANA
(U.S.A.) (HAGÅTÑA)

Ulithi   *Fais*
Yap ○
Yap

FEDERATED STATES
OF MICRONESIA

Ngulu
Sorol

PALAU
□ KOROR

Eauripik Atoll

8

7

6

5

Karakelong *Talaud Is*

e b e s
e a

Sangir Is

Manadao
Tondano

Gorontalo

Togian Is

Peleng
Togian Is
Togori Gulf
Banggai Is

Kendari
Kolaka
Muna *Buton*
Baubau
Kabaena

Sea
Flores
Endeh
Maumere
Sawu Sea
Sawu

Roti

Sula Is
Taliabu

Buru

Molucca Sea

Morotai
Tobelo
Ternate   *Halmahera*

Waigeo

Bacan
Misoöl
Obi   *Dampir Str.* Mt Kwoka
Sorong   ▲3000
Misoöl   Doberai
Peninsula
Manokwari Biak
Biak

Seram Sea

Namlea   ▲3019
Ambon   Bula
Seram
Banda Is

Wowoni

Tukangbesi Is

B a n d a   S e a

I A

Wetar   *Roma*
Alor
DILI □ EAST   *Leti Is*   *Babar Is*
TIMOR
▲2960   *Timor*
Mutis
▲2427 *Timor*
Kupang

Berau Gulf
Babo
Fakfak
Kaimana

Adi

Kai Is
Aru Is Wokam
Kobroör

Trangan

Damar

Tanimbar Is
Saumlakki
Selaru

Cenderawasih Gulf
Serui

Yapen

Maoke Range
▲5030
Pk Jaya

A r a f u r a
Sea

Jayapura
Vanimo

Aitape

Mamberamo

NEW
Pk Mandala▲
4700

Amamapare

IRIAN
JAYA

GUINEA

C. Vals

Merauke

Balimo

Fly

Morehead

Ninigo
Group
Wuvulu I.

Equator

Wewak
Sepik

Central Ra.

Wabag
Tari

Balimo

Daru

4

3

0°

5°

2

G 125° H 130° I 135° J 140° K

Melville I.
(Aust.)

Croker I.
(Aust.)

C. Wessel

Prince of Wales I.
AUSTRALIA

C. York
Bamaga

10°

1

145° L 150°

A 130° B 135° C 140° D E

5 Fangzheng
Wanda Shan
Dongjianghong
Hulin
Iman
Amgu
La Pérouse Strait
Sea of Okhotsk
Wakkanai
Iturup
45°
Shangzhi
Linkou
Jixi
Muling
Sikhote-Alin Range
Monbetsu
Abashin
Kitami
Kunashir
Shikotan-to
Mudanjiang
Lake Khanka
Spassk Dal'niy
Ussuri
Hokkaido
Asahikawa 2290
Asahi-dake
Hidaka-sanmyaku
CHINA
Pipa Dingzi 1397
RUSSIAN
Iwanai
Otaru
Bibai
Yubari
Obihiro
Kushiro
Dunhua
Wangqing
Hunchun
FEDERATION
Ussuriysk
Sapporo
Yakumo
Tomakomai
Yanji
Tumen
Helong
Hunchun
Vladivostok
Suchan
Muroran
Samani
4 Zengfeng Shan 1677
Unggi
Nakhodka
Rudnaya Pristan'
Mori
Hakodate
45°
2541
Kambo Ho
Najin
Chongjin
Tsugaru-kaikyo
Mutsu
Kimchaek
Goshogawara
Aomori
NORTH
Towada
Hachinohe
Hirosaki
40° KOREA
Noshiro
Odate
40°
Akita
Morioka
Miyako
Sea
Kamaishi
of Japan
Hanamaki
Kesennuma
Sakata
Ichinoseki
Ishinomaki
3 Kangnung
Nogwak-san 1321
Ullung-do
Sadoga-shima
Ryotsu
Yamagata
Tendo
Sendai
35°
Niigata
Fukushima
Ulchin
Tok-to (Take-shima)
Suzu
Nanao
Toyama-wan
Kashiwazaki
Joetsu
Agano
Aizu-wakamatsu
Koriyama
Iwaki
SOUTH
Takaoka
Toyama
Nagaoka
Nagano
Hitachi
KOREA
Changgi Gap
Oki-shoto
Kanazawa
Yariga-take 3180
Matsumoto
Ueda
Utsunomiya
Mito
Honshu
P'ohang
Komatsu
Okaya
Maebashi
Oyama
Tsuchiura
Ulsan
Fukui
Tsuruga
Kofu
Urawa
Sakura
Choshi
Pusan
Matsue
Tottori
Yonago
Maizuru
Ogaki
Gifu
Ichinomiya
TOKYO
Funabashi
35° Korea Strait
Chugoku-sanchi
Biwa-ko
Nagoya
Toyota
Yokohama
Chiba
35°
Tsushima
Masuda
Okayama
Kobe
Kyoto
Osaka
Suzuka
Tsu
Fuji-san 3776
Kawasaki
Yokosuka
Higashi-suido
Hiroshima
Sakai
Matsusaka
Ise
Numazu
Shizuoka
Hamamatsu
Iki-shima
Shimonoseki
Seto-naikai
Matsuyama
Takamatsu
Wakayama
Izu-shoto
Kita-Kyushu
Shikoku-sanchi 1955
Tokushima
Sasebo
Fukuoka
Kurume 1981
Kochi
Shingu
Arao
Oita 1788
Yawatahama
Uwajima 1229
Shikoku
Hachijo-jima
Nagasaki
1759
Kumamoto
Nobeoka
PACIFIC
OCEAN
1739
Miyazaki
2 Kagoshima 1700
Kyushu
Osumi-kaikyo
Yaku-shima
Tanega-shima
30°
30°
1 Tokara-retto

135° C 140° D 145° E

A 130° B

SCALE 1 : 7 500 000

0   100   200   300   400 km

Albers Equal Area Conic projection

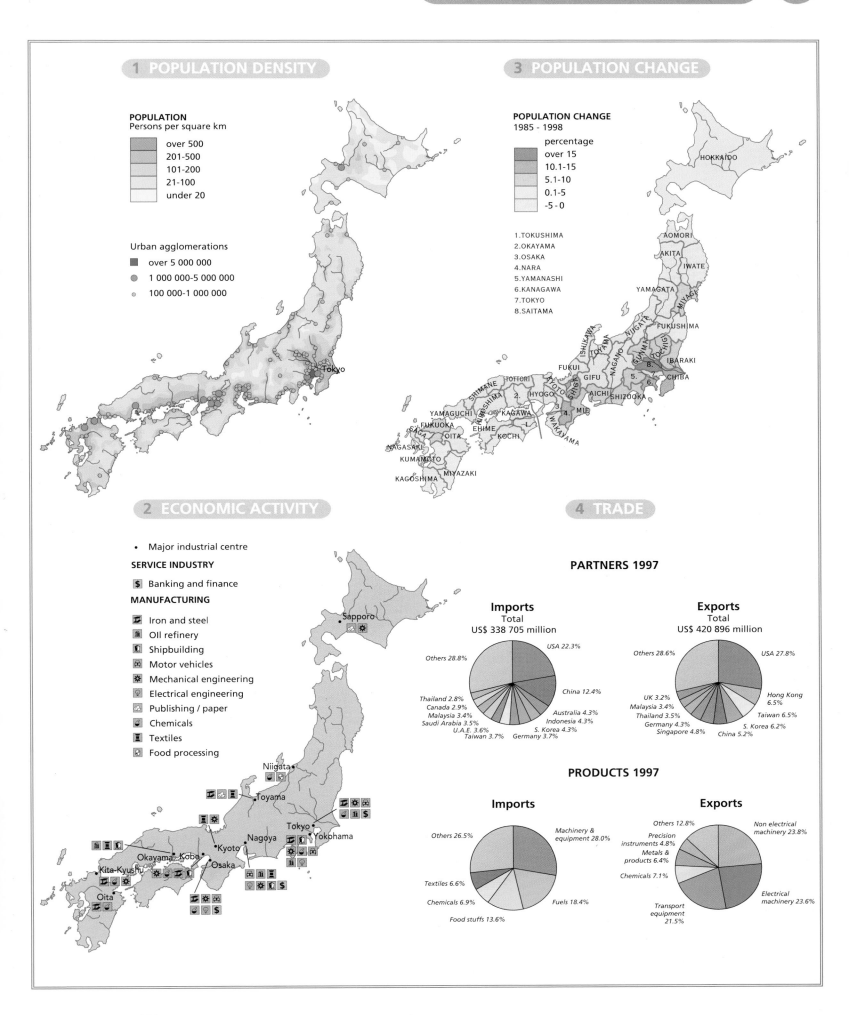

1 POPULATION DENSITY

POPULATION
Persons per square km

over 500
201-500
101-200
21-100
under 20

Urban agglomerations
over 5 000 000
1 000 000-5 000 000
100 000-1 000 000

Tokyo

3 POPULATION CHANGE

POPULATION CHANGE
1985 - 1998

percentage
over 15
10.1-15
5.1-10
0.1-5
-5 - 0

1.TOKUSHIMA
2.OKAYAMA
3.OSAKA
4.NARA
5.YAMANASHI
6.KANAGAWA
7.TOKYO
8.SAITAMA

HOKKAIDO
AOMORI
AKITA
IWATE
YAMAGATA
MIYAGI
ISHIKAWA
TOYAMA
NIIGATA
FUKUSHIMA
NAGANO
GUNMA
TOCHIGI
IBARAKI
FUKUI
GIFU
CHIBA
SHIGA
AICHI
SHIZUOKA
TOTTORI
KYOTO
HYOGO
MIE
SHIMANE
HIROSHIMA
OKAYAMA
KAGAWA
WAKAYAMA
YAMAGUCHI
EHIME
KOCHI
FUKUOKA
SAGA
OITA
NAGASAKI
KUMAMOTO
MIYAZAKI
KAGOSHIMA

2 ECONOMIC ACTIVITY

• Major industrial centre

SERVICE INDUSTRY
$ Banking and finance

MANUFACTURING
Iron and steel
Oil refinery
Shipbuilding
Motor vehicles
Mechanical engineering
Electrical engineering
Publishing / paper
Chemicals
Textiles
Food processing

Sapporo
Niigata
Toyama
Tokyo
Yokohama
Nagoya
Kyoto
Okayama Kobe
Kita-Kyushu
Osaka
Oita

4 TRADE

PARTNERS 1997

Imports
Total
US$ 338 705 million

USA 22.3%
China 12.4%
Australia 4.3%
Indonesia 4.3%
S. Korea 4.3%
Germany 3.7%
Taiwan 3.7%
U.A.E. 3.6%
Saudi Arabia 3.5%
Malaysia 3.4%
Canada 2.9%
Thailand 2.8%
Others 28.8%

Exports
Total
US$ 420 896 million

USA 27.8%
Hong Kong 6.5%
Taiwan 6.5%
S. Korea 6.2%
China 5.2%
Singapore 4.8%
Germany 4.3%
Thailand 3.5%
Malaysia 3.4%
UK 3.2%
Others 28.6%

PRODUCTS 1997

Imports

Machinery & equipment 28.0%
Fuels 18.4%
Food stuffs 13.6%
Chemicals 6.9%
Textiles 6.6%
Others 26.5%

Exports

Non electrical machinery 23.8%
Electrical machinery 23.6%
Transport equipment 21.5%
Chemicals 7.1%
Metals & products 6.4%
Precision instruments 4.8%
Others 12.8%

SCALE 1 : 15 000 000

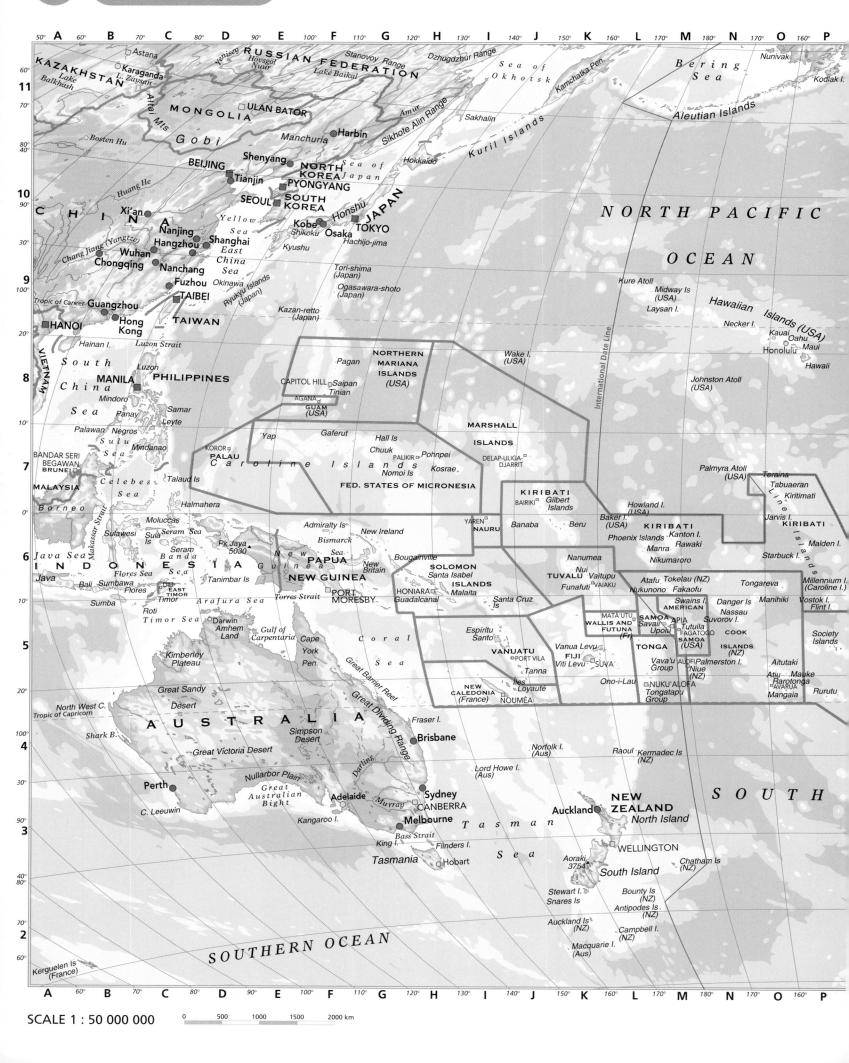

SCALE 1 : 50 000 000

0   500   1000   1500   2000 km

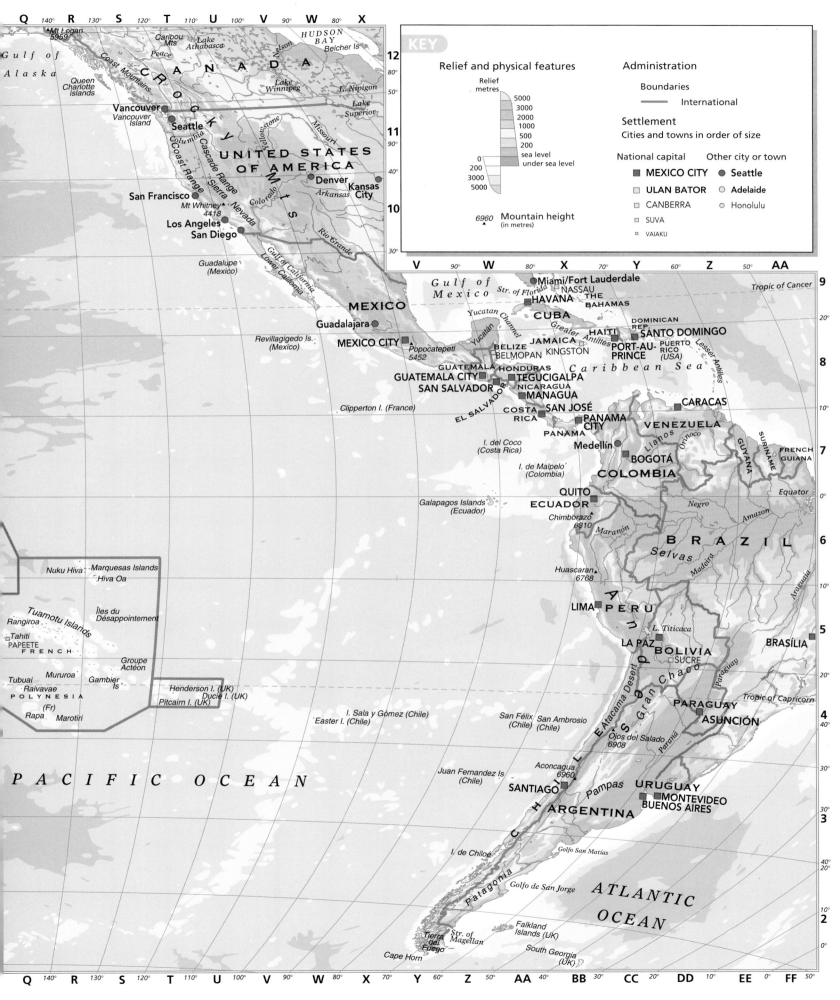

Hammer - Aitoff projection

KEY

Relief and physical features

Relief metres
5000
3000
2000
1000
500
200
sea level
0
under sea level
200
4000
6000

▲ 3754 Mountain height (in metres)

Water features

～ River
┅ Intermittent river
◯ Lake / Reservoir
◯ Intermittent lake
Marsh
Coral reef

Communications

━ Railway
━ Road
⊕ Main airport

Administration

Boundaries
━━ International
── Internal

Settlement

Cities and towns in order of size

National capital
□ CANBERRA
□ SUVA

Other city or town
● Sydney
◉ Adelaide
◎ Newcastle
◦ Darwin

SCALE 1 : 20 000 000

0   200   400   600   800 km

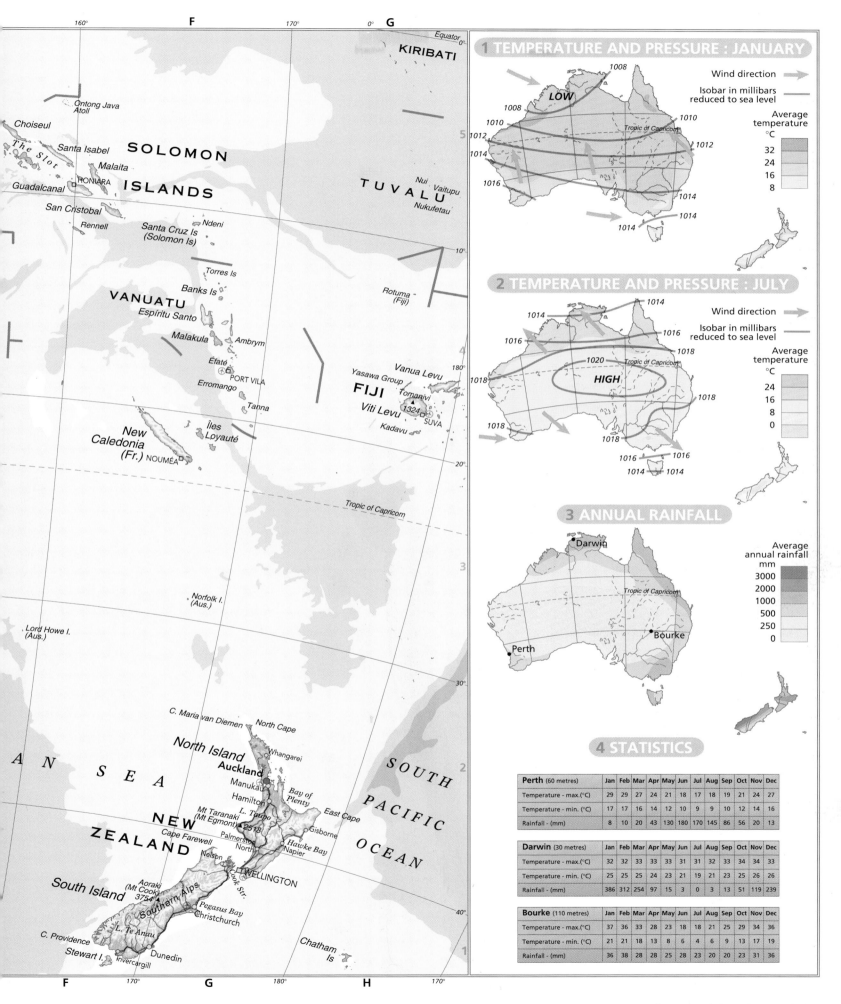

1 TEMPERATURE AND PRESSURE : JANUARY

Wind direction →
Isobar in millibars reduced to sea level —

Average temperature °C
32
24
16
8

2 TEMPERATURE AND PRESSURE : JULY

Wind direction →
Isobar in millibars reduced to sea level —

Average temperature °C
24
16
8
0

3 ANNUAL RAINFALL

Average annual rainfall mm
3000
2000
1000
500
250
0

4 STATISTICS

Perth (60 metres)	Jan	Feb	Mar	Apr	May	Jun	Jul	Aug	Sep	Oct	Nov	Dec
Temperature - max.(°C)	29	29	27	24	21	18	17	18	19	21	24	27
Temperature - min. (°C)	17	17	16	14	12	10	9	9	10	12	14	16
Rainfall - (mm)	8	10	20	43	130	180	170	145	86	56	20	13

Darwin (30 metres)	Jan	Feb	Mar	Apr	May	Jun	Jul	Aug	Sep	Oct	Nov	Dec
Temperature - max.(°C)	32	32	33	33	33	31	31	32	33	34	34	33
Temperature - min. (°C)	25	25	25	24	23	21	19	21	23	25	26	26
Rainfall - (mm)	386	312	254	97	15	3	0	3	13	51	119	239

Bourke (110 metres)	Jan	Feb	Mar	Apr	May	Jun	Jul	Aug	Sep	Oct	Nov	Dec
Temperature - max.(°C)	37	36	33	28	23	18	18	21	25	29	34	36
Temperature - min. (°C)	21	21	18	13	8	6	4	6	9	13	17	19
Rainfall - (mm)	36	38	28	28	25	28	23	20	20	23	31	36

Lambert Azimuthal Equal Area projection

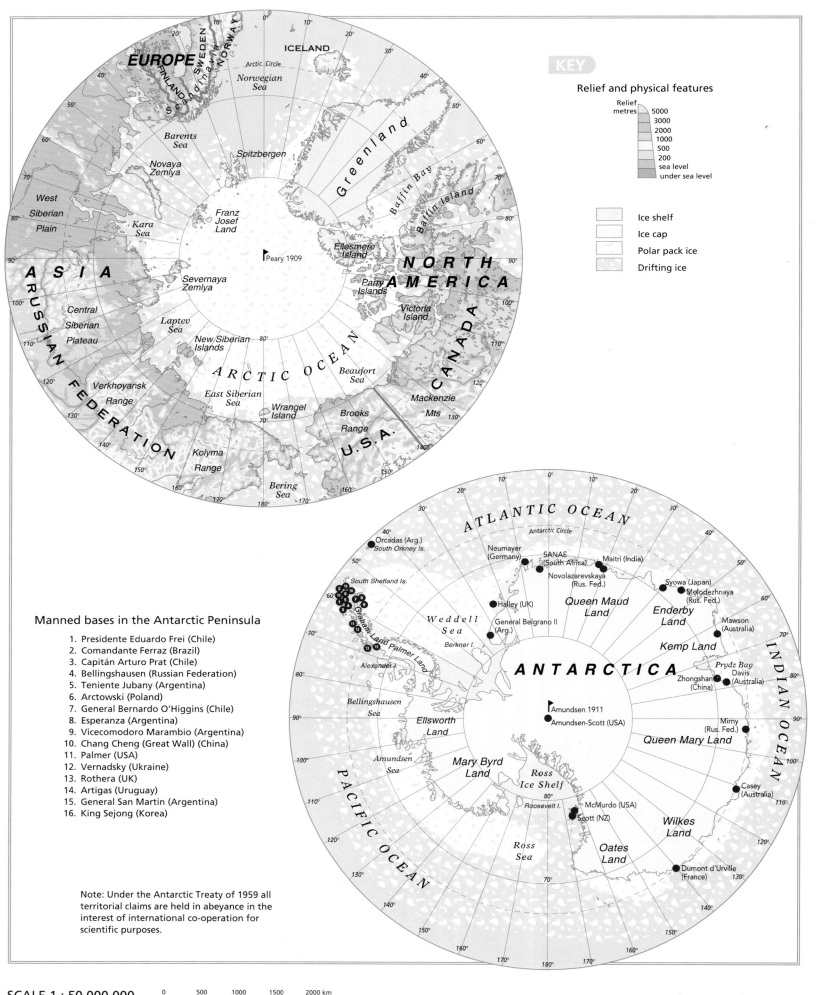

KEY

Relief and physical features

Relief metres	
	5000
	3000
	2000
	1000
	500
	200
	sea level
	under sea level

Ice shelf

Ice cap

Polar pack ice

Drifting ice

Arctic Ocean map labels

EUROPE
ICELAND
NORWAY
SWEDEN
FINLAND
Scandinavia
Arctic Circle
Norwegian Sea
Barents Sea
Spitzbergen
Novaya Zemlya
West Siberian Plain
Kara Sea
Franz Josef Land
Greenland
Baffin Bay
Baffin Island
Ellesmere Island
NORTH AMERICA
Peary 1909
ASIA
RUSSIAN FEDERATION
Severnaya Zemlya
Central Siberian Plateau
Laptev Sea
New Siberian Islands
Parry Islands
Victoria Island
CANADA
Verkhoyansk Range
East Siberian Sea
ARCTIC OCEAN
Beaufort Sea
Mackenzie Mts
Wrangel Island
Brooks Range
U.S.A.
Kolyma Range
Bering Sea

Manned bases in the Antarctic Peninsula

1. Presidente Eduardo Frei (Chile)
2. Comandante Ferraz (Brazil)
3. Capitán Arturo Prat (Chile)
4. Bellingshausen (Russian Federation)
5. Teniente Jubany (Argentina)
6. Arctowski (Poland)
7. General Bernardo O'Higgins (Chile)
8. Esperanza (Argentina)
9. Vicecomodoro Marambio (Argentina)
10. Chang Cheng (Great Wall) (China)
11. Palmer (USA)
12. Vernadsky (Ukraine)
13. Rothera (UK)
14. Artigas (Uruguay)
15. General San Martin (Argentina)
16. King Sejong (Korea)

Note: Under the Antarctic Treaty of 1959 all territorial claims are held in abeyance in the interest of international co-operation for scientific purposes.

Antarctica map labels

ATLANTIC OCEAN
Antarctic Circle
Orcadas (Arg.)
South Orkney Is.
Neumayer (Germany)
SANAE (South Africa)
Maitri (India)
Novolazarevskaya (Rus. Fed.)
Syowa (Japan)
Molodezhnaya (Rus. Fed.)
South Shetland Is.
Halley (UK)
Queen Maud Land
Enderby Land
Mawson (Australia)
Weddell Sea
General Belgrano II (Arg.)
Berkner I.
Graham Land / Palmer Land
Kemp Land
Alexander I.
Prydz Bay
Davis (Australia)
Zhongshan (China)
ANTARCTICA
Bellingshausen Sea
Amundsen 1911
Amundsen-Scott (USA)
Mirny (Rus. Fed.)
Ellsworth Land
Queen Mary Land
Amundsen Sea
Mary Byrd Land
Ross Ice Shelf
Casey (Australia)
Roosevelt I.
McMurdo (USA)
Scott (NZ)
Wilkes Land
PACIFIC OCEAN
Ross Sea
Oates Land
Dumont d'Urville (France)
INDIAN OCEAN

SCALE 1 : 50 000 000

0   500   1000   1500   2000 km

Polar Stereographic projection

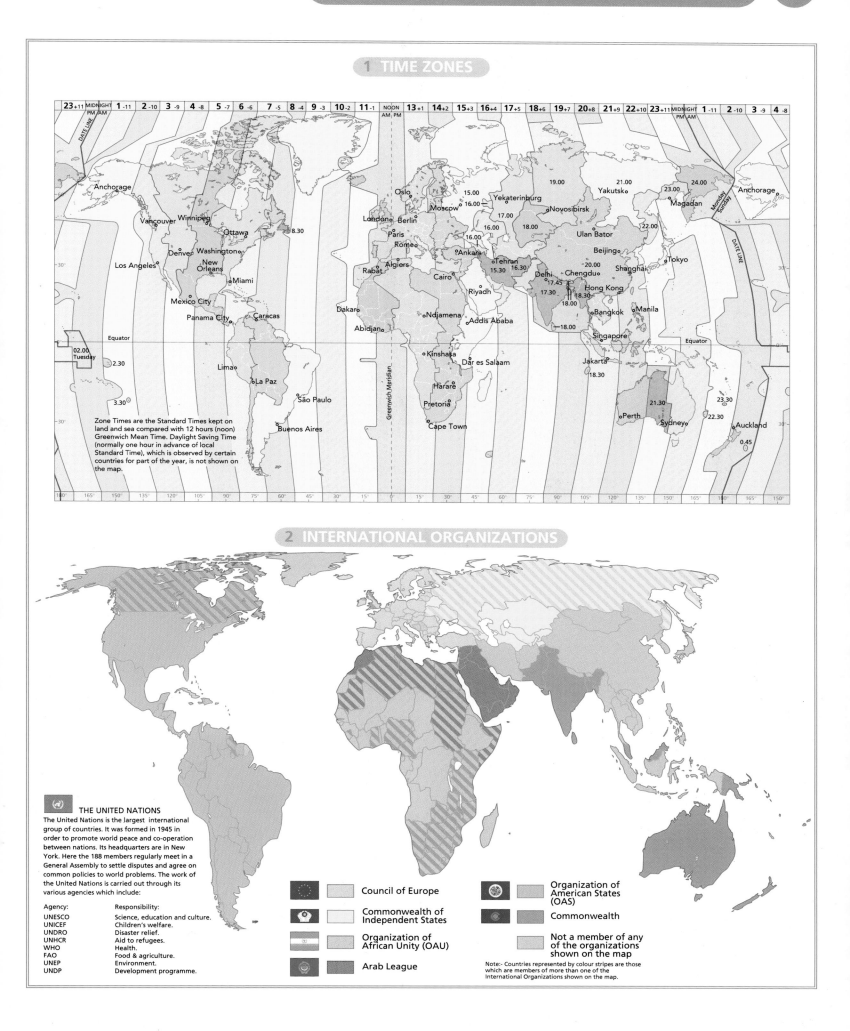

1 TIME ZONES

23 +11 MIDNIGHT PM AM	1 -11	2 -10	3 -9	4 -8	5 -7	6 -6	7 -5	8 -4	9 -3	10 -2	11 -1	NOON AM, PM	13 +1	14 +2	15 +3	16 +4	17 +5	18 +6	19 +7	20 +8	21 +9	22 +10	23 +11 MIDNIGHT PM AM	1 -11	2 -10	3 -9	4 -8

Anchorage
Oslo
19.00
Yakutsk
21.00
23.00
24.00
Anchorage
15.00
Moscow
Yekaterinburg
Novosibirsk
Magadan
Monday Sunday
London
Berlin
16.00
17.00
22.00
Paris
16.00
18.00
Ulan Bator
Vancouver
Winnipeg
Ottawa
8.30
Rome
Ankara
16.00
Beijing
Denver
Washington
New Orleans
Algiers
Tehran
-20.00
Tokyo
Los Angeles
Rabat
Cairo
15.30
16.30
Chengdu
Shanghai
Miami
Riyadh
Delhi
17.45
Hong Kong
17.30
18.30
Mexico City
Dakar
18.00
Bangkok
Manila
Panama City
Caracas
Ndjamena
Addis Ababa
Abidjan
-18.00
Singapore
Equator
Equator
02.00
Tuesday
2.30
Kinshasa
Dar es Salaam
Jakarta
Lima
18.30
La Paz
Harare
3.30
São Paulo
Pretoria
21.30
23.30
Perth
Sydney
22.30
Buenos Aires
Cape Town
Auckland
0.45

Zone Times are the Standard Times kept on land and sea compared with 12 hours (noon) Greenwich Mean Time. Daylight Saving Time (normally one hour in advance of local Standard Time), which is observed by certain countries for part of the year, is not shown on the map.

180° 165° 150° 135° 120° 105° 90° 75° 60° 45° 30° 15° 0° 15° 30° 45° 60° 75° 90° 105° 120° 135° 150° 165° 180° 165° 150°

2 INTERNATIONAL ORGANIZATIONS

THE UNITED NATIONS
The United Nations is the largest international group of countries. It was formed in 1945 in order to promote world peace and co-operation between nations. Its headquarters are in New York. Here the 188 members regularly meet in a General Assembly to settle disputes and agree on common policies to world problems. The work of the United Nations is carried out through its various agencies which include:

Agency:	Responsibility:
UNESCO	Science, education and culture.
UNICEF	Children's welfare.
UNDRO	Disaster relief.
UNHCR	Aid to refugees.
WHO	Health.
FAO	Food & agriculture.
UNEP	Environment.
UNDP	Development programme.

Council of Europe

Commonwealth of Independent States

Organization of African Unity (OAU)

Arab League

Organization of American States (OAS)

Commonwealth

Not a member of any of the organizations shown on the map

Note:- Countries represented by colour stripes are those which are members of more than one of the International Organizations shown on the map.

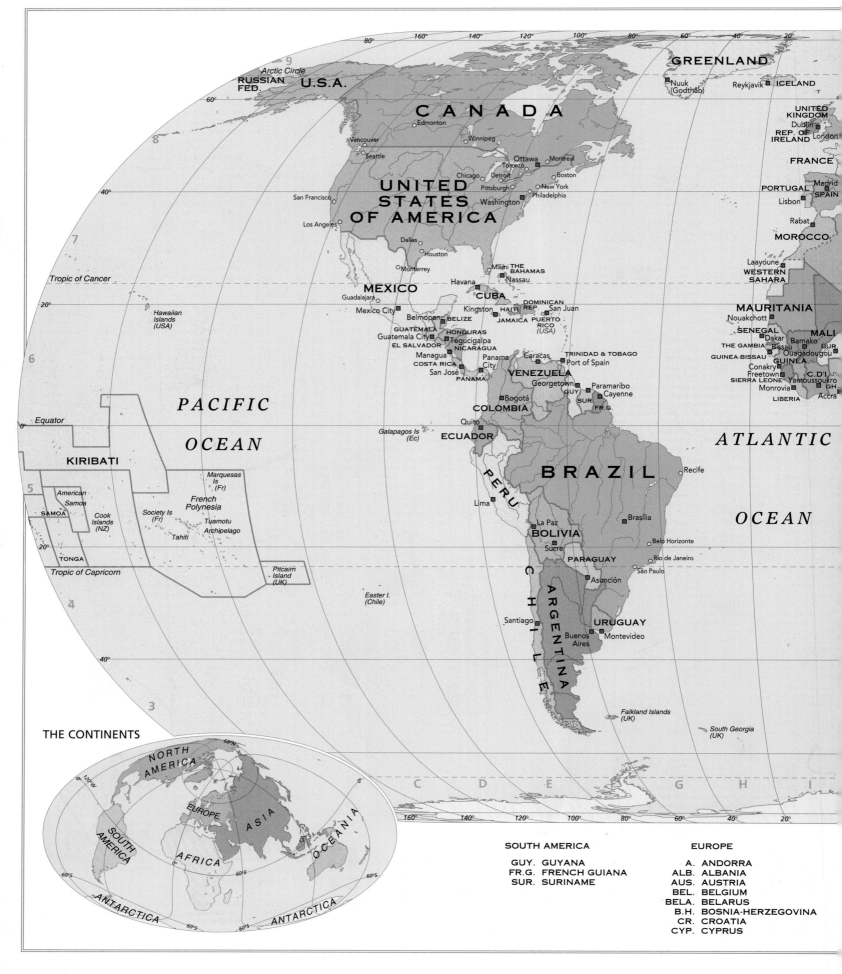

GREENLAND

Nuuk (Godthåb)   Reykjavik   ICELAND

RUSSIAN FED.   U.S.A.

Arctic Circle

C A N A D A

UNITED KINGDOM

Dublin

REP. OF IRELAND   London

Edmonton

FRANCE

Vancouver   Winnipeg

Seattle   Ottawa   Montreal

PORTUGAL   Madrid   SPAIN

Chicago   Detroit   Boston   Lisbon

UNITED   Toronto   New York

STATES   Pittsburgh   Washington   Philadelphia   Rabat

San Francisco   OF AMERICA   MOROCCO

Los Angeles

Dallas   Laayoune

Houston   WESTERN

Monterrey   Miami   THE   SAHARA

BAHAMAS

MEXICO   Havana   Nassau   MAURITANIA

Guadalajara   CUBA   Nouakchott

Mexico City   Kingston   HAITI   DOMINICAN   SENEGAL   MALI

REP.   San Juan   Dakar   Bamako   BUR.

Belmopan   BELIZE   JAMAICA   PUERTO   THE GAMBIA   Bissau   Ouagadougou

GUATEMALA   RICO   GUINEA-BISSAU   GUINEA

Guatemala City   HONDURAS   (USA)   Conakry   C.D'I.

EL SALVADOR   Tegucigalpa   Freetown   Yamoussoukro   GH.

Managua   NICARAGUA   SIERRA LEONE   Monrovia   Accra

COSTA RICA   Caracas   TRINIDAD & TOBAGO   LIBERIA

San José   Panama   Port of Spain

PANAMA   City   VENEZUELA

Georgetown   Paramaribo

GUY.   Cayenne

Bogotá   SUR.

COLOMBIA   FR.G.

Quito

PACIFIC   Galapagos Is   ECUADOR

(Ec)

OCEAN   ATLANTIC

Equator

KIRIBATI   B R A Z I L   Recife

Marquesas

Is   OCEAN

(Fr)   PERU

American   French   Lima

Samoa   Polynesia

SAMOA   Cook   Society Is   La Paz   Brasília

Islands   (Fr)   Tuamotu

(NZ)   Tahiti   Archipelago   BOLIVIA   Belo Horizonte

Sucre   Rio de Janeiro

TONGA   PARAGUAY   São Paulo

Tropic of Capricorn   Pitcairn   Asunción

Island

(UK)

Easter I.

(Chile)

ARGENTINA

URUGUAY

Santiago   Buenos   Montevideo

Aires

Falkland Islands

(UK)

South Georgia

(UK)

THE CONTINENTS

NORTH AMERICA

EUROPE   ASIA

SOUTH AMERICA

AFRICA   OCEANIA

ANTARCTICA   ANTARCTICA

SOUTH AMERICA	EUROPE
GUY. GUYANA	A. ANDORRA
FR.G. FRENCH GUIANA	ALB. ALBANIA
SUR. SURINAME	AUS. AUSTRIA
	BEL. BELGIUM
	BELA. BELARUS
	B.H. BOSNIA-HERZEGOVINA
	CR. CROATIA
	CYP. CYPRUS

SCALE 1 : 77 500 000

0   800   1600   2400   3200 km

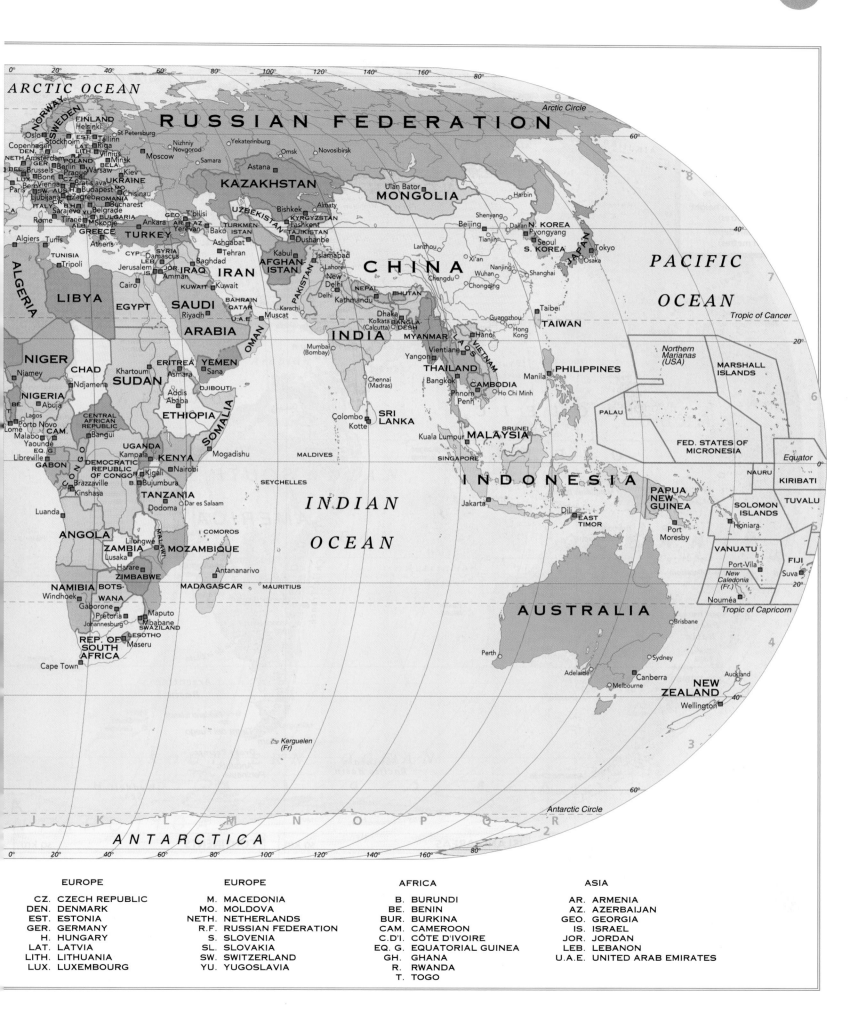

EUROPE		EUROPE		AFRICA		ASIA	
CZ.	CZECH REPUBLIC	M.	MACEDONIA	B.	BURUNDI	AR.	ARMENIA
DEN.	DENMARK	MO.	MOLDOVA	BE.	BENIN	AZ.	AZERBAIJAN
EST.	ESTONIA	NETH.	NETHERLANDS	BUR.	BURKINA	GEO.	GEORGIA
GER.	GERMANY	R.F.	RUSSIAN FEDERATION	CAM.	CAMEROON	IS.	ISRAEL
H.	HUNGARY	S.	SLOVENIA	C.D'I.	CÔTE D'IVOIRE	JOR.	JORDAN
LAT.	LATVIA	SL.	SLOVAKIA	EQ. G.	EQUATORIAL GUINEA	LEB.	LEBANON
LITH.	LITHUANIA	SW.	SWITZERLAND	GH.	GHANA	U.A.E.	UNITED ARAB EMIRATES
LUX.	LUXEMBOURG	YU.	YUGOSLAVIA	R.	RWANDA		
				T.	TOGO		

Eckert IV projection

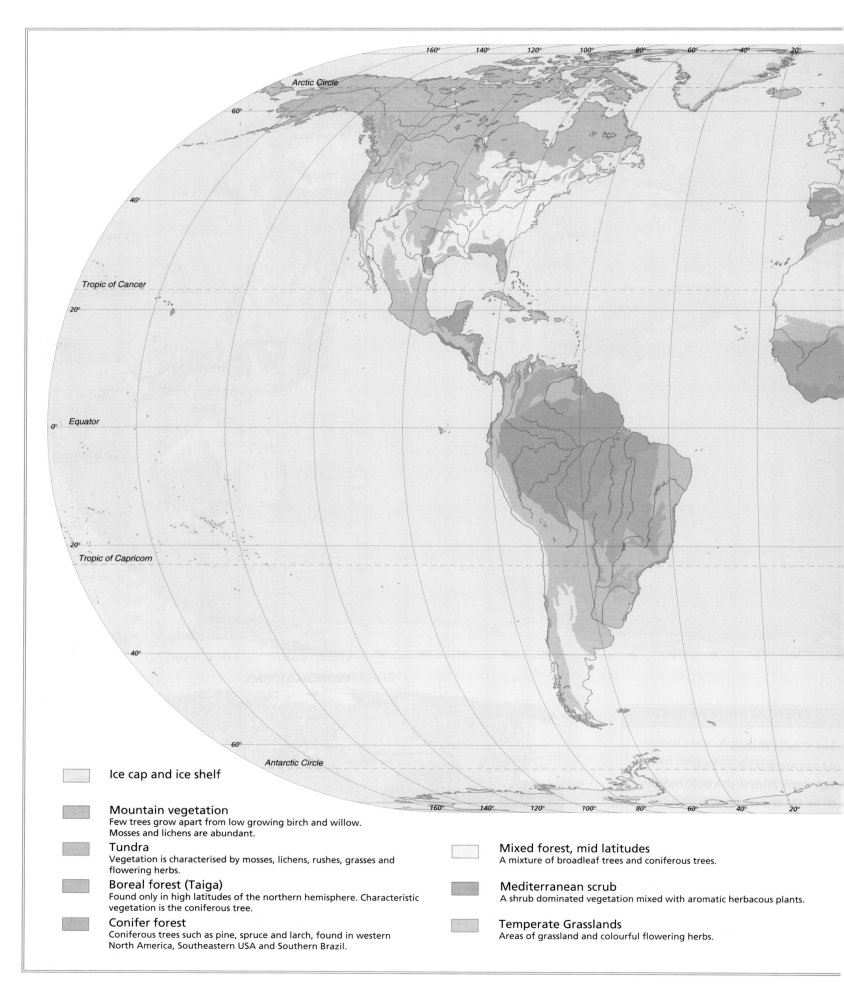

Ice cap and ice shelf

Mountain vegetation
Few trees grow apart from low growing birch and willow.
Mosses and lichens are abundant.

Tundra
Vegetation is characterised by mosses, lichens, rushes, grasses and
flowering herbs.

Boreal forest (Taiga)
Found only in high latitudes of the northern hemisphere. Characteristic
vegetation is the coniferous tree.

Conifer forest
Coniferous trees such as pine, spruce and larch, found in western
North America, Southeastern USA and Southern Brazil.

Mixed forest, mid latitudes
A mixture of broadleaf trees and coniferous trees.

Mediterranean scrub
A shrub dominated vegetation mixed with aromatic herbacous plants.

Temperate Grasslands
Areas of grassland and colourful flowering herbs.

SCALE 1 : 80 000 000

0   800   1600   2400   3200 km

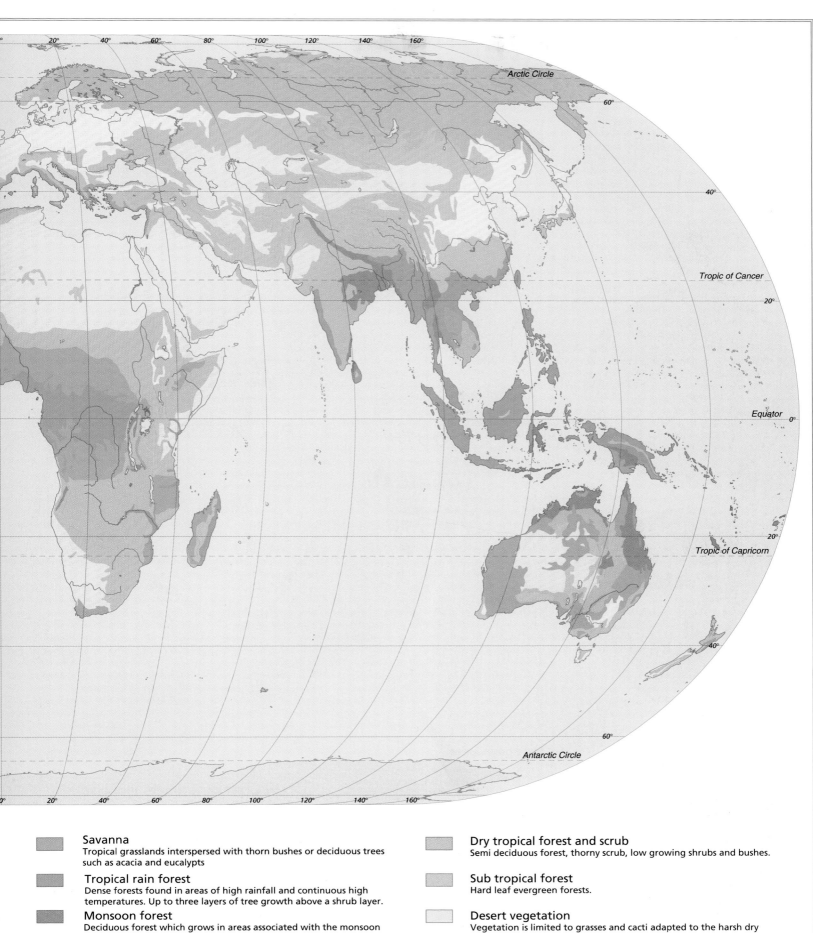

Savanna
Tropical grasslands interspersed with thorn bushes or deciduous trees such as acacia and eucalypts

Tropical rain forest
Dense forests found in areas of high rainfall and continuous high temperatures. Up to three layers of tree growth above a shrub layer.

Monsoon forest
Deciduous forest which grows in areas associated with the monsoon climate.

Dry tropical forest and scrub
Semi deciduous forest, thorny scrub, low growing shrubs and bushes.

Sub tropical forest
Hard leaf evergreen forests.

Desert vegetation
Vegetation is limited to grasses and cacti adapted to the harsh dry conditions of desert areas.

Eckert IV projection

CONTINENTAL DRIFT

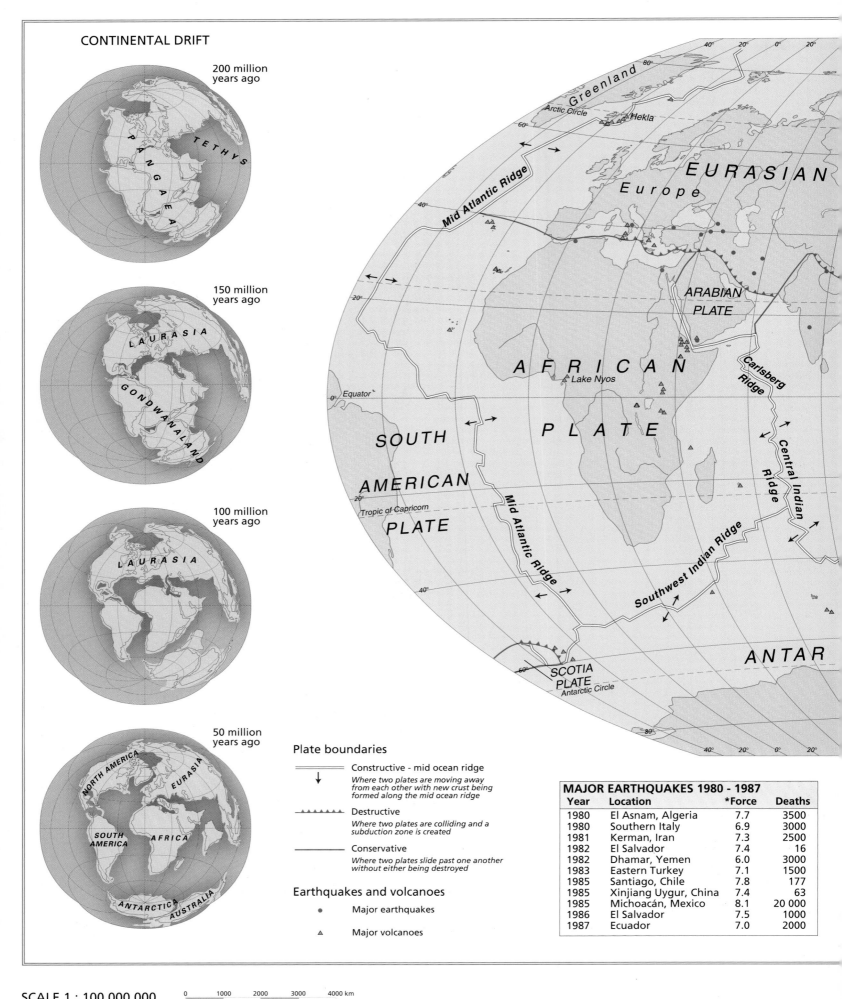

200 million years ago

PANGAEA

TETHYS

150 million years ago

LAURASIA

GONDWANALAND

100 million years ago

LAURASIA

50 million years ago

NORTH AMERICA

EURASIA

SOUTH AMERICA

AFRICA

ANTARCTICA

AUSTRALIA

Greenland

Arctic Circle

Hekla

Mid Atlantic Ridge

EURASIAN

Europe

ARABIAN PLATE

AFRICAN

Carlsberg Ridge

Lake Nyos

PLATE

Central Indian Ridge

SOUTH

Equator

AMERICAN

Tropic of Capricorn

PLATE

Mid Atlantic Ridge

Southwest Indian Ridge

ANTAR

SCOTIA PLATE

Antarctic Circle

Plate boundaries

Constructive - mid ocean ridge
Where two plates are moving away from each other with new crust being formed along the mid ocean ridge

Destructive
Where two plates are colliding and a subduction zone is created

Conservative
Where two plates slide past one another without either being destroyed

Earthquakes and volcanoes

●   Major earthquakes

▲   Major volcanoes

MAJOR EARTHQUAKES 1980 - 1987			
Year	Location	*Force	Deaths
1980	El Asnam, Algeria	7.7	3500
1980	Southern Italy	6.9	3000
1981	Kerman, Iran	7.3	2500
1982	El Salvador	7.4	16
1982	Dhamar, Yemen	6.0	3000
1983	Eastern Turkey	7.1	1500
1985	Santiago, Chile	7.8	177
1985	Xinjiang Uygur, China	7.4	63
1985	Michoacán, Mexico	8.1	20 000
1986	El Salvador	7.5	1000
1987	Ecuador	7.0	2000

SCALE 1 : 100 000 000

0   1000   2000   3000   4000 km

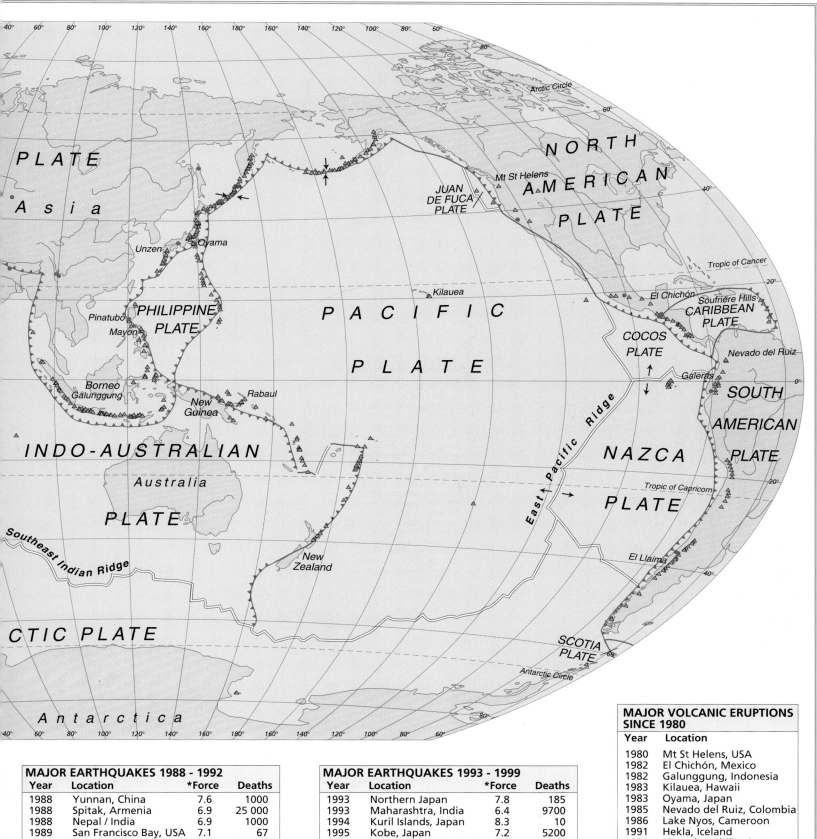

PLATE

Asia

PLATE

NORTH

AMERICAN

PLATE

Arctic Circle

Mt St Helens

JUAN
DE FUCA
PLATE

Tropic of Cancer

Unzen

Oyama

Kilauea

El Chichón

Soufriére Hills

CARIBBEAN
PLATE

PACIFIC

PHILIPPINE
PLATE

Pinatubo

Mayon

PLATE

Nevado del Ruiz

COCOS
PLATE

Galeras

SOUTH

Borneo
Gálunggung

New
Guinea

Rabaul

AMERICAN

East Pacific Ridge

NAZCA

PLATE

INDO-AUSTRALIAN

Australia

PLATE

Tropic of Capricorn

PLATE

El Llaima

Southeast Indian Ridge

New
Zealand

CTIC PLATE

SCOTIA
PLATE

Antarctic Circle

Antarctica

MAJOR VOLCANIC ERUPTIONS SINCE 1980

Year	Location
1980	Mt St Helens, USA
1982	El Chichón, Mexico
1982	Galunggung, Indonesia
1983	Kilauea, Hawaii
1983	Oyama, Japan
1985	Nevado del Ruiz, Colombia
1986	Lake Nyos, Cameroon
1991	Hekla, Iceland
1991	Pinatubo, Philippines
1991	Unzen, Japan
1993	Mayon, Philippines
1993	Galeras, Colombia
1994	El Llaima, Chile
1994	Rabaul, PNG
1997	Soufriére Hills, Montserrat

MAJOR EARTHQUAKES 1988 - 1992

Year	Location	*Force	Deaths
1988	Yunnan, China	7.6	1000
1988	Spitak, Armenia	6.9	25 000
1988	Nepal / India	6.9	1000
1989	San Francisco Bay, USA	7.1	67
1990	Manjil, Iran	7.7	50 000
1990	Luzon, Philippines	7.7	1600
1991	Georgia	7.1	114
1991	Uttar Pradesh, India	6.1	1600
1992	Flores, Indonesia	7.5	2500
1992	Erzincan, Turkey	6.8	500
1992	Cairo, Egypt	5.9	550

MAJOR EARTHQUAKES 1993 - 1999

Year	Location	*Force	Deaths
1993	Northern Japan	7.8	185
1993	Maharashtra, India	6.4	9700
1994	Kuril Islands, Japan	8.3	10
1995	Kobe, Japan	7.2	5200
1995	Sakhalin, Russian Fed	7.6	2500
1996	Yunnan, China	7.0	251
1997	Quae'n, Iran	7.1	2400
1999	Izmit, Turkey	7.4	15 657

* Earthquake force measured on the Richter scale

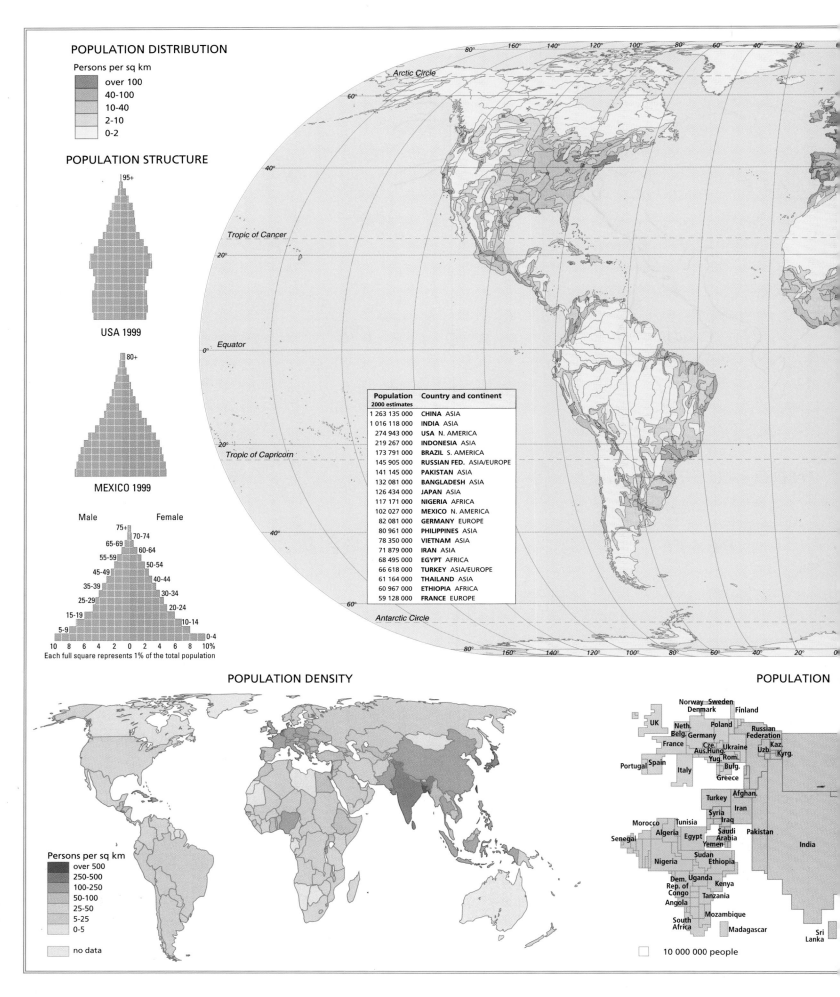

POPULATION DISTRIBUTION

Persons per sq km

- over 100
- 40-100
- 10-40
- 2-10
- 0-2

POPULATION STRUCTURE

USA 1999

95+

MEXICO 1999

80+

Male   Female

75+
70-74
65-69   60-64
55-59   50-54
45-49   40-44
35-39   30-34
25-29   20-24
15-19   10-14
5-9   0-4

10 8 6 4 2 0 2 4 6 8 10%

Each full square represents 1% of the total population

Population 2000 estimates	Country and continent
1 263 135 000	CHINA ASIA
1 016 118 000	INDIA ASIA
274 943 000	USA N. AMERICA
219 267 000	INDONESIA ASIA
173 791 000	BRAZIL S. AMERICA
145 905 000	RUSSIAN FED. ASIA/EUROPE
141 145 000	PAKISTAN ASIA
132 081 000	BANGLADESH ASIA
126 434 000	JAPAN ASIA
117 171 000	NIGERIA AFRICA
102 027 000	MEXICO N. AMERICA
82 081 000	GERMANY EUROPE
80 961 000	PHILIPPINES ASIA
78 350 000	VIETNAM ASIA
71 879 000	IRAN ASIA
68 495 000	EGYPT AFRICA
66 618 000	TURKEY ASIA/EUROPE
61 164 000	THAILAND ASIA
60 967 000	ETHIOPIA AFRICA
59 128 000	FRANCE EUROPE

POPULATION DENSITY

Persons per sq km

- over 500
- 250-500
- 100-250
- 50-100
- 25-50
- 5-25
- 0-5

no data

POPULATION

10 000 000 people

SCALE 1 : 100 000 000

0   1000   2000   3000   4000 km

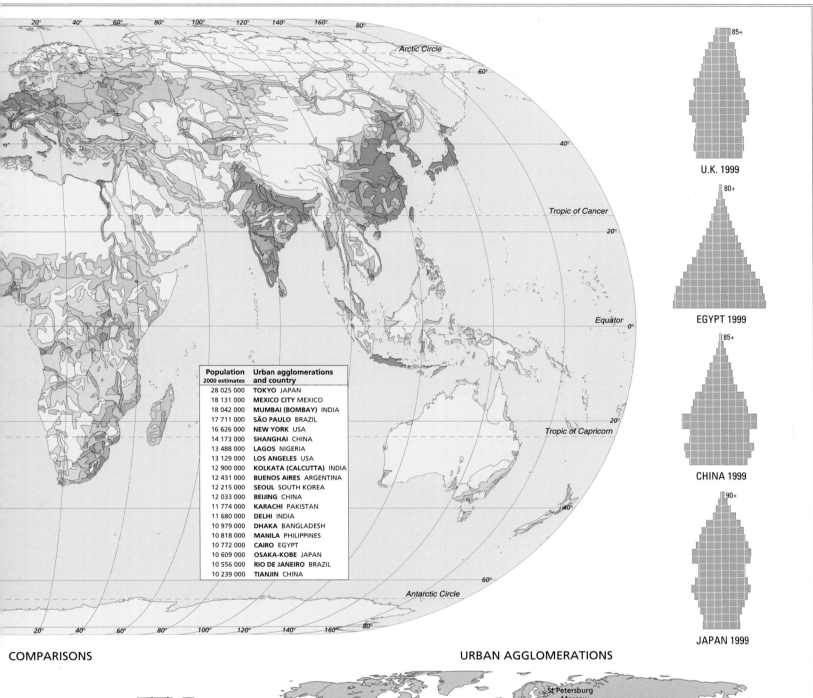

Population 2000 estimates	Urban agglomerations and country
28 025 000	**TOKYO** JAPAN
18 131 000	**MEXICO CITY** MEXICO
18 042 000	**MUMBAI (BOMBAY)** INDIA
17 711 000	**SÃO PAULO** BRAZIL
16 626 000	**NEW YORK** USA
14 173 000	**SHANGHAI** CHINA
13 488 000	**LAGOS** NIGERIA
13 129 000	**LOS ANGELES** USA
12 900 000	**KOLKATA (CALCUTTA)** INDIA
12 431 000	**BUENOS AIRES** ARGENTINA
12 215 000	**SEOUL** SOUTH KOREA
12 033 000	**BEIJING** CHINA
11 774 000	**KARACHI** PAKISTAN
11 680 000	**DELHI** INDIA
10 979 000	**DHAKA** BANGLADESH
10 818 000	**MANILA** PHILIPPINES
10 772 000	**CAIRO** EGYPT
10 609 000	**OSAKA-KOBE** JAPAN
10 556 000	**RIO DE JANEIRO** BRAZIL
10 239 000	**TIANJIN** CHINA

U.K. 1999

EGYPT 1999

CHINA 1999

JAPAN 1999

COMPARISONS

URBAN AGGLOMERATIONS

Urban agglomerations 2000
- over 15 million population
- 10-15 million population
- 5-10 million population

Eckert IV projection

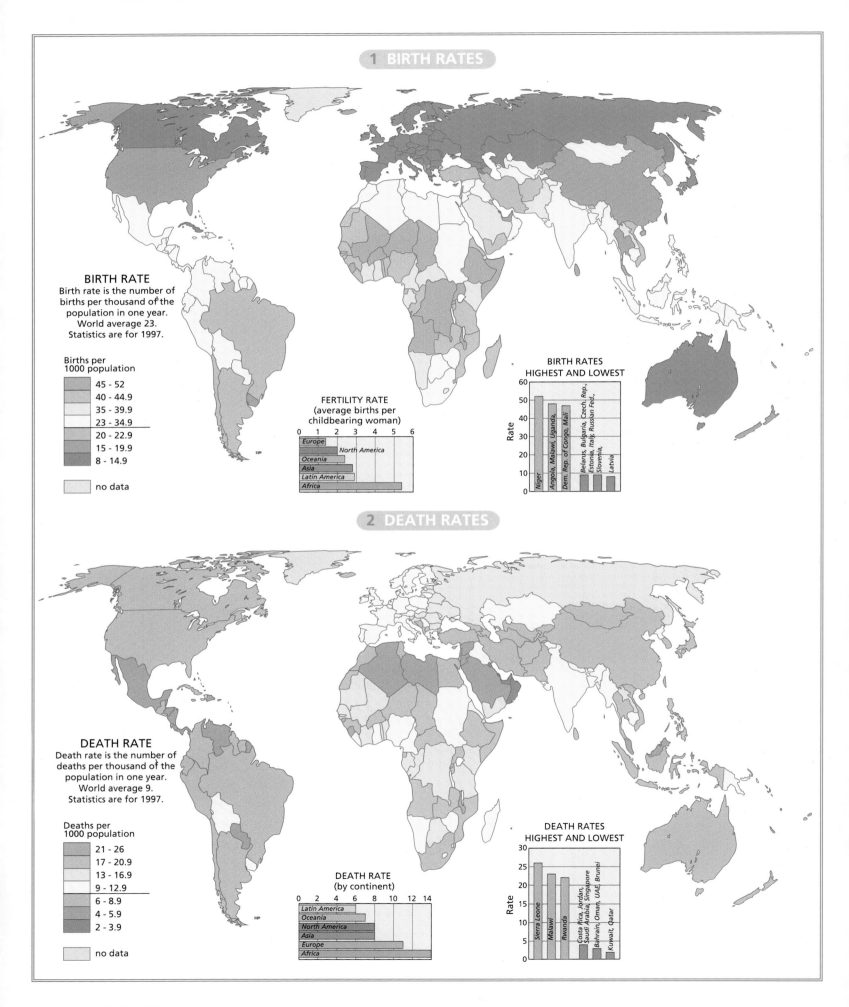

1 BIRTH RATES

BIRTH RATE
Birth rate is the number of births per thousand of the population in one year. World average 23. Statistics are for 1997.

Births per 1000 population
- 45 - 52
- 40 - 44.9
- 35 - 39.9
- 23 - 34.9
- 20 - 22.9
- 15 - 19.9
- 8 - 14.9

no data

FERTILITY RATE
(average births per childbearing woman)

Europe
North America
Oceania
Asia
Latin America
Africa

0 1 2 3 4 5 6

BIRTH RATES HIGHEST AND LOWEST

Niger
Angola, Malawi, Uganda,
Dem. Rep. of Congo, Mali
Belarus, Bulgaria, Czech. Rep.,
Estonia, Italy, Russian Fed.,
Slovenia,
Latvia

Rate — 0, 10, 20, 30, 40, 50, 60

2 DEATH RATES

DEATH RATE
Death rate is the number of deaths per thousand of the population in one year. World average 9. Statistics are for 1997.

Deaths per 1000 population
- 21 - 26
- 17 - 20.9
- 13 - 16.9
- 9 - 12.9
- 6 - 8.9
- 4 - 5.9
- 2 - 3.9

no data

DEATH RATE
(by continent)

Latin America
Oceania
North America
Asia
Europe
Africa

0 2 4 6 8 10 12 14

DEATH RATES HIGHEST AND LOWEST

Sierra Leone
Malawi
Rwanda
Costa Rica, Jordan,
Saudi Arabia, Singapore
Bahrain, Oman, UAE, Brunei
Kuwait, Qatar

Rate — 0, 5, 10, 15, 20, 25, 30

SCALE 1 : 140 000 000

Eckert IV projection

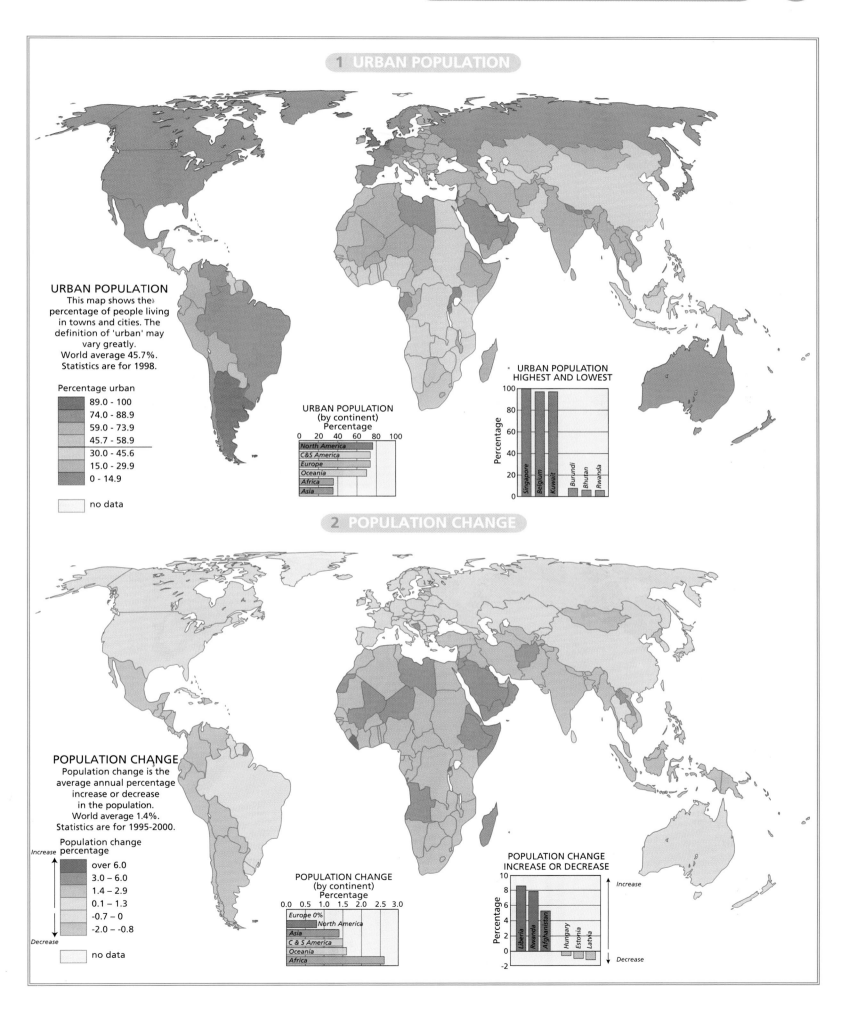

1 URBAN POPULATION

URBAN POPULATION
This map shows the percentage of people living in towns and cities. The definition of 'urban' may vary greatly.
World average 45.7%.
Statistics are for 1998.

Percentage urban
- 89.0 - 100
- 74.0 - 88.9
- 59.0 - 73.9
- 45.7 - 58.9
- 30.0 - 45.6
- 15.0 - 29.9
- 0 - 14.9

no data

URBAN POPULATION
(by continent)
Percentage
0   20   40   60   80   100
North America
C&S America
Europe
Oceania
Africa
Asia

URBAN POPULATION
HIGHEST AND LOWEST
Percentage
100
80
60
40
20
0
Singapore Belgium Kuwait Burundi Bhutan Rwanda

2 POPULATION CHANGE

POPULATION CHANGE
Population change is the average annual percentage increase or decrease in the population.
World average 1.4%.
Statistics are for 1995-2000.

Population change percentage
Increase
- over 6.0
- 3.0 – 6.0
- 1.4 – 2.9
- 0.1 – 1.3
- -0.7 – 0
- -2.0 – -0.8
Decrease

no data

POPULATION CHANGE
(by continent)
Percentage
0.0   0.5   1.0   1.5   2.0   2.5   3.0
Europe 0%
North America
Asia
C & S America
Oceania
Africa

POPULATION CHANGE
INCREASE OR DECREASE
Percentage
10
8
6
4
2
0
-2
Liberia Rwanda Afghanistan Hungary Estonia Latvia
Increase
Decrease

SCALE 1 : 140 000 000

Eckert IV projection

THREATS TO THE ENVIRONMENT

- *Forest
- Severe marine pollution
- Partial marine pollution
- River pollution
- Forest areas under threat
- Forest above global average of deforestation
- Current nuclear test site
- Former nuclear test site
- Major city with air pollution problem due to industry and vehicle exhaust
- Offshore oil production

*Includes Tropical rain forest, Monsoon forest and Dry tropical scrub. See World Natural Vegetation pp120-121.

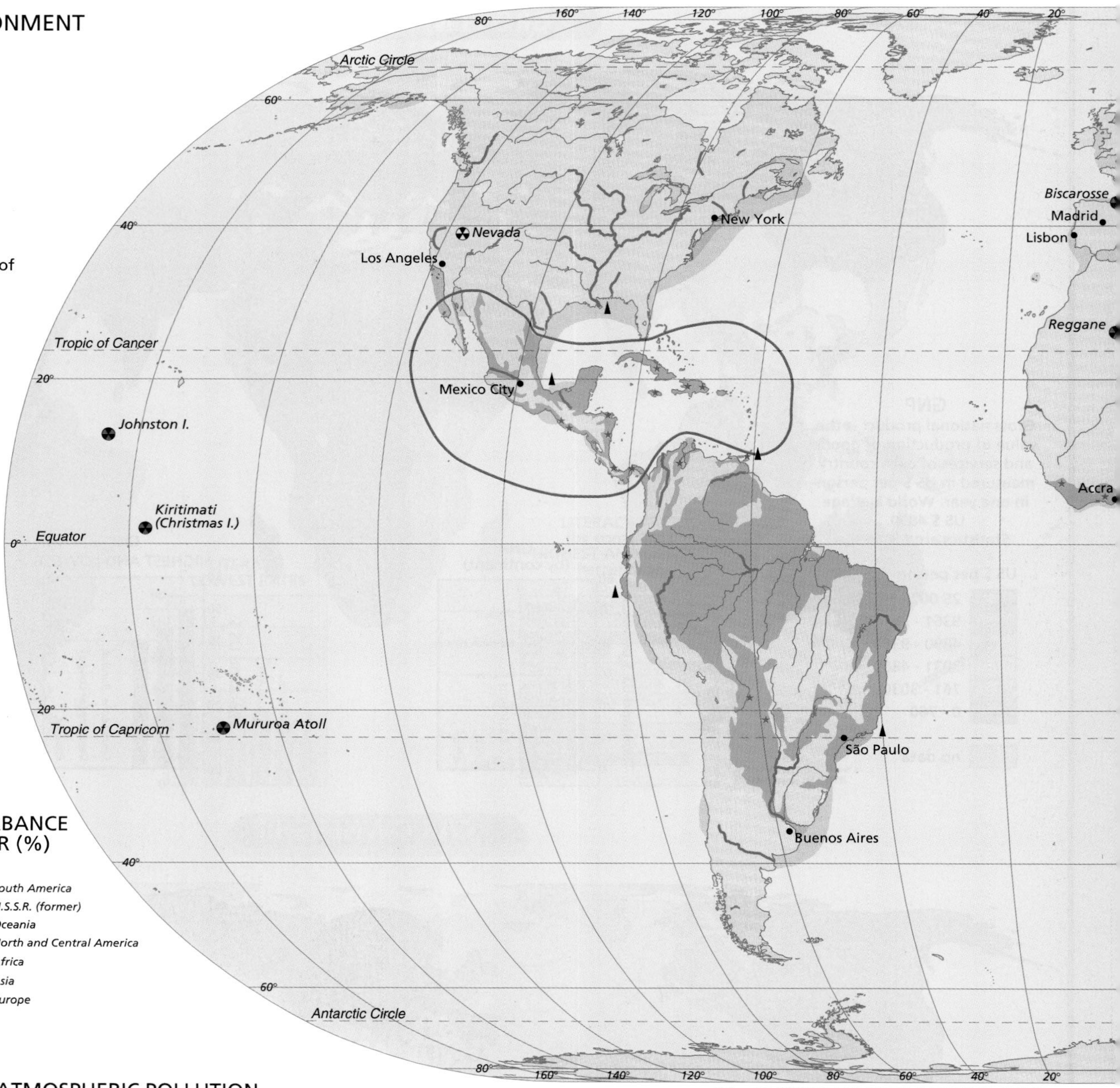
Arctic Circle
Nevada
New York
Los Angeles
Biscarosse
Madrid
Lisbon
Reggane
Tropic of Cancer
Mexico City
Johnston I.
Accra
Kiritimati (Christmas I.)
Equator
Tropic of Capricorn
Mururoa Atoll
São Paulo
Buenos Aires
Antarctic Circle

DEGREE OF HUMAN DISTURBANCE TO NATURAL LAND COVER (%)

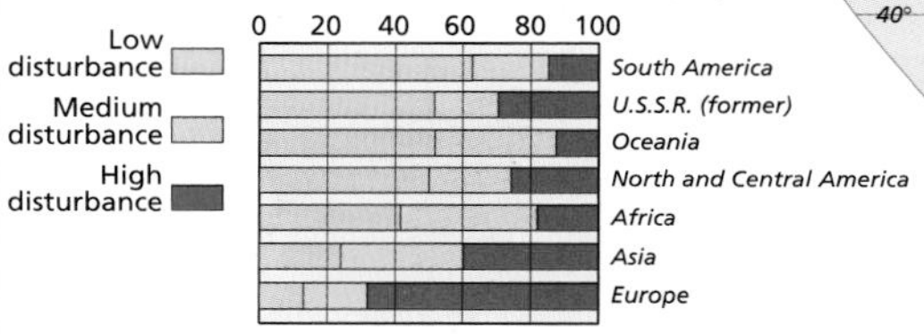
Low disturbance
Medium disturbance
High disturbance
0 20 40 60 80 100
South America
U.S.S.R. (former)
Oceania
North and Central America
Africa
Asia
Europe

ATMOSPHERIC POLLUTION (Carbon Dioxide emissions, 1997)

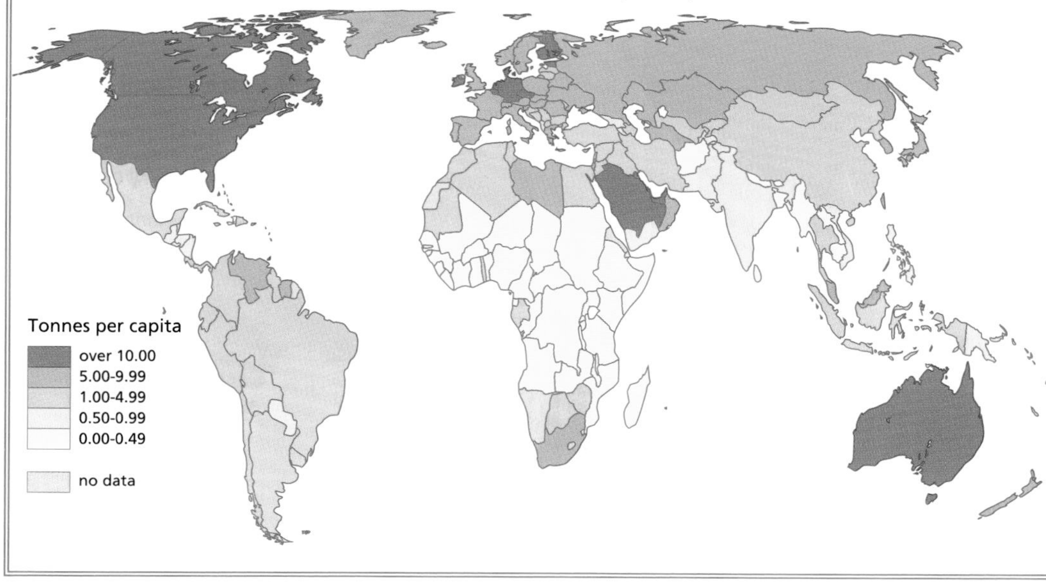
Tonnes per capita
over 10.00
5.00-9.99
1.00-4.99
0.50-0.99
0.00-0.49
no data

CO_2 EMISSIONS FROM FOSSIL FUEL CONSUMPTION 1955-1997

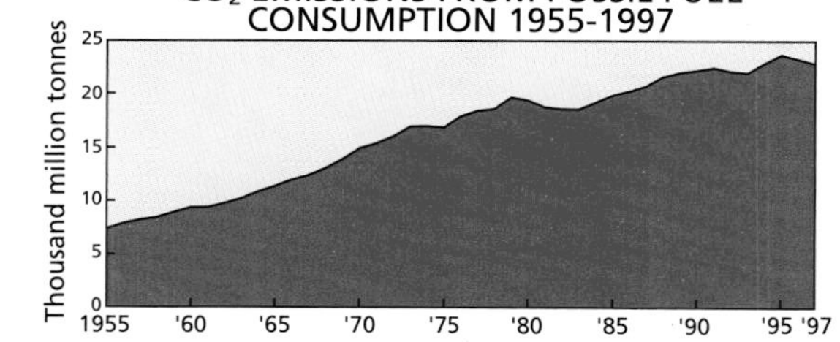
Thousand million tonnes
0
5
10
15
20
25
1955 '60 '65 '70 '75 '80 '85 '90 '95 '97

CO_2 EMISSIONS, 1997

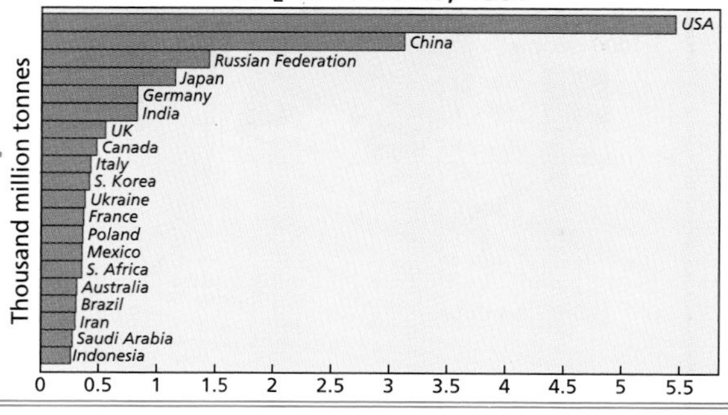
Thousand million tonnes
USA
China
Russian Federation
Japan
Germany
India
UK
Canada
Italy
S. Korea
Ukraine
France
Poland
Mexico
S. Africa
Australia
Brazil
Iran
Saudi Arabia
Indonesia
0 0.5 1 1.5 2 2.5 3 3.5 4 4.5 5 5.5

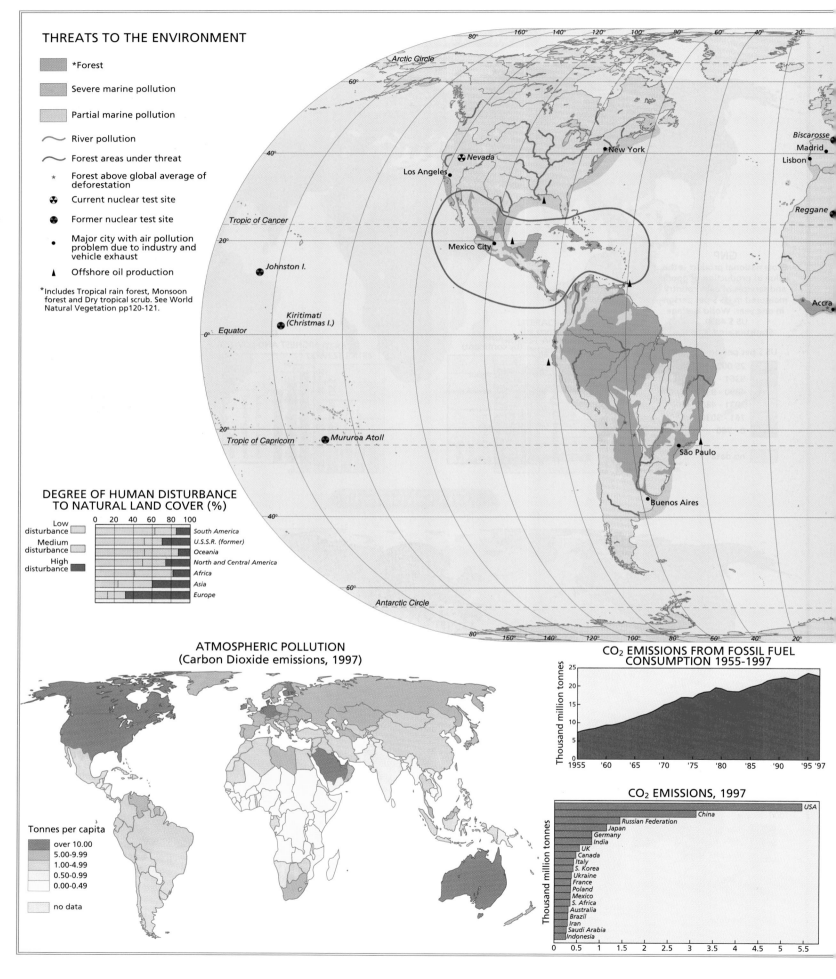

SCALE 1 : 100 000 000

0 1000 2000 3000 4000 km

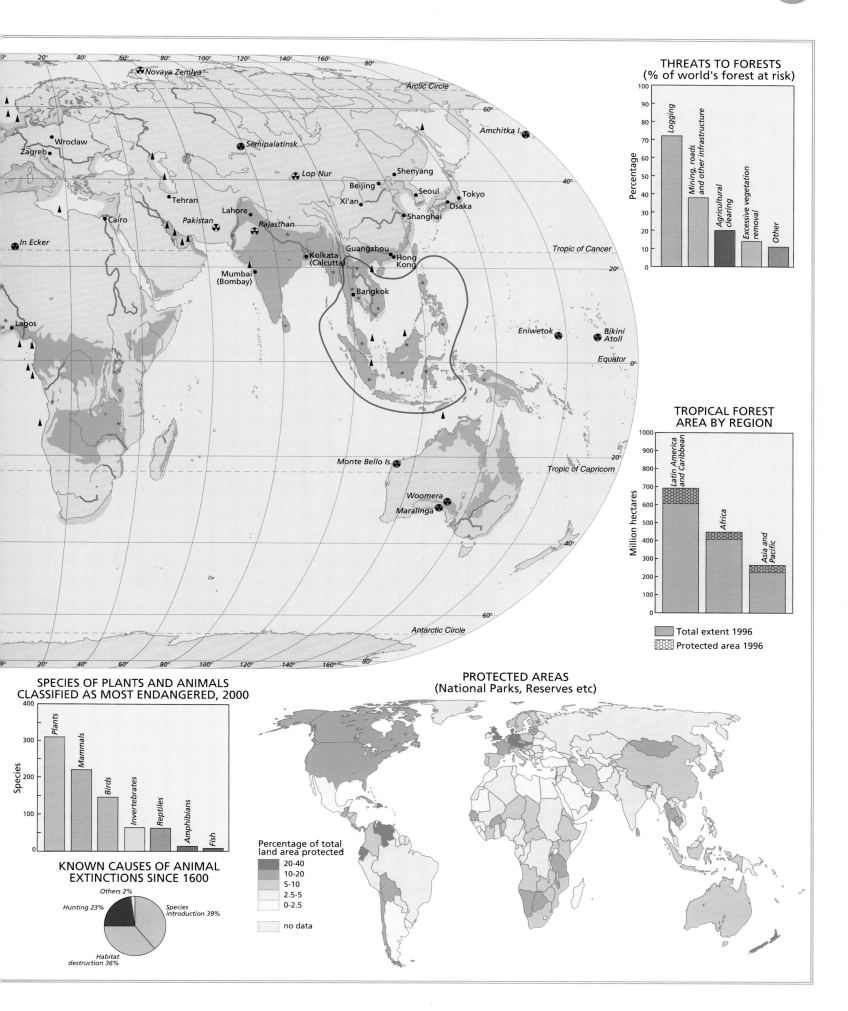

THREATS TO FORESTS
(% of world's forest at risk)

TROPICAL FOREST AREA BY REGION

- Total extent 1996
- Protected area 1996

SPECIES OF PLANTS AND ANIMALS CLASSIFIED AS MOST ENDANGERED, 2000

KNOWN CAUSES OF ANIMAL EXTINCTIONS SINCE 1600

Others 2%
Hunting 23%
Species introduction 39%
Habitat destruction 36%

PROTECTED AREAS
(National Parks, Reserves etc)

Percentage of total land area protected
- 20-40
- 10-20
- 5-10
- 2.5-5
- 0-2.5
- no data

Eckert IV projection

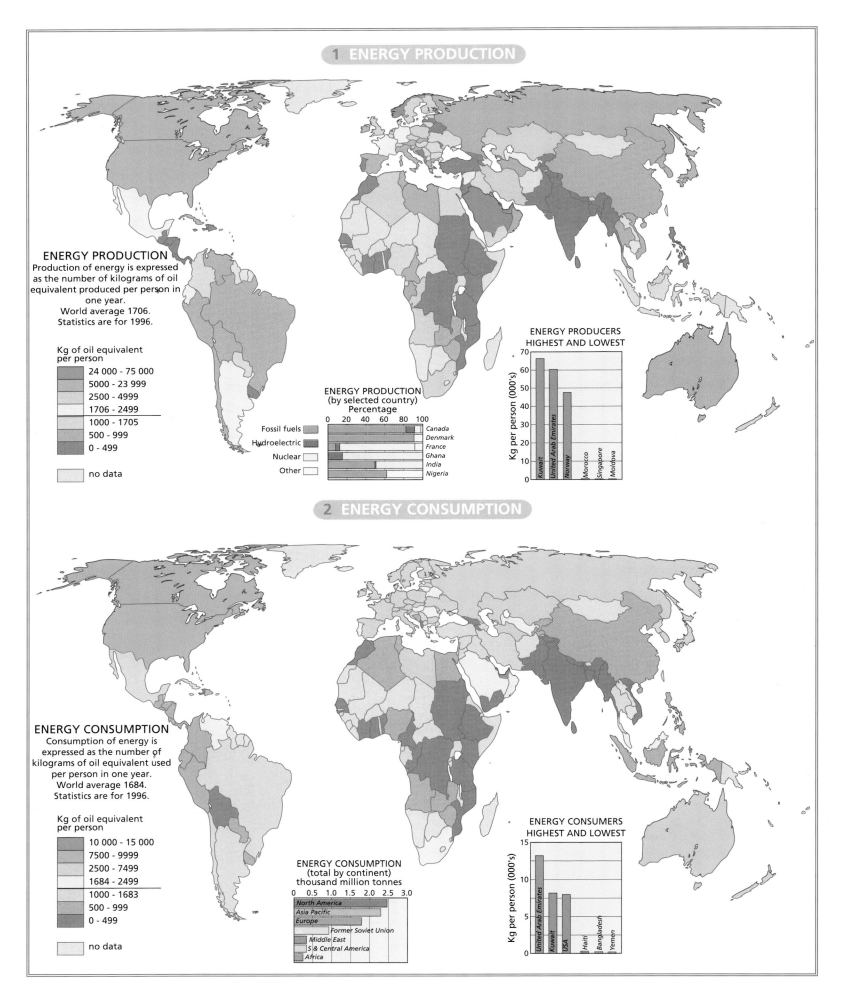

1 ENERGY PRODUCTION

ENERGY PRODUCTION
Production of energy is expressed as the number of kilograms of oil equivalent produced per person in one year.
World average 1706.
Statistics are for 1996.

Kg of oil equivalent per person

- 24 000 - 75 000
- 5000 - 23 999
- 2500 - 4999
- 1706 - 2499
- 1000 - 1705
- 500 - 999
- 0 - 499

no data

ENERGY PRODUCTION
(by selected country)
Percentage

0 20 40 60 80 100

- Fossil fuels
- Hydroelectric
- Nuclear
- Other

Canada
Denmark
France
Ghana
India
Nigeria

ENERGY PRODUCERS
HIGHEST AND LOWEST

Kg per person (000's)

70
60
50
40
30
20
10
0

Kuwait
United Arab Emirates
Norway
Morocco
Singapore
Moldova

2 ENERGY CONSUMPTION

ENERGY CONSUMPTION
Consumption of energy is expressed as the number of kilograms of oil equivalent used per person in one year.
World average 1684.
Statistics are for 1996.

Kg of oil equivalent per person

- 10 000 - 15 000
- 7500 - 9999
- 2500 - 7499
- 1684 - 2499
- 1000 - 1683
- 500 - 999
- 0 - 499

no data

ENERGY CONSUMPTION
(total by continent)
thousand million tonnes

0 0.5 1.0 1.5 2.0 2.5 3.0

- North America
- Asia Pacific
- Europe
- Former Soviet Union
- Middle East
- S & Central America
- Africa

ENERGY CONSUMERS
HIGHEST AND LOWEST

Kg per person (000's)

15
10
5
0

United Arab Emirates
Kuwait
USA
Haiti
Bangladesh
Yemen

SCALE 1 : 140 000 000

Eckert IV projection

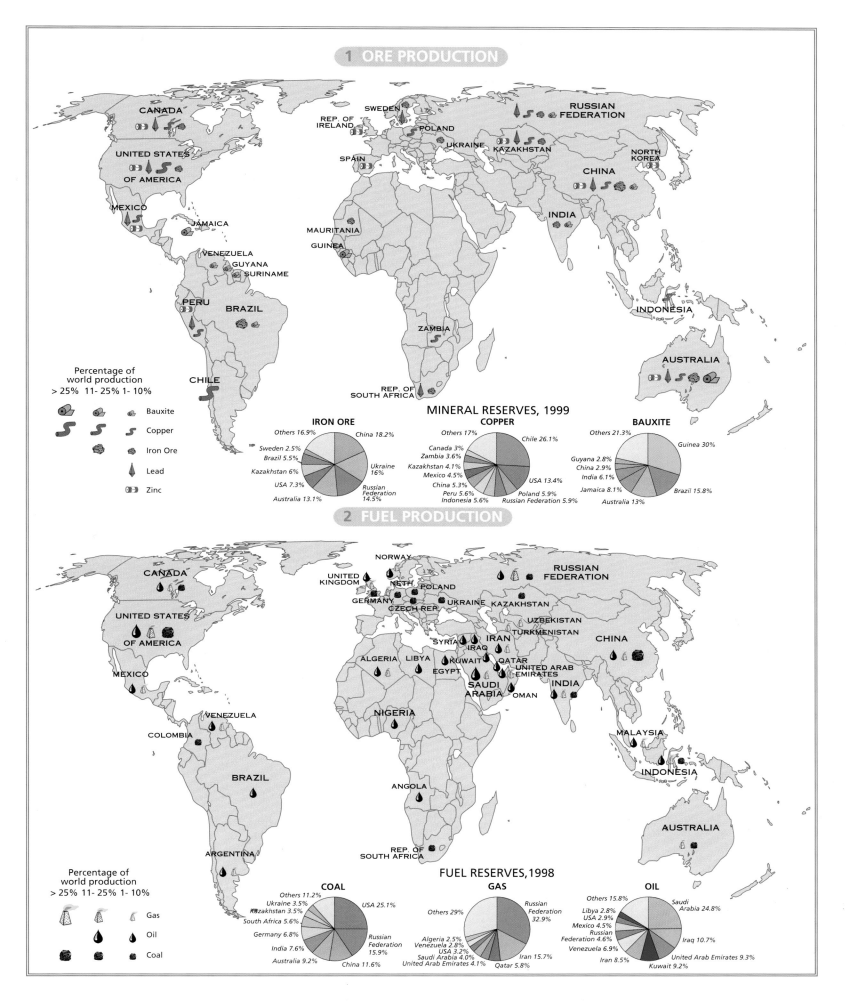

1 ORE PRODUCTION

CANADA
UNITED STATES OF AMERICA
MEXICO
JAMAICA
VENEZUELA
GUYANA
SURINAME
PERU
BRAZIL
CHILE
REP. OF IRELAND
SWEDEN
SPAIN
POLAND
UKRAINE
KAZAKHSTAN
RUSSIAN FEDERATION
NORTH KOREA
CHINA
INDIA
MAURITANIA
GUINEA
ZAMBIA
REP. OF SOUTH AFRICA
INDONESIA
AUSTRALIA

Percentage of world production
> 25% 11- 25% 1- 10%

Bauxite
Copper
Iron Ore
Lead
Zinc

MINERAL RESERVES, 1999

IRON ORE
China 18.2%
Others 16.9%
Sweden 2.5%
Brazil 5.5%
Kazakhstan 6%
USA 7.3%
Australia 13.1%
Ukraine 16%
Russian Federation 14.5%

COPPER
Chile 26.1%
Others 17%
Canada 3%
Zambia 3.6%
Kazakhstan 4.1%
Mexico 4.5%
China 5.3%
Peru 5.6%
Indonesia 5.6%
USA 13.4%
Poland 5.9%
Russian Federation 5.9%

BAUXITE
Guinea 30%
Others 21.3%
Guyana 2.8%
China 2.9%
India 6.1%
Jamaica 8.1%
Australia 13%
Brazil 15.8%

2 FUEL PRODUCTION

CANADA
UNITED STATES OF AMERICA
MEXICO
VENEZUELA
COLOMBIA
BRAZIL
ARGENTINA
NORWAY
UNITED KINGDOM
NETH.
GERMANY
POLAND
CZECH REP.
UKRAINE
KAZAKHSTAN
UZBEKISTAN
TURKMENISTAN
SYRIA
IRAQ
IRAN
ALGERIA
LIBYA
EGYPT
KUWAIT
QATAR
UNITED ARAB EMIRATES
SAUDI ARABIA
OMAN
NIGERIA
ANGOLA
RUSSIAN FEDERATION
CHINA
INDIA
MALAYSIA
INDONESIA
AUSTRALIA
REP. OF SOUTH AFRICA

Percentage of world production
> 25% 11- 25% 1- 10%

Gas
Oil
Coal

FUEL RESERVES, 1998

COAL
USA 25.1%
Others 11.2%
Ukraine 3.5%
Kazakhstan 3.5%
South Africa 5.6%
Germany 6.8%
India 7.6%
Australia 9.2%
China 11.6%
Russian Federation 15.9%

GAS
Russian Federation 32.9%
Others 29%
Algeria 2.5%
Venezuela 2.8%
USA 3.2%
Saudi Arabia 4.0%
United Arab Emirates 4.1%
Qatar 5.8%
Iran 15.7%

OIL
Saudi Arabia 24.8%
Others 15.8%
Libya 2.8%
USA 2.9%
Mexico 4.5%
Russian Federation 4.6%
Venezuela 6.9%
Iran 8.5%
Kuwait 9.2%
United Arab Emirates 9.3%
Iraq 10.7%

SCALE 1 : 140 000 000

Eckert IV projection

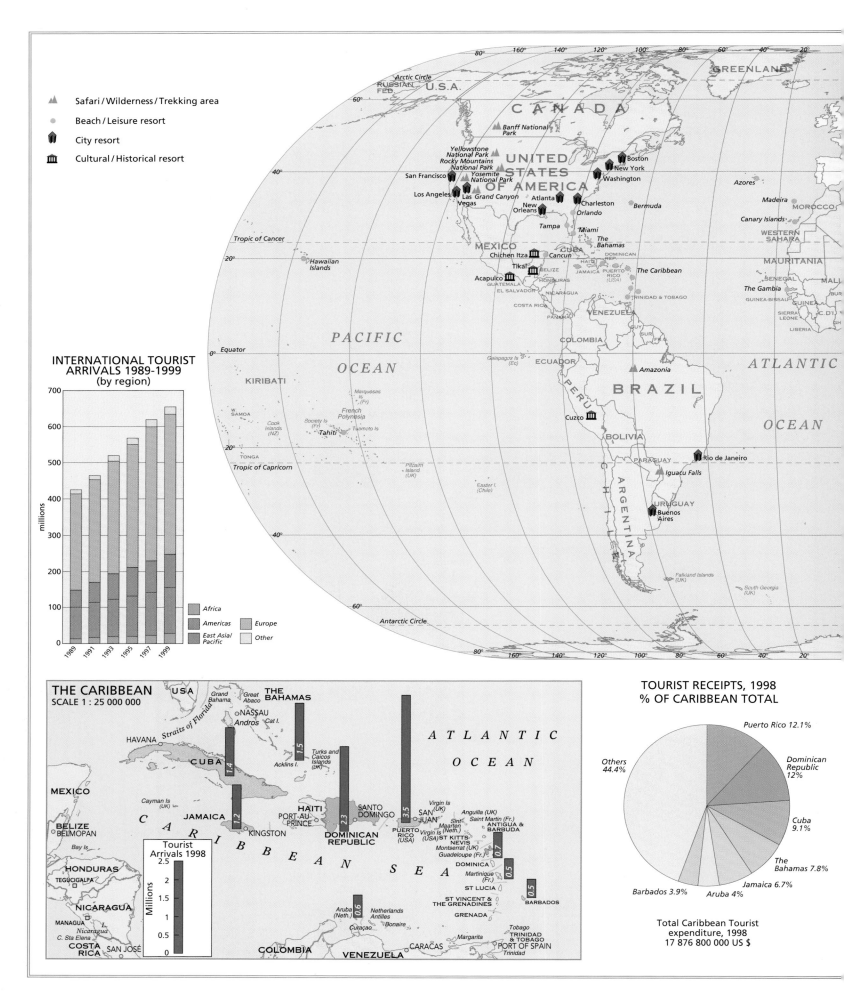

▲▲ Safari / Wilderness / Trekking area

● Beach / Leisure resort

⬠ City resort

🏛 Cultural / Historical resort

INTERNATIONAL TOURIST ARRIVALS 1989-1999
(by region)

millions

700
600
500
400
300
200
100
0

1989 1991 1993 1995 1997 1999

Africa
Americas
East Asia/Pacific
Europe
Other

THE CARIBBEAN
SCALE 1 : 25 000 000

USA
Grand Bahama
Great Abaco
THE BAHAMAS
NASSAU
Andros
Cat I.

HAVANA
Straits of Florida
CUBA 1.4
Turks and Caicos Islands (UK)
Acklins I.
1.5

MEXICO

Cayman Is (UK)

BELIZE
BELMOPAN

JAMAICA 1.2
KINGSTON

HAITI
PORT-AU-PRINCE

SANTO DOMINGO
DOMINICAN REPUBLIC 2.3

PUERTO RICO (USA) 3.5
SAN JUAN

Virgin Is (UK)
Virgin Is (USA)
Sint Maarten (Neth.)

ATLANTIC OCEAN

Anguilla (UK)
Saint Martin (Fr.)
ANTIGUA & BARBUDA 0.7
ST KITTS-NEVIS
Montserrat (UK)
Guadeloupe (Fr.)

DOMINICA 0.5
Martinique (Fr.)

ST LUCIA 0.5

Bay Is

HONDURAS
TEGUCIGALPA

Tourist Arrivals 1998

Millions
2.5
2
1.5
1
0.5
0

CARIBBEAN SEA

ST VINCENT & THE GRENADINES

GRENADA

BARBADOS 0.5

NICARAGUA
MANAGUA
L. Nicaragua
C. Sta Elena

COSTA RICA
SAN JOSÉ

Aruba (Neth.) 0.6
Curaçao
Netherlands Antilles
Bonaire

COLOMBIA
VENEZUELA
CARACAS

Margarita
Tobago
TRINIDAD & TOBAGO
PORT OF SPAIN
Trinidad

TOURIST RECEIPTS, 1998
% OF CARIBBEAN TOTAL

Puerto Rico 12.1%

Dominican Republic 12%

Cuba 9.1%

The Bahamas 7.8%

Jamaica 6.7%

Aruba 4%

Barbados 3.9%

Others 44.4%

Total Caribbean Tourist expenditure, 1998
17 876 800 000 US $

SCALE 1 : 100 000 000   0   1000   2000   3000   4000 km

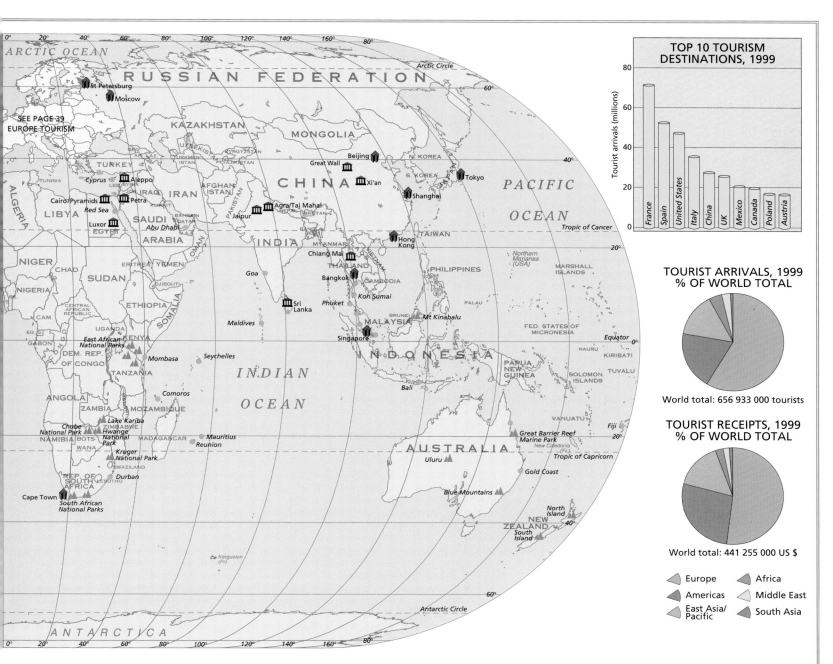

TOP 10 TOURISM DESTINATIONS, 1999

Tourist arrivals (millions)

France, Spain, United States, Italy, China, UK, Mexico, Canada, Poland, Austria

TOURIST ARRIVALS, 1999 % OF WORLD TOTAL

World total: 656 933 000 tourists

TOURIST RECEIPTS, 1999 % OF WORLD TOTAL

World total: 441 255 000 US $

- Europe
- Africa
- Americas
- Middle East
- East Asia/Pacific
- South Asia

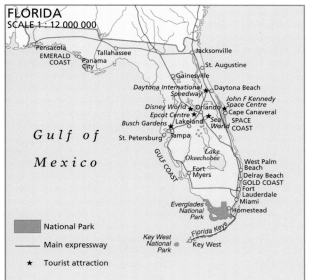

FLORIDA
SCALE 1 : 12 000 000

Pensacola
EMERALD COAST
Panama City
Tallahassee
Jacksonville
St. Augustine
Gainesville
Daytona International Speedway
Daytona Beach
John F Kennedy Space Centre
Disney World ★ Orlando
Epcot Centre ★
Cape Canaveral
Busch Gardens ★ Lakeland
Sea World ★
SPACE COAST
St. Petersburg
Tampa

Gulf of Mexico

GULF COAST
Lake Okeechobee
West Palm Beach
Fort Myers
Delray Beach
GOLD COAST
Fort Lauderdale
Miami
GOLD COAST
Everglades National Park
Homestead
Florida Keys
Key West National Park
Key West

- National Park
- Main expressway
- ★ Tourist attraction

GROWTH IN TOURISM - Tourist Arrivals (1000's)

Country	1995	1996	1997	1998	1999	Annual change 1998/1999
Australia	3726	4165	4318	4167	4326	3.8
China	20 034	22 765	23 770	25 073	27 047	7.9
Egypt	2872	3528	3657	3213	4489	39.7
France	60 033	62 406	67 310	70 000	71 400	2.0
India	2124	2288	2374	2359	2384	1.1
Mexico	20 241	21 405	19 351	19 810	20 216	2.0
Morocco	2602	2693	3072	3243	3950	21.8
Spain	38 803	40 541	43 252	47 749	51 958	8.8
Thailand	6952	7244	7294	7843	8280	5.6
Tunisia	4120	3885	4263	4718	4880	3.4
United Kingdom	23 537	25 163	25 515	25 745	25 740	0.0
United States	43 318	46 489	47 752	46 395	46 983	1.3

Eckert IV projection

	KEY INFORMATION		POPULATION						
FLAG	COUNTRY	CAPITAL CITY	TOTAL 1998	DENSITY persons per sq km 1998	BIRTH RATE per 1000 population 1997	DEATH RATE per 1000 population 1997	LIFE EXPEC-TANCY in years 1997	POP. CHANGE average % per annum 1995 - 2000	URBAN POP. % 1998
	Afghanistan	Kabul	21 354 000	32.7	43	18	46	5.3	20
	Albania	Tiranë	3 119 000	108.5	19	7	72	0.6	40
	Algeria	Algiers	30 081 000	12.6	27	5	70	2.3	59
	Angola	Luanda	12 092 000	9.7	48	19	46	3.3	33
	Argentina	Buenos Aires	36 123 000	13.1	20	8	73	1.3	89
	Armenia	Yerevan	3 536 000	118.7	12	6	74	0.2	69
	Australia	Canberra	18 520 000	2.4	14	7	78	1.1	85
	Austria	Vienna	8 140 000	97.1	10	10	77	0.6	65
	Azerbaijan	Baku	7 669 000	88.6	18	6	71	0.8	57
	Bahamas, The	Nassau	296 000	21.2	22	6	73	1.6	87
	Bahrain	Manama	595 000	861.1	22	3	69	2.1	91
	Bangladesh	Dhaka	124 774 000	866.5	28	10	58	1.6	23
	Barbados	Bridgetown	268 000	623.3	14	9	75	0.3	48
	Belarus	Minsk	10 315 000	49.7	9	13	68	-0.1	71
	Belgium	Brussels	10 141 000	332.3	11	10	77	0.3	97
	Belize	Belmopan	230 000	10.0	30	4	72	2.5	46
	Benin	Porto Novo	5 781 000	51.3	43	13	53	2.8	41
	Bhutan	Thimbu	2 004 000	43.0	40	9	66	2.8	6
	Bolivia	La Paz/Sucre	7 957 000	7.2	33	9	61	2.3	61
	Bosnia-Herzegovina	Sarajevo	3 675 000	71.9	13	7	73	3.9	42
	Botswana	Gaborone	1 570 000	2.7	34	15	47	2.2	49
	Brazil	Brasília	165 851 000	19.5	21	7	67	1.2	80
	Brunei	Bandar Seri Begawan	315 000	54.6	25	3	71	2.1	70
	Bulgaria	Sofia	8 336 000	75.1	9	14	71	-0.5	69
	Burkina	Ouagadougou	11 305 000	41.2	45	19	44	2.8	17
	Burundi	Bujumbura	6 457 000	232.0	43	20	42	2.8	8
	Cambodia	Phnom Penh	10 716 000	59.2	34	12	54	2.2	15
	Cameroon	Yaoundé	14 305 000	30.1	39	11	57	2.7	47
	Canada	Ottawa	30 563 000	3.1	12	7	79	0.9	77
	Cape Verde	Praia	408 000	101.2	36	8	68	2.5	56
	Central African Republic	Bangui	3 485 000	5.6	37	19	45	2.1	40
	Chad	Ndjamena	7 270 000	5.7	45	17	49	2.8	23
	Chile	Santiago	14 824 000	19.6	20	6	75	1.4	85
	China	Beijing	1 262 817 000	132.0	17	8	70	0.9	31
	Colombia	Bogotá	40 803 000	35.7	25	6	70	1.7	73
	Comoros	Moroni	658 000	353.4	38	10	59	3.1	31
	Congo	Brazzaville	2 785 000	8.1	44	16	48	2.8	61
	Congo, Dem. Rep. of	Kinshasa	49 139 000	21.0	47	15	51	2.6	30
	Costa Rica	San José	3 841 000	75.2	23	4	77	2.1	47
	Côte d'Ivoire	Yamoussoukro	14 292 000	44.3	37	16	47	2.0	45
	Croatia	Zagreb	4 481 000	79.3	10	13	72	-0.1	57
	Cuba	Havana	11 116 000	100.3	14	7	76	0.4	75
	Cyprus	Nicosia	771 000	83.3	14	8	77	1.3	55
	Czech Republic	Prague	10 282 000	130.4	9	11	74	-0.1	75
	Denmark	Copenhagen	5 270 000	122.3	13	11	75	0.2	85

LAND			EDUCATION AND HEALTH				DEVELOPMENT				
AREA sq km	CULTIV-ATED AREA '000s sq km 1997	FOREST '000s sq km 1997	ADULT LITERACY % 1998	*SCHOOL ENROL-MENT Secondary, gross % 1997	DOCTORS PER 100 000 PERSONS 1990-1998	FOOD INTAKE calories per capita per day 1997	ENERGY CONSU-MPTION million tonnes of oil equivalent 1997	TRADE BALANCE millions US $ 1996	GNP PER CAPITA US $ 1998	COUNTRY	TIME ZONES + OR - GMT
652 225	81	-	35.0	22	-	1747	-	-300	-	Afghanistan	+4½
28 748	7	10	83.7	35	141	2961	1.0	-960	810	Albania	+1
2 381 741	80	19	61.5	63	83	2853	26.5	2200	1550	Algeria	+1
1 246 700	35	222	42.0	12	-	1903	6.8	1500	340	Angola	+1
2 766 889	272	339	97.0	77	268	3093	61.7	-2220	8970	Argentina	-3
29 800	6	3	98.0	98	312	2371	1.8	-540	480	Armenia	+4
7 682 300	531	409	100.0	153	250	3224	101.6	-12 850	20 300	Australia	+8 to +10½
83 855	15	39	100.0	102	327	3536	27.8	-5810	26 850	Austria	+1
86 600	19	10	99.0	77	390	2236	12.0	-400	490	Azerbaijan	+4
13 939	-	-	95.0	86	141	2499	-	-230	-	Bahamas, The	-5
691	-	-	86.0	94	11	-	-	340	7660	Bahrain	+3
143 998	82	10	40.1	19	18	2086	24.3	-3410	350	Bangladesh	+6
430	-	-	97.0	98	113	3176	-	-760	7890	Barbados	-4
207 600	63	74	99.5	93	379	3226	25.1	-530	2200	Belarus	+2
30 520	8	7	99.0	147	365	3619	57.1	14 760	25 380	Belgium	+1
22 965	1	-	80.0	50	47	2907	-	-180	2610	Belize	-6
112 620	16	46	38.3	17	7	2487	2.1	-250	380	Benin	+1
46 620	2	-	42.0	-	20	-	-	0	390	Bhutan	+6
1 098 581	21	483	84.5	40	51	2174	4.3	-650	1000	Bolivia	-4
51 130	7	27	86.0	-	-	2266	1.8	-	-	Bosnia-Herzegovina	+1
581 370	3	139	75.6	68	20	2183	-	70	3600	Botswana	+2
8 511 965	653	5511	84.0	50	134	2974	172.0	-3760	4570	Brazil	-2 to -5
5 765	-	-	91.0	77	-	2857	-	90	24 000	Brunei	+8
110 994	45	32	98.5	77	333	2686	20.6	-1400	1230	Bulgaria	+2
274 200	34	43	22.4	9	<5**	2121	-	-430	240	Burkina	GMT
27 835	11	3	45.8	8	6	1685	-	-90	140	Burundi	+2
181 000	38	98	37.6	28	58	2048	-	-360	280	Cambodia	+7
475 442	72	196	73.5	26	7	2111	5.8	190	610	Cameroon	+1
9 970 610	457	2446	97.0	108	221	3119	238.0	18 190	20 020	Canada	-3½ to 8
4 033	-	0	73.0	10	29	3015	-	-230	1060	Cape Verde	-1
622 436	20	299	44.2	10	6	2016	-	-120	300	Central African Republic	+1
1 284 000	33	110	39.9	10	2	2032	-	10	230	Chad	+1
756 945	23	79	95.5	75	108	2796	23.0	510	4810	Chile	-4
9 562 000	1354	1333	83.3	71	115	2897	1113.1	29 210	750	China	+8
1 141 748	44	530	91.0	72	105	2597	30.5	910	2600	Colombia	-5
1 862	1	-	58.0	24	10	1858	-	-40	370	Comoros	+3
342 000	2	195	78.3	52	27	2144	1.2	1130	690	Congo	+1
2 345 410	79	-	71.0	30	10	1755	14.5	280	110	Congo, Dem. Rep. of	+1 to +2
51 100	5	12	95.0	50	126	2649	-	250	2780	Costa Rica	-6
322 463	74	55	44.7	24	10	2610	5.6	1150	700	Côte d'Ivoire	GMT
56 538	14	18	98.0	84	201	2445	7.7	-3500	4520	Croatia	+1
110 860	45	18	96.0	77	518	2480	14.3	-2360	-	Cuba	-5
9 251	1	-	97.0	-	231	3429	-	-2700	-	Cyprus	+2
78 864	33	26	100.0	103	293	3244	40.6	-2020	5040	Czech Republic	+1
43 075	24	4	100.0	123	283	3407	21.1	4520	33 260	Denmark	+1

*Total enrolment in secondary level of education, regardless of age, is expressed as a percentage of the official secondary school-age population.
**Estimate.

	KEY INFORMATION		POPULATION						
FLAG	COUNTRY	CAPITAL CITY	TOTAL 1998	DENSITY persons per sq km 1998	BIRTH RATE per 1000 population 1997	DEATH RATE per 1000 population 1997	LIFE EXPEC- TANCY in years 1997	POP. CHANGE average % per annum 1995 - 2000	URBAN POP. % 1998
	Djibouti	Djibouti	623 000	26.9	39	16	48	2.7	82
	Dominica	Roseau	71 000	94.7	19	8	78	0.1	70
	Dominican Republic	Santo Domingo	8 232 000	169.9	26	5	71	1.7	64
	Ecuador	Quito	12 175 000	44.8	25	6	70	2.0	63
	Egypt	Cairo	65 978 000	66.0	25	7	66	1.9	45
	El Salvador	San Salvador	6 032 000	286.7	28	6	69	2.2	46
	Equatorial Guinea	Malabo	431 000	15.4	44	18	48	2.5	43
	Eritrea	Asmara	3 577 000	30.5	41	12	51	3.7	18
	Estonia	Tallinn	1 429 000	31.6	9	13	70	-1.0	69
	Ethiopia	Addis Ababa	59 649 000	52.6	46	20	43	3.2	17
	Fiji	Suva	796 000	43.4	24	6	63	1.6	41
	Finland	Helsinki	5 154 000	15.2	12	10	77	0.3	66
	France	Paris	58 683 000	107.9	12	9	78	0.3	75
	Gabon	Libreville	1 167 000	4.4	37	16	52	2.8	79
	Gambia, The	Banjul	1 229 000	108.8	43	13	53	2.3	31
	Georgia	T'bilisi	5 059 000	72.6	10	7	73	-0.1	60
	Germany	Berlin	82 133 000	229.5	10	10	77	0.3	87
	Ghana	Accra	19 162 000	80.3	36	9	60	2.8	37
	Greece	Athens	10 600 000	80.3	10	10	78	0.3	60
	Guatemala	Guatemala City	10 801 000	99.2	34	7	64	2.8	39
	Guinea	Conakry	7 337 000	29.8	41	17	46	1.4	31
	Guinea-Bissau	Bissau	1 161 000	32.1	42	21	44	2.0	23
	Guyana	Georgetown	850 000	4.0	24	7	66	1.0	36
	Haiti	Port-au-Prince	7 952 000	286.6	32	13	54	1.9	34
	Honduras	Tegucigalpa	6 147 000	54.8	34	5	69	2.8	51
	Hungary	Budapest	10 116 000	108.7	10	14	71	-0.6	64
	Iceland	Reykjavík	276 000	2.7	15	7	79	1.0	92
	India	New Delhi	982 223 000	298.8	27	9	63	1.6	28
	Indonesia	Jakarta	206 338 000	107.5	24	8	65	1.5	39
	Iran	Tehran	65 758 000	39.9	22	6	69	2.2	61
	Iraq	Baghdad	21 800 000	49.7	33	10	58	2.8	71
	Ireland, Republic of	Dublin	3 681 000	52.4	14	9	76	0.2	59
	Israel	Jerusalem	5 984 000	288.1	21	6	77	1.9	91
	Italy	Rome	57 369 000	190.4	9	10	78	0.0	67
	Jamaica	Kingston	2 538 000	230.9	24	6	75	0.9	55
	Japan	Tokyo	126 281 000	334.3	10	7	80	0.2	79
	Jordan	Amman	6 304 000	70.7	31	4	71	3.3	73
	Kazakhstan	Astana	16 319 000	6.0	14	10	65	0.1	56
	Kenya	Nairobi	29 008 000	49.8	37	13	52	2.2	31
	Kuwait	Kuwait	1 811 000	101.6	22	2	76	3.0	97
	Kyrgyzstan	Bishkek	4 643 000	23.4	22	7	67	0.4	34
	Laos	Vientiane	5 163 000	21.8	38	14	53	3.1	22
	Latvia	Riga	2 424 000	38.1	8	13	69	-1.1	69
	Lebanon	Beirut	3 191 000	305.3	22	6	70	1.8	89
	Lesotho	Maseru	2 062 000	67.9	35	12	56	2.5	26

LAND			EDUCATION AND HEALTH				DEVELOPMENT				
AREA sq km	CULTIV-ATED AREA '000s sq km 1997	FOREST '000s sq km 1997	ADULT LITERACY % 1998	*SCHOOL ENROL-MENT Secondary, gross % 1997	DOCTORS PER 100 000 PERSONS 1990-1998	FOOD INTAKE calories per capita per day 1997	ENERGY CONSU-MPTION million tonnes of oil equivalent 1997	TRADE BALANCE millions US $ 1996	GNP PER CAPITA US $ 1998	COUNTRY	TIME ZONES + OR - GMT
23 200	-	-	46.0	14	20	2084	-	-300	-	Djibouti	+3
750	-	-	90.0	-	46	3059	-	-80	3010	Dominica	-4
48 442	15	16	83.0	50	77	2288	5.5	-4270	1770	Dominican Republic	-4
272 045	30	111	90.5	10	111	2679	8.5	1350	1530	Ecuador	-5
1 000 250	33	0	53.7	75	202	3287	39.6	-12 720	1290	Egypt	+2
21 041	8	1	77.9	33	91	2562	4.1	-2080	1850	El Salvador	-6
28 051	2	-	81.0	-	21	-	-	-20	1500	Equatorial Guinea	+1
117 400	4	3	51.9	21	2	1622	-	-	200	Eritrea	+3
45 200	11	136	98.0	103	312	2849	5.6	-1180	3390	Estonia	+2
1 133 880	105	-	36.0	12	4	1858	17.1	-890	100	Ethiopia	+3
18 330	3	-	92.0	70	38	2865	-	-300	2110	Fiji	+12
338 145	21	200	100.0	117	269	3100	33.1	10 100	24 110	Finland	+2
543 965	195	150	99.0	111	280	3518	247.5	9600	24 940	France	+1
267 667	5	179	63.3	-	19	2556	1.6	1190	3950	Gabon	+1
11 295	2	1	34.5	25	2	2350	-	-220	340	Gambia, The	GMT
69 700	11	30	99.0	73	436	2614	2.3	-360	930	Georgia	+4
357 868	121	107	100.0	102	319	3382	347.3	67 890	25 850	Germany	+1
238 537	46	90	68.9	31	4	2611	6.9	-140	390	Ghana	GMT
131 957	39	65	96.5	98	387	3649	25.6	19 090	11 650	Greece	+2
108 890	19	38	67.6	25	90	2339	5.6	-1850	1640	Guatemala	-6
245 857	15	64	36.0	12	15	2232	-	-290	540	Guinea	GMT
36 125	4	23	36.7	11	18	2430	-	-50	160	Guinea-Bissau	GMT
214 969	5	-	98.0	76	33	2530	-	-90	770	Guyana	-4
27 750	9	0	48.0	23	16	1869	1.8	-730	410	Haiti	-5
112 088	20	41	73.0	32	22	2403	3.2	-1650	730	Honduras	-6
93 030	50	17	99.0	103	337	3313	25.3	-3000	4510	Hungary	+1
102 820	-	-	100.0	104	-	3117	-	-510	28 010	Iceland	GMT
3 287 263	1699	650	53.3	49	48	2496	461.0	-8030	430	India	+5½
1 919 445	310	1098	85.5	52	12	2886	138.8	24 570	680	Indonesia	+7 to +9
1 648 000	194	15	74.5	74	90	2836	108.3	2550	1770	Iran	+3½
438 317	55	1	53.3	41	51	2619	27.1	2900	-	Iraq	+3
70 282	13	6	99.0	117	167	3565	12.5	24 050	18 340	Ireland, Republic of	GMT
20 770	4	1	96.0	88	459	3278	17.6	-7880	15 940	Israel	+2
301 245	109	65	98.5	91	550	3507	163.3	14 790	20 250	Italy	+1
10 991	3	2	86.0	71	57	2553	4.0	-1670	1680	Jamaica	-5
377 727	43	251	100.0	106	177	2932	514.9	108 740	32 380	Japan	+9
89 206	4	0	88.7	-	158	3014	4.8	-1970	1520	Jordan	+2
2 717 300	301	105	100.0	85	360	3085	38.4	1680	1310	Kazakhstan	+4 to +6
582 646	45	13	80.5	24	15	1977	14.1	-1170	330	Kenya	+3
17 818	-	0	80.6	65	178	3096	16.2	4500	-	Kuwait	+3
198 500	14	7	97.0	80	310	2447	2.8	-120	350	Kyrgyzstan	+5
236 800	9	-	45.8	29	20	2108	-	-180	330	Laos	+7
63 700	18	29	100.0	84	303	2864	4.5	-1180	2430	Latvia	+2
10 452	3	1	84.9	82	191	3277	5.2	-5490	3560	Lebanon	+2
30 355	3	0	82.2	29	5	2244	-	-710	570	Lesotho	+2

*Total enrolment in secondary level of education, regardless of age, is expressed as a percentage of the official secondary school-age population.
**Estimate.

KEY INFORMATION			POPULATION						
FLAG	COUNTRY	CAPITAL CITY	TOTAL 1998	DENSITY persons per sq km 1998	BIRTH RATE per 1000 population 1997	DEATH RATE per 1000 population 1997	LIFE EXPEC- TANCY in years 1997	POP. CHANGE average % per annum 1995 - 2000	URBAN POP. % 1998
	Liberia	Monrovia	2 666 000	23.9	42	12	59	8.6	46
	Libya	Tripoli	5 339 000	3.0	29	5	70	3.3	87
	Lithuania	Vilnius	3 694 000	56.7	10	12	71	-0.3	68
	Luxembourg	Luxembourg	422 000	163.2	13	9	77	1.1	90
	Macedonia	Skopje	1 999 000	77.7	16	8	72	0.7	61
	Madagascar	Antananarivo	15 057 000	25.6	42	11	57	3.1	28
	Malawi	Lilongwe	10 346 000	87.3	48	23	43	2.5	22
	Malaysia	Kuala Lumpur	21 410 000	64.3	26	5	72	2.0	56
	Maldives	Male	271 000	909.4	26	5	69	3.4	27
	Mali	Bamako	10 694 000	8.6	47	16	50	3.0	29
	Malta	Valletta	384 000	1215.2	13	8	78	0.6	90
	Mauritania	Nouakchott	2 529 000	2.5	41	14	53	2.5	55
	Mauritius	Port Louis	1 141 000	559.3	17	7	71	1.1	41
	Mexico	Mexico City	95 831 000	48.6	25	5	72	1.6	74
	Micronesia, Fed. States of	Pohnpei	114 000	162.6	33	8	66	2.8	28
	Moldova	Chişinău	4 378 000	129.9	11	10	67	0.1	46
	Mongolia	Ulan Bator	2 579 000	1.6	23	7	66	2.1	62
	Morocco	Rabat	27 377 000	61.3	26	7	67	1.8	55
	Mozambique	Maputo	18 880 000	23.6	41	20	45	2.5	38
	Myanmar	Yangon	44 497 000	65.8	27	10	60	1.8	27
	Namibia	Windhoek	1 660 000	2.0	36	12	56	2.4	30
	Nepal	Kathmandu	22 847 000	155.2	34	11	57	2.5	11
	Netherlands	Amsterdam/The Hague	15 678 000	377.5	12	9	78	0.5	89
	New Zealand	Wellington	3 796 000	14.0	15	7	77	1.1	86
	Nicaragua	Managua	4 807 000	37.0	32	5	68	2.6	55
	Niger	Niamey	10 078 000	8.0	52	18	47	3.3	20
	Nigeria	Abuja	106 409 000	115.2	40	12	54	2.8	42
	North Korea	Pyonyang	23 348 000	193.7	21	9	63	1.6	60
	Norway	Oslo	4 419 000	13.6	14	10	78	0.4	75
	Oman	Muscat	2 382 000	7.7	30	3	73	4.2	81
	Pakistan	Islamabad	148 166 000	184.3	36	8	62	2.7	36
	Panama	Panama City	2 767 000	35.9	23	5	74	1.6	56
	Papua New Guinea	Port Moresby	4 600 000	9.9	32	10	58	2.2	17
	Paraguay	Asunción	5 222 000	12.8	31	5	70	2.6	55
	Peru	Lima	24 797 000	19.3	27	6	69	1.7	72
	Philippines	Manila	72 944 000	243.1	29	6	68	2.0	57
	Poland	Warsaw	38 718 000	123.8	11	10	73	0.1	65
	Portugal	Lisbon	9 869 000	111.0	11	11	75	-0.1	61
	Qatar	Doha	579 000	50.6	19	2	72	1.8	92
	Romania	Bucharest	22 474 000	94.6	10	13	69	-0.2	56
	Russian Federation	Moscow	147 434 000	8.6	9	14	67	-0.3	77
	Rwanda	Kigali	6 604 000	250.7	46	22	40	7.9	6
	Samoa	Apia	174 000	61.5	29	5	65	1.1	21
	São Tomé and Príncipe	São Tomé	141 000	146.0	43	9	64	2.0	44
	Saudi Arabia	Riyadh	20 181 000	9.2	35	4	71	3.4	85

AREA sq km	CULTIVATED AREA '000s sq km 1997	FOREST '000s sq km 1997	ADULT LITERACY % 1998	*SCHOOL ENROLMENT Secondary, gross % 1997	DOCTORS PER 100 000 PERSONS 1990-1998	FOOD INTAKE calories per capita per day 1997	ENERGY CONSUMPTION million tonnes of oil equivalent 1997	TRADE BALANCE millions US $ 1996	GNP PER CAPITA US $ 1998	COUNTRY	TIME ZONES + OR - GMT
111 369	3	-	51.0	14	-	2044	-	100	-	Liberia	GMT
1 759 540	21	4	78.0	100	137	3289	15.1	3660	-	Libya	+2
65 200	30	20	99.0	86	399	3261	8.8	-1790	2440	Lithuania	+1
2 586	-	-	99.0	-	213	3619	-	14 760	45 570	Luxembourg	+1
25 713	7	10	89.0	58	219	2664	-	-	1290	Macedonia	+1
587 041	31	151	65.0	13	24	2022	-	-510	260	Madagascar	+3
118 484	17	33	58.3	17	2	2043	-	-120	200	Malawi	+2
332 965	76	155	86.6	62	43	2977	48.5	19 030	3600	Malaysia	+8
298	-	-	96.0	63	19	2485	-	-340	1230	Maldives	+5
1 240 140	47	116	38.4	11	4	2030	-	-190	250	Mali	GMT
316	-	-	91.0	86	250	3398	-	-860	9440	Malta	+1
1 030 700	5	6	41.4	16	11	2622	-	130	410	Mauritania	GMT
2 040	1	0	83.5	65	85	2917	-	-570	3700	Mauritius	+4
1 972 545	273	554	91.0	63	107	3097	141.5	-11 530	3970	Mexico	-6 to -8
701	-	-	81.0	-	46	-	-	-	1800	Micronesia	+10 to +11
33 700	22	4	98.5	79	356	2567	4.4	-90	410	Moldova	+2
1 565 000	13	94	61.5	56	268	1917	-	-100	400	Mongolia	+8
446 550	96	38	47.0	39	34	3078	9.3	-3350	1250	Morocco	GMT
799 380	32	169	42.0	7	<5**	1832	7.7	-1210	210	Mozambique	+2
676 577	102	272	84.0	35	28	2862	13.0	-1600	-	Myanmar	+6½
824 292	8	124	81.0	61	23	2183	-	-150	1940	Namibia	+1
147 181	30	48	39.7	37	5	2366	7.2	-780	210	Nepal	+5¾
41 526	9	3	100.0	140	260	3284	74.9	15 130	24 760	Netherlands	+1
270 534	33	79	100.0	120	210	3395	16.7	-1910	14 700	New Zealand	+12 to 12¾
130 000	27	56	67.5	47	82	2186	2.6	-1330	390	Nicaragua	-6
1 267 000	50	26	14.4	7	3	2097	-	-80	190	Niger	+1
923 768	307	138	60.9	34	21	2735	88.7	930	300	Nigeria	+1
120 538	20	62	95.0	-	-	1837	-	-320	-	North Korea	+9
323 878	9	81	100.0	119	250	3357	24.2	11 100	34 330	Norway	+1
309 500	-	0	68.2	66	120	-	6.8	2310	-	Oman	+4
803 940	216	17	44.0	30	52	2476	56.8	-1770	480	Pakistan	+5
77 082	7	28	91.1	69	119	2430	2.3	-2620	3080	Panama	-5
462 840	7	369	63.2	14	18	2224	-	680	890	Papua New Guinea	+10
406 752	23	115	92.5	44	67	2566	4.2	-1570	1760	Paraguay	-4
1 285 216	42	676	89.0	70	73	2302	15.1	-1950	2460	Peru	-5
300 000	95	68	97.5	79	11	2366	38.3	2320	1050	Philippines	+8
312 683	144	87	99.0	97	230	3366	105.2	-18 100	3900	Poland	+1
88 940	29	29	91.4	116	291	3667	20.4	-14 110	10 690	Portugal	GMT
11 437	-	-	80.0	78	143	-	-	1500	-	Qatar	+3
237 500	99	62	98.0	78	176	3253	44.1	-1740	1390	Romania	+2
17 075 400	1280	7635	99.0	86	380	2904	592.0	33 200	2300	Russian Federation	+2 to +12
26 338	12	3	63.9	13	<5**	2056	-	-230	230	Rwanda	+2
2 831	1	-	98.0	62	38	2828	-	-80	1020	Samoa	-11
964	-	-	-	-	32	2138	-	-20	280	São Tomé and Príncipe	GMT
2 200 000	38	2	66.5	61	166	2783	98.4	20 550	-	Saudi Arabia	+3

*Total enrolment in secondary level of education, regardless of age, is expressed as a percentage of the official secondary school-age population.
**Estimate.

KEY INFORMATION			POPULATION						
FLAG	COUNTRY	CAPITAL CITY	TOTAL 1998	DENSITY persons per sq km 1998	BIRTH RATE per 1000 population 1997	DEATH RATE per 1000 population 1997	LIFE EXPEC-TANCY in years 1997	POP. CHANGE average % per annum 1995 - 2000	URBAN POP. % 1998
	Senegal	Dakar	9 003 000	45.8	40	13	52	2.7	46
	Seychelles	Victoria	76 000	167.0	19	8	70	1.0	35
	Sierra Leone	Freetown	4 568 000	63.7	46	26	37	3.0	34
	Singapore	Singapore	3 476 000	5439.7	13	4	76	1.5	100
	Slovakia	Bratislava	5 377 000	109.7	11	10	73	0.1	57
	Slovenia	Ljubljana	1 993 000	98.4	9	10	75	-0.1	50
	Solomon Islands	Honiara	417 000	14.7	37	4	70	3.2	18
	Somalia	Mogadishu	9 237 000	14.5	47	19	46	3.9	26
	South Africa, Republic of	Pretoria/Cape Town	39 357 000	32.3	25	8	65	2.2	53
	South Korea	Seoul	46 109 000	464.5	15	6	72	0.9	80
	Spain	Madrid	39 628 000	78.5	9	10	78	0.1	77
	Sri Lanka	Colombo	18 455 000	281.3	19	6	73	1.0	23
	St Lucia	Castries	150 000	244.0	22	7	70	1.3	38
	St Vincent and the Grenadines	Kingstown	112 000	288.0	21	7	73	0.9	50
	Sudan	Khartoum	28 292 000	11.3	33	12	55	2.2	34
	Suriname	Paramaribo	414 000	2.5	24	6	70	1.2	50
	Swaziland	Mbabane	952 000	54.8	42	10	39	2.8	32
	Sweden	Stockholm	8 875 000	19.7	10	11	79	0.3	83
	Switzerland	Bern	7 299 000	176.8	11	9	79	0.7	68
	Syria	Damascus	15 333 000	82.8	29	5	69	2.5	54
	Taiwan	Taibei	21 908 135	605.5	12	6	75	-	-
	Tajikistan	Dushanbe	6 015 000	42.0	23	6	68	1.9	28
	Tanzania	Dodoma	32 102 000	34.0	41	16	48	2.3	31
	Thailand	Bangkok	60 300 000	117.5	17	7	69	0.8	21
	Togo	Lomé	4 397 000	77.4	41	16	49	2.7	32
	Tonga	Nuku'alofa	98 000	131.0	24	7	72	0.4	42
	Trinidad and Tobago	Port of Spain	1 283 000	250.1	16	7	73	0.8	73
	Tunisia	Tunis	9 335 000	56.9	23	7	70	1.8	64
	Turkey	Ankara	64 479 000	82.7	22	7	69	1.6	73
	Turkmenistan	Ashgabat	4 309 000	8.8	24	7	66	1.9	45
	Tuvalu	Funafuti	11 000	440.0	24	9	-	-	48
	Uganda	Kampala	20 554 000	85.3	48	20	42	2.6	14
	Ukraine	Kiev	50 861 000	84.2	9	15	67	-0.4	68
	United Arab Emirates	Abu Dhabi	2 377 453	28.4	18	3	75	2.0	85
	United Kingdom	London	58 649 000	240.3	12	11	77	0.1	89
	United States of America	Washington	274 028 000	27.9	15	8	76	0.8	77
	Uruguay	Montevideo	3 289 000	18.7	18	10	74	0.6	91
	Uzbekistan	Tashkent	23 574 000	52.7	27	6	69	1.9	38
	Vanuatu	Port Vila	182 000	14.9	35	7	65	2.5	19
	Venezuela	Caracas	23 242 000	25.5	25	5	73	2.0	86
	Vietnam	Hanoi	77 562 000	235.3	21	7	68	1.8	20
	Yemen	Sana	16 887 000	32.0	40	13	54	3.7	24
	Yugoslavia	Belgrade	10 635 000	104.1	13	11	72	0.5	52
	Zambia	Lusaka	8 781 000	11.7	42	19	43	2.5	39
	Zimbabwe	Harare	11 377 000	29.1	31	12	52	2.1	34

AREA sq km	CULTIV-ATED AREA '000s sq km 1997	FOREST '000s sq km 1997	ADULT LITERACY % 1998	*SCHOOL ENROL-MENT Secondary, gross % 1997	DOCTORS PER 100 000 PERSONS 1990-1998	FOOD INTAKE calories per capita per day 1997	ENERGY CONSU-MPTION million tonnes of oil equivalent 1997	TRADE BALANCE millions US $ 1996	GNP PER CAPITA US $ 1998	COUNTRY	TIME ZONES + OR - GMT
196 720	23	74	35.5	16	7	2418	2.8	-420	530	Senegal	GMT
455	-	-	84.0	-	104	2487	-	-330	6450	Seychelles	+4
71 740	5	13	31.5	17	10	2035	-	-90	140	Sierra Leone	GMT
639	-	0	92.0	73	147	-	26.9	3630	30 060	Singapore	+8
49 035	16	20	100.0	94	325	2984	17.2	-1000	3700	Slovakia	+1
20 251	3	11	99.0	93	219	3101	6.4	-1320	9760	Slovenia	+1
28 370	1	-	62.0	18	-	2122	-	-30	750	Solomon Islands	+11
637 657	11	-	24.0	5	4	1566	-	0	-	Somalia	+3
1 219 080	163	85	84.5	84	59	2990	107.2	-120	2880	South Africa, Republic of	+2
99 274	19	76	97.5	102	127	3155	176.4	-24 510	7970	South Korea	+9
504 782	192	84	97.0	132	400	3310	107.3	-35 530	14 080	Spain	+1
65 610	19	18	91.0	75	23	2302	7.2	-1290	810	Sri Lanka	+6
616	-	-	-	-	35	2734	-	-290	3410	St Lucia	-4
389	-	-	-	-	46	2472	-	-140	2420	St Vincent & the Grenadines	-4
2 505 813	169	416	55.5	20	10	2395	11.5	-1110	290	Sudan	+2
163 820	-	-	93.0	-	40	2665	-	-160	1660	Suriname	-3
17 364	2	-	78.0	52	-	2483	-	-290	1400	Swaziland	+2
449 964	28	244	100.0	139	299	3194	51.9	16 360	25 620	Sweden	+1
41 293	4	11	100.0	-	301	3223	26.2	450	40 080	Switzerland	+1
185 180	55	2	72.7	42	109	3352	14.6	-740	1020	Syria	+2
36 179	-	-	94.0	-	-	-	72.7	10 610	-	Taiwan	+8
143 100	9	4	99.0	76	210	2001	3.4	60	350	Tajikistan	+5
945 087	40	325	73.4	5	4	1995	14.3	-770	210	Tanzania	+3
513 115	204	116	95.0	57	24	2360	80.0	7870	2200	Thailand	+7
56 785	24	12	54.9	27	6	2469	-	-400	330	Togo	GMT
748	-	-	99.0	-	-	-	-	-70	1690	Tonga	+13
5 130	1	2	93.5	72	90	2661	8.2	-750	4430	Trinidad and Tobago	-4
164 150	49	6	68.6	66	67	3283	6.8	-2590	2050	Tunisia	+1
779 452	292	89	81.4	63	103	3525	71.3	-13 050	3160	Turkey	+2
488 100	17	38	98.0	111	353	2306	12.2	520	640	Turkmenistan	+5
25	-	-	99.0	-	89	-	-	-10	-	Tuvalu	+12
241 038	68	61	65.0	12	4	2085	-	-910	320	Uganda	+3
603 700	341	92	99.5	93	429	2795	150.1	-60	850	Ukraine	+2
83 600	1	1	74.3	80	168	3390	30.9	630	18 220	United Arab Emirates	+4
244 082	64	24	100.0	133	164	3276	228.0	-52 300	21 400	United Kingdom	GMT
9 809 386	1790	2125	99.0	97	245	3699	2162.2	-364 850	29 340	United States	-5 to -10
176 215	13	8	96.0	85	309	2816	2.9	-560	6180	Uruguay	-3
447 400	49	91	88.0	93	335	2433	42.6	-250	870	Uzbekistan	+5
12 190	1	-	64.0	21	-	2700	-	-60	1270	Vanuatu	+11
912 050	35	440	92.0	40	194	2321	57.5	3660	3500	Venezuela	-4
329 565	72	91	93.0	41	40	2484	39.3	-80	330	Vietnam	+7
527 968	16	0	45.0	34	26	2051	3.4	-120	300	Yemen	+3
102 173	41	18	98.0	64	200	3031	-	-	-	Yugoslavia	+1
752 614	53	314	76.4	29	10	1970	6.0	40	330	Zambia	+2
390 759	32	87	87.5	48	14	2145	10.0	660	610	Zimbabwe	+2

*Total enrolment in secondary level of education, regardless of age, is expressed as a percentage of the official secondary school-age population.
**Estimate.

How to use the Index

All the names on the maps in this atlas, except some of those on the special topic maps, are included in the index.

The names are arranged in **alphabetical order.** Where the name has more than one word the separate words are considered as one to decide the position of the name in the index:

Thetford
Thetford Mines
The Trossachs
The Wash
The Weald
Thiers

Where there is more than one place with the same name, the country name is used to decide the order:

London Canada
London England

If both places are in the same country, the county or state name is also used:

Avon *r.* Bristol England
Avon *r.* Dorset England

Each entry in the index starts with the name of the place or feature, followed by the name of the country or region in which it is located. This is followed by the number of the most appropriate page on which the name appears, usually the largest scale map. Next comes the alphanumeric reference followed by the latitude and longitude.

Names of physical features such as rivers, capes, mountains etc are followed by a description. The descriptions are usually shortened to one or two letters, these abbreviations are keyed below. Town names are followed by a description only when the name may be confused with that of a physical feature:

Big Spring *town*

To help to distinguish the different parts of each entry, different styles of type are used:

place name   country name   alphanumeric
              or   grid reference
              region name

description   page   latitude/
(if any)   number longitude

Thames *r.* England **15** **C2** 51.30N 0.05E

To use the **alphanumeric grid reference** to find a feature on the map, first find the correct page and then look at the black letters printed outside the blue frame along the top and bottom of the map and the black numbers printed outside the blue frame at the sides of the map. When you have found the correct letter and number follow the grid boxes up and along until you find the correct grid box in which the feature appears. You must then search the grid box until you find the name of the feature.

The **latitude and longitude reference** gives a more exact description of the position of the feature.

Page 6 of the atlas describes lines of latitude and lines of longitude, and explains how they are numbered and divided into degrees and minutes. Each name in the index has a different latitude and longitude reference, so the feature can be located accurately. The lines of latitude and lines of longitude shown on each map are numbered in degrees. These numbers are printed in black along the top, bottom and sides of the map frame.

The drawing above shows part of the map on page 20 and the lines of latitude and lines of longitude.

The index entry for Wexford is given as follows

Wexford Rep. of Ire. **20 E2** 52.20N 6.28W

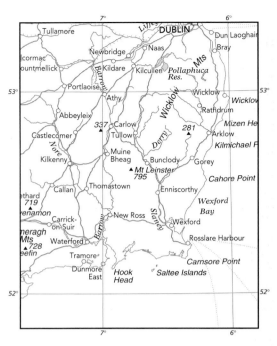

To locate Wexford, first find latitude 52N and estimate 20 minutes north from 52 degrees to find 52.20N, then find longitude 6W and estimate 28 minutes west from 6 degrees to find 6.28W. The symbol for the town of Wexford is where latitude 52.20N and longitude 6.28W meet.

On maps at a smaller scale than the map of Ireland, it is not possible to show every line of latitude and longitude. Only every 5 or 10 degrees of latitude and longitude may be shown. On these maps you must estimate the degrees and minutes to find the exact location of a feature.

Abbreviations

A. and B.	Argyll and Bute
Afghan.	Afghanistan
Ala.	Alabama
b., **B.**	bay, Bay
Bangla.	Bangladesh
Bosnia.	Bosnia-Herzegovina
B.V.Is.	British Virgin Islands
c., **C.**	cape, Cape
Cambs.	Cambridgeshire
C.A.R.	Central African Republic
Carib. Sea	Caribbean Sea
Colo.	Colorado
Czech Rep.	Czech Republic
d.	internal division eg. county, state
D. and G.	Dumfries and Galloway
Del.	Delaware
Derbys.	Derbyshire
des.	desert
Dom. Rep.	Dominican Republic
Equat. Guinea	Equatorial Guinea
E. Sussex	East Sussex
E. Yorks.	East Riding of Yorkshire
est.	estuary
f.	physical feature eg. valley, plain, geographic district
Fla.	Florida
g., **G.**	Gulf
Ga.	Georgia
Glos.	Gloucestershire
Hants.	Hampshire
Herts.	Hertfordshire
High.	Highland

i., **I.**, *is.*, **Is.**	island, Island, islands, Islands
Ill.	Illinois
I.o.M.	Isle of Man
I.o.W.	Isle of Wight
l., **L.**	lake, Lake
La.	Louisiana
Lancs.	Lancashire
Leics.	Leicestershire
Liech.	Liechtenstein
Lincs.	Lincolnshire
Lux.	Luxembourg
Man.	Manitoba
Med. Sea	Mediterranean Sea
Miss.	Mississippi
Mt.	Mount
mtn., **Mtn.**	mountain, Mountain
mts., **Mts.**	mountains, Mountains
N. Ayr.	North Ayrshire
N.C.	North Carolina
N. Cal.	New Caledonia
Neth.	Netherlands
Neth. Ant.	Netherlands Antilles
Nev.	Nevada
Nfld.	Newfoundland
N. Korea	North Korea
N. Mex.	New Mexico
Northum.	Northumberland
Notts.	Nottinghamshire
N.Y.	New York
Oc.	Ocean
Oreg.	Oregon
Oxon.	Oxfordshire
P. and K.	Perth and Kinross

Pem.	Pembrokeshire
pen., **Pen.**	peninsula, Peninsula
Phil.	Philadelphia
P.N.G.	Papua New Guinea
Pt.	Point
r., **R.**	river, River
Rep.of Ire.	Republic of Ireland
Resr.	Reservoir
R.S.A.	Republic of South Africa
Russian Fed.	Russian Federation
Sd.	Sound
S.C.	South Carolina
Shrops.	Shropshire
S. Korea	South Korea
Staffs.	Staffordshire
str., **Str.**	strait, Strait
Switz.	Switzerland
Tex.	Texas
U.A.E.	United Arab Emirates
U.K.	United Kingdom
U.S.A.	United States of America
U.S. V.Is.	United States Virgin Islands
Va.	Virginia
Warwicks.	Warwickshire
W. Isles	Western Isles
W. Sahara	Western Sahara
W. Sussex	West Sussex
W. Va.	West Virginia
Wilts.	Wiltshire
Wyo.	Wyoming
Yugo.	Yugoslavia

A

Aachen Germany 48 C4 ....50.46N 6.06E
Aalen Germany 48 E3 ....48.50N 10.05E
Aalst Belgium 42 D2 ....50.57N 4.03E
Abadan Iran 95 G5 ....30.21N 48.15E
Abadeh Iran 95 H5 ....31.10N 52.40E
Abadla Algeria 84 D5 ....31.01N 2.45W
Abakan Russian Fed. 59 L3 ....53.43N 91.25E
Abancay Peru 76 C3 ....13.35S 72.55W
Abashiri Japan 106 D4 ....44.02N 144.17E
Abaya, L. Ethiopia 85 H2 ....6.20N 38.00E
Abaza Russian Fed. 102 G8 ....52.44N 90.12E
Abbeville France 44 D7 ....50.06N 1.51E
Abbeyfeale Rep. of Ire. 20 B2 ....52.24N 9.18W
Abbey Head Scotland 17 F2 ....54.45N 3.58W
Abbeyleix Rep. of Ire. 20 D2 ....52.55N 7.20W
Abbottabad Pakistan 95 L5 ....34.12N 73.15E
Abéché Chad 85 G3 ....13.49N 20.49E
Åbenrå Denmark 43 B1 ....55.03N 9.26E
Abeokuta Nigeria 84 E2 ....7.10N 3.26E
Aberaeron Wales 12 C4 ....52.15N 4.16W
Aberchirder Scotland 19 G2 ....57.33N 2.38W
Aberdare Wales 12 D3 ....51.43N 3.27W
Aberdare Range mts. Kenya 87 B2 ....0.20S 36.07E
Aberdeen Scotland 19 G2 ....57.08N 2.07W
Aberdeen U.S.A. 64 G6 ....45.28N 98.30W
Aberdeen City d. Scotland 8 D5 ....57.08N 2.07W
Aberdeenshire d. Scotland 8 D5 ....57.22N 2.35W
Aberfeldy Scotland 17 F4 ....56.38N 3.52W
Aberford England 15 F2 ....53.51N 1.20W
Aberfoyle Scotland 16 E4 ....56.11N 4.23W
Abergavenny Wales 12 D3 ....51.49N 3.01W
Abergele Wales 12 D5 ....53.17N 3.34W
Aberporth Wales 12 C4 ....52.08N 4.33W
Abersoch Wales 12 C4 ....52.50N 4.31W
Abertillery Wales 12 D3 ....51.44N 3.09W
Aberystwyth Wales 12 C4 ....52.25N 4.06W
Abha Saudi Arabia 94 F3 ....18.13N 42.30E
Abidjan Côte d'Ivoire 84 D2 ....5.19N 4.01W
Abilene U.S.A. 64 G3 ....32.27N 99.45W
Abingdon England 10 D2 ....51.40N 1.17W
Abington Scotland 17 F3 ....55.29N 3.42W
Abitibi, L. Canada 65 K6 ....48.42N 79.45W
Aboyne Scotland 19 G2 ....57.05N 2.49W
Abqaiq Saudi Arabia 95 G4 ....25.55N 49.40E
Abu' Arīsh Saudi Arabia 94 F2 ....16.58N 42.50E
Abu Dhabi U.A.E. 95 H3 ....24.27N 54.23E
Abu Hamed Sudan 85 H3 ....19.32N 33.20E
Abuja Nigeria 84 E2 ....9.12N 7.11E
Abu Matariq Sudan 94 C1 ....10.58N 26.17E
Abunã r. Brazil 76 D4 ....9.41S 65.20W
Abu Simbel Egypt 94 D3 ....22.18N 31.40E
Abyad Sudan 94 C1 ....13.46N 26.28E
Acapulco Mexico 70 E4 ....16.51N 99.56W
Acarigua Venezuela 71 K2 ....9.35N 69.12W
Accra Ghana 84 D2 ....5.33N 0.15W
Accrington England 15 E2 ....53.46N 2.22W
Acheloös r. Greece 56 E3 ....38.20N 21.04E
Achill I. Rep. of Ire. 20 A3 ....53.57N 10.00W
Achinsk Russian Fed. 59 L3 ....56.10N 90.10E
A'Chralaig mtn. Scotland 18 D2 ....57.11N 5.09W
Acklins I. The Bahamas 71 J5 ....22.30N 74.10W
Acle England 11 G3 ....52.38N 1.33E
Aconcagua mtn. Argentina 75 B3 ....32.37S 70.00W
A Coruña Spain 46 A5 ....43.22N 8.24W
Acre r. Brazil 76 D4 ....8.45S 67.23W
Acre d. Brazil 76 C4 ....8.50S 71.30W
Actéon, Groupe is. French Polynesia 109 Q4
............22.00S 136.00W
Adaja r. Spain 46 C4 ....41.32N 4.52W
Adamawa Highlands Nigeria/Cameroon 84 F2
............7.05N 12.00E
Adana Turkey 57 L2 ....37.00N 35.19E
Adapazari Turkey 57 J4 ....40.45N 30.23E
Adda r. Italy 50 C6 ....45.08N 9.55E
Ad Dahna des. Saudi Arabia 95 G2 ....26.00N 47.00E
Ad Dakhla W. Sahara 84 C4 ....23.43N 15.57W
Ad Dammam Saudi Arabia 95 H4 ....26.23N 50.08E
Adderbury England 10 D3 ....52.01N 1.19W
Ad Dir'īyah Saudi Arabia 95 G4 ....24.45N 46.32E
Addis Ababa Ethiopia 85 H2 ....9.03N 38.42E
Ad Dīwānīyah Iraq 94 F5 ....31.59N 44.57E
Adelaide Australia 110 C2 ....34.56S 138.36E
Aden Yemen 94 F1 ....12.50N 45.00E
Aden, G. of Indian Oc. 85 I3 ....13.00N 50.00E
Adi i. Indonesia 105 I3 ....4.10S 133.10E
Adī Ärk'ay Ethiopia 94 E1 ....13.35N 37.57E
Adige r. Italy 50 E6 ....45.08N 12.05E
Ādīgrat Ethiopia 94 E1 ....14.18N 39.31E
Adilang Uganda 87 A3 ....2.44N 33.28E
Adi Ugri Eritrea 94 E1 ....14.55N 38.53E
Adıyaman Turkey 57 N2 ....37.46N 38.15E
Admiralty Is. P.N.G. 108 J6 ....2.30S 147.20E
Adour r. France 44 C3 ....43.28N 1.35W
Adriatic Sea Med. Sea 50 F5 ....42.30N 16.00E
Adwa Ethiopia 85 H3 ....14.12N 38.56E
Aegean Sea Med. Sea 56 G3 ....39.00N 25.00E
Afghanistan Asia 95 K5 ....33.00N 65.30E
Afmadow Somalia 87 C3 ....0.27N 42.05E
Africa 82
Afyon Turkey 57 J3 ....38.46N 30.32E
Agadez Niger 84 E3 ....17.00N 7.56E
Agadir Morocco 84 D5 ....30.30N 9.40W
Agana Guam 105 K6 ....13.28N 144.45E
Agano r. Japan 106 C3 ....38.25N 139.02E
Agartala India 97 I5 ....23.49N 91.15E
Agde France 44 E3 ....43.19N 3.28E
Agen France 44 D4 ....44.12N 0.38E
Ägere Maryam Ethiopia 87 B4 ....5.40N 38.11E
Aghla Mtn. Rep. of Ire. 16 B4 ....54.50N 8.10W
Agios Efstratios i. Greece 56 G3 ....39.30N 25.00E
Agirwat Hills Sudan 94 E2 ....17.00N 35.10E
Agra India 97 F6 ....27.09N 78.00E
Ağri Turkey 94 F6 ....39.44N 43.04E
Agrigento Italy 50 E2 ....37.19N 13.36E
Agrihan i. N. Mariana Is. 105 L7 ....18.44N 145.39E
Aguascalientes Mexico 70 D5 ....21.51N 102.18W
Aguascalientes d. Mexico 70 D4 ....22.00N 102.18W
Aguilar de Campóo Spain 46 C5 ....42.55N 4.15W
Aguilas Spain 46 E2 ....37.25N 1.35W
Agulhas, C. R.S.A. 86 B1 ....34.50S 20.00E
Agulhas Negras mtn. Brazil 72 F4 ....22.20S 44.43W

Ahar Iran 95 G6 ....38.25N 47.07E
Ahaus Germany 42 G4 ....52.04N 7.01E
Ahmadabad India 96 E5 ....23.03N 72.40E
Ahmadnagar India 96 E4 ....19.08N 74.48E
Ahmadpur East Pakistan 95 L4 ....29.09N 71.16E
Ahmadpur Sial Pakistan 95 L5 ....30.41N 71.46E
Ahvaz Iran 95 G5 ....31.17N 48.44E
Aigina i. Greece 56 F2 ....37.43N 23.30E
Ailsa Craig i. Scotland 16 D3 ....55.15N 5.07W
Aïn Beïda Algeria 52 E4 ....35.50N 7.27E
Aïn Sefra Algeria 84 D5 ....32.45N 0.35W
Aïr mts. Niger 84 E3 ....18.30N 8.30E
Airdrie Canada 62 G3 ....51.20N 114.00W
Airdrie Scotland 17 F3 ....55.52N 3.59W
Aisne r. France 44 E6 ....49.27N 2.51E
Aitape P.N.G. 105 K3 ....3.10S 142.17E
Aitutaki i. Cook Is. 108 P5 ....18.52S 159.46W
Aix-en-Provence France 44 F3 ....43.31N 5.27E
Aizu-wakamatsu Japan 106 C3 ....37.30N 139.58E
Ajaccio France 44 H2 ....41.55N 8.43E
Ajdabiya Libya 85 G5 ....30.48N 20.15E
Akhdar, Al Jabal al mts. Libya 85 G5 ....32.10N 22.00E
Akhdar, Jabal mts. Oman 95 I3 ....23.10N 57.25E
Akhisar Turkey 57 H3 ....38.54N 27.49E
Akimiski I. Canada 63 J3 ....53.00N 81.20W
Akita Japan 106 D3 ....39.44N 140.05E
Akkajaure l. Sweden 43 D4 ....67.40N 17.30E
Akobo r. Sudan/Ethiopia 82 G5 ....8.30N 33.15E
Akordat Eritrea 85 H3 ....15.35N 37.55E
Akpatok I. Canada 63 L4 ....60.30N 68.30W
Akranes Iceland 43 X2 ....64.19N 22.05W
Akron U.S.A. 65 J5 ....41.04N 81.31W
Aksaray Turkey 94 D5 ....38.22N 34.02E
Akşehir Turkey 57 J3 ....38.22N 31.24E
Aksu China 102 E6 ....42.10N 80.00E
Aksum Ethiopia 94 E1 ....14.08N 38.48E
Aktau Kazakhstan 58 H2 ....43.37N 51.11E
Aktogay Kazakhstan 102 D7 ....46.59N 79.42E
Aktyubinsk Kazakhstan 58 H3 ....50.16N 57.13E
Akureyri Iceland 43 Y2 ....65.41N 18.04W
Alabama d. U.S.A. 65 I3 ....31.05N 87.55W
Alabama r. U.S.A. 65 I3 ....33.00N 87.00W
Alagoas d. Brazil 77 G4 ....9.30S 37.00W
Alagoinhas Brazil 77 G3 ....12.09S 38.21W
Al Ahmadi Kuwait 95 G4 ....29.05N 48.04E
Alakol', L. Kazakhstan 102 E7 ....46.00N 81.40E
Alakurtti Russian Fed. 43 G4 ....67.00N 30.23E
Alamagan i. N. Mariana Is. 105 L7 ....17.35N 145.50E
Åland is. Finland 43 E3 ....60.20N 20.00E
Alanya Turkey 57 J2 ....36.32N 32.02E
Al Artawīyah Saudi Arabia 95 G4 ....26.31N 45.21E
Alaska d. U.S.A. 62 C4 ....65.00N 153.00W
Alaska, G. of U.S.A. 62 D3 ....58.45N 145.00W
Alaska Pen. U.S.A. 62 C3 ....56.00N 160.00W
Alaska Range mts. U.S.A. 62 C4 ....62.10N 152.00W
Alausí Ecuador 76 C4 ....2.00S 78.50W
Alavus Finland 43 E3 ....62.35N 23.37E
Alaw Resr. Wales 12 C5 ....53.20N 4.25W
Albacete Spain 46 E3 ....39.00N 1.52W
Alba Iulia Romania 56 F7 ....46.04N 23.33E
Albania Europe 56 E4 ....41.00N 20.00E
Albany Australia 110 A2 ....34.57S 117.54E
Albany r. Canada 63 J3 ....52.10N 82.00W
Albany U.S.A. 65 J3 ....31.37N 84.10W
Albany N.Y. U.S.A. 65 L5 ....42.40N 73.49W
Al Basrah Iraq 95 G5 ....30.33N 47.50E
Al Bayda' Libya 85 G5 ....32.50N 21.50E
Albenga Italy 50 C6 ....44.03N 8.13E
Alberche r. Spain 46 C4 ....40.00N 4.45W
Albert France 44 E6 ....50.00N 2.40E
Albert, L. Africa 86 C5 ....1.45N 31.00E
Alberta d. Canada 62 G3 ....55.00N 115.00W
Albert Lea U.S.A. 65 H5 ....43.39N 93.22W
Albert Nile r. Uganda 85 H2 ....3.30N 32.00E
Albi France 44 E3 ....43.56N 2.08E
Al Biyadh f. Saudi Arabia 95 G3 ....22.00N 47.00E
Alboran, Isla de i. Spain 46 D1 ....35.55N 3.10W
Ålborg Denmark 43 B2 ....57.03N 9.56E
Al Bu Kamal Syria 94 F5 ....34.27N 40.55E
Albuquerque U.S.A. 64 E4 ....35.05N 106.38W
Al Buraymi U.A.E. 95 I3 ....24.15N 55.45E
Albury Australia 110 D2 ....36.03S 146.53E
Alcalá de Henares Spain 46 D4 ....40.28N 3.22W
Alcalá la Real Spain 46 D2 ....37.28N 3.55W
Alcañiz Spain 46 E4 ....41.03N 0.09W
Alcázar de San Juan Spain 46 D3 ....39.24N 3.12W
Alcester England 10 D3 ....52.13N 1.52W
Alcoy Spain 46 E3 ....38.42N 0.29W
Alcúdia Spain 46 G3 ....39.51N 3.09E
Aldabra Is. Indian Oc. 86 D4 ....9.00S 47.00E
Aldan Russian Fed. 59 O3 ....58.44N 125.22E
Aldan r. Russian Fed. 59 P4 ....63.30N 130.00E
Aldbrough England 15 G2 ....53.50N 0.07W
Aldeburgh England 11 G3 ....52.09N 1.35E
Alderley Edge England 15 E2 ....53.18N 2.15W
Alderney i. Channel Is. 13 Z9 ....49.42N 2.11W
Aldershot England 10 E2 ....51.15N 0.47W
Aldingham England 14 D3 ....54.08N 3.08W
Aldridge England 10 D3 ....52.36N 1.55W
Aleksandrovsk-Sakhalinskiy Russian Fed. 59 Q3
............50.55N 142.12E
Aleksin Russian Fed. 55 O6 ....54.31N 37.07E
Alençon France 44 D6 ....48.25N 0.05E
Aleppo Syria 94 E6 ....36.14N 37.10E
Alès France 44 F4 ....44.08N 4.05E
Alessandria Italy 50 C6 ....44.54N 8.37E
Ålesund Norway 43 A3 ....62.28N 6.11E
Aleutian Is. U.S.A. 108 N12 ....52.00N 176.00W
Aleutian Range mts. U.S.A. 62 C3 ....58.00N 156.00W
Alexander Archipelago is. U.S.A. 62 E3
............56.30N 134.30W
Alexander I. Antarctica 112 ....72.00S 70.00W
Alexander, C. South Georgia 75 F1 ....54.05S 37.58W
Alexandria Egypt 94 C5 ....31.13N 29.55E
Alexandria Scotland 16 E3 ....55.59N 4.35W
Alexandria La. U.S.A. 65 H3 ....31.19N 92.29W
Alexandria Va. U.S.A. 65 K4 ....38.48N 77.03W
Alexandroupoli Greece 56 G4 ....40.50N 25.53E
Aleysk Russian Fed. 102 E8 ....52.32N 82.17E
Al Farwānīyah Kuwait 95 G4 ....29.04N 47.50E
Alford England 15 H2 ....53.17N 0.11E
Alfreton England 15 F2 ....53.06N 1.22W
Algarve f. Portugal 46 A2 ....37.20N 8.00W
Algeciras Spain 46 C2 ....36.08N 5.27W
Algeria Africa 84 E4 ....28.00N 2.00E
Al Ghaydah Yemen 95 H2 ....16.12N 52.16E

Alghero Italy 50 C4 ....40.33N 8.20E
Algiers Algeria 84 E5 ....36.50N 3.00E
Al Hamad des. Asia 94 E5 ....31.45N 39.00E
Al Hamadah al Hamra' f. Libya 52 F2 ....29.00N 12.00E
Al Hasakah Syria 94 F6 ....36.29N 40.45E
Al Hibak f. Saudi Arabia 95 H3 ....21.00N 53.30E
Al Hillah Iraq 94 F5 ....32.28N 44.29E
Al Hoceima Morocco 46 D1 ....35.15N 3.55W
Aliakmonas r. Greece 56 F4 ....40.30N 22.38E
Ali Bayramli Azerbaijan 95 G6 ....39.56N 48.55E
Alicante Spain 46 E3 ....38.21N 0.29W
Alice Springs town Australia 110 C3 ....23.42S 133.52E
Alingsås Sweden 43 C2 ....57.55N 12.30E
Al Jaghbub Libya 85 G4 ....29.42N 24.38E
Al Jaharah Kuwait 95 G4 ....29.20N 47.41E
Al Jauf Saudi Arabia 94 E4 ....29.49N 39.52E
Al Jawf Libya 85 G4 ....24.09N 23.19E
Al Jawsh Libya 52 F3 ....32.00N 11.40E
Al Jubayl Saudi Arabia 95 G4 ....26.59N 49.40E
Al Khaburah Oman 95 I3 ....23.58N 57.10E
Al Khasab Oman 95 I4 ....26.14N 56.15E
Al Khums Libya 53 F3 ....32.39N 14.15E
Alkmaar Neth. 42 D4 ....52.37N 4.44E
Al Kut Iraq 95 G5 ....32.30N 45.51E
Allahabad India 97 G6 ....25.57N 81.50E
Allegheny Mts. U.S.A. 65 K5 ....38.00N 81.00W
Allendale Town England 15 E3 ....54.54N 2.15W
Allen, Lough Rep. of Ire. 20 C4 ....54.07N 8.04W
Allentown U.S.A. 65 K5 ....40.37N 75.30W
Alleppey India 96 F2 ....9.30N 76.22E
Aller r. Germany 48 D5 ....52.57N 9.11E
Alliance U.S.A. 64 F5 ....42.08N 103.00W
Allier r. France 44 E5 ....46.58N 3.04E
Al Lith Saudi Arabia 94 F3 ....20.09N 40.16E
Al Mudawwara Jordan 94 E5 ....29.20N 36.00E
Al Mukha Yemen 94 F1 ....13.19N 43.15E
Almuñécar Spain 46 D2 ....36.44N 3.41W
Al Mahrah f. Yemen 95 H2 ....15.30N 51.00E
Almansa Spain 46 E3 ....38.52N 1.06W
Almanzor mtn. Spain 46 C4 ....40.20N 5.22W
Almaty Kazakhstan 102 D6 ....43.19N 76.55E
Almeirim Brazil 77 E4 ....1.30S 52.35W
Almelo Neth. 42 F4 ....52.21N 6.40E
Almería Spain 46 D2 ....36.50N 2.26W
Almina, Punta c. Morocco 46 C1 ....35.54N 5.17W
Al Mish'ab Saudi Arabia 95 G4 ....28.00N 48.48E
Almodôvar Portugal 46 A2 ....37.31N 8.03W
Almond r. Scotland 17 F4 ....56.25N 3.28W
Al Nu'ayriyah Saudi Arabia 95 G4 ....27.27N 48.17E
Alnwick England 15 F4 ....55.25N 1.41W
Alofi Niue 108 N5 ....19.03S 169.54W
Alor i. Indonesia 105 G2 ....8.20S 124.30E
Alpes Maritimes mts. France 44 G4 ....44.07N 7.08E
Alpine U.S.A. 64 F3 ....30.22N 103.40W
Alps mts. Europe 34 E2 ....46.00N 7.30E
Al Qa'amiyat f. Saudi Arabia 95 G2 ....18.30N 49.00E
Al Qaddahiyah Libya 53 G3 ....31.24N 15.12E
Al Qāmishli Syria 94 F6 ....37.05N 41.11E
Al Qunfidhah Saudi Arabia 94 F2 ....19.08N 41.15E
Alsager England 15 E2 ....53.07N 2.20W
Alston England 15 E3 ....54.48N 2.26W
Alta r. Norway 43 E5 ....70.00N 23.15E
Altai Mts. Mongolia 102 G7 ....46.30N 93.30E
Altamira Brazil 77 E4 ....3.12S 52.12W
Altamura Italy 50 G4 ....40.50N 16.32E
Altay China 102 F7 ....47.48N 88.07E
Altay Mongolia 102 H7 ....46.20N 97.00E
Altenburg Germany 48 F4 ....50.59N 12.27E
Altiplano f. Bolivia 76 D3 ....18.00S 67.30W
Altiplano Mexicano mts. N. America 60 I4
............24.00N 105.00W
Alton England 10 E2 ....51.08N 0.59W
Altoona U.S.A. 65 K5 ....40.32N 78.23W
Altrincham England 15 E2 ....53.25N 2.21W
Altun Shan mts. China 102 F5 ....38.10N 87.50E
Al'Uqaylah Libya 53 G3 ....30.15N 19.12E
Alur Setar Malaysia 104 C5 ....6.06N 100.23E
Al'Uthmanīyah Saudi Arabia 95 G4 ....25.16N 49.24E
Al'Uwaynat Libya 94 B3 ....21.53N 24.51E
Alva U.S.A. 64 G4 ....36.48N 98.40W
Älvdalen Sweden 43 C3 ....61.14N 14.05E
Alveley England 10 C3 ....52.28N 2.20W
Älvsbyn Sweden 43 E4 ....65.41N 21.00E
Al Wajh Saudi Arabia 94 E4 ....26.16N 36.28E
Alwen Resr. Wales 12 D5 ....53.05N 3.35W
Al Widyan f. Iraq/Saudi Arabia 94 F5 ....31.00N 42.00E
Alyth Scotland 17 F4 ....56.38N 3.14W
Alytus Lithuania 55 I6 ....54.24N 24.03E
Amadeus, L. Australia 110 C3 ....24.50S 130.45E
Amadjuak L. Canada 63 K4 ....65.00N 71.00W
Amadora Portugal 46 A3 ....38.45N 9.13W
Åmål Sweden 43 C2 ....59.04N 12.41E
Amamapare Indonesia 105 J3 ....4.56S 136.43E
Amapá d. Brazil 77 E5 ....2.00N 52.00W
Amarillo U.S.A. 64 F4 ....35.14N 101.50W
Amasya Turkey 57 L4 ....40.37N 35.50E
Amazon r. Brazil 77 E4 ....2.00S 50.00W
Amazonas d. Brazil 77 D4 ....4.50S 64.00W
Amazon Delta f. Brazil 77 F5 ....0.00 50.00W
Ambarchik Russian Fed. 59 S4 ....69.39N 162.27E
Ambato Ecuador 76 C4 ....1.18S 78.36W
Ambergate England 15 F2 ....53.03N 1.29W
Ambergris Cay i. Belize 70 G4 ....18.00N 87.58W
Amble England 15 F4 ....55.20N 1.34W
Ambleside England 14 E3 ....54.26N 2.58W
Ambon Indonesia 105 H3 ....3.50S 128.10E
Amboseli Nat. Park Kenya 87 B2 ....2.40S 37.10E
Ambrym i. Vanuatu 111 F4 ....16.15S 168.10E
Ameland i. Neth. 42 E5 ....53.28N 5.48E
American Samoa is. Pacific Oc. 108 O5
............14.20S 170.00W
Amersfoort Neth. 42 E4 ....52.10N 5.23E
Amersham England 11 E2 ....51.40N 0.38W
Amesbury England 10 D2 ....51.10N 1.46W
Amga r. Russian Fed. 59 P4 ....62.40N 135.20E
Amgu Russian Fed. 106 C5 ....45.48N 137.36E
Amgun r. Russian Fed. 59 P3 ....53.10N 139.47E
Amiens France 44 E6 ....49.54N 2.18E
Amino Ethiopia 87 C3 ....4.25N 41.52E
Amlwch Wales 12 C5 ....53.24N 4.21W
Amman Jordan 94 E5 ....31.57N 35.56E
Ammanford Wales 12 C3 ....51.48N 4.00W
Amol Iran 95 H6 ....36.26N 52.24E
Amorgos i. Greece 56 G2 ....36.49N 25.54E
Amos Canada 63 K2 ....48.35N 78.05W

Ampthill England 11 E3 ....52.03N 0.30W
Amravati India 97 F5 ....20.58N 77.50E
Amritsar India 96 E7 ....31.35N 74.56E
Amstelveen Neth. 42 D4 ....52.18N 4.51E
Amsterdam Neth. 42 D4 ....52.22N 4.54E
Amu Darya r. Asia 90 H7 ....43.50N 59.00E
Amund Ringnes I. Canada 63 I5 ....78.00N 96.00W
Amundsen G. Canada 62 F5 ....70.30N 122.00W
Amundsen Sea Antarctica 112 ....70.00S 110.00W
Amuntai Indonesia 104 F3 ....2.24S 115.14E
Amur r. Russian Fed. 59 P3 ....53.17N 140.00E
Anabar r. Russian Fed. 59 N5 ....72.40N 113.30E
Anadyr Russian Fed. 59 T4 ....64.40N 177.32E
Anadyr r. Russian Fed. 59 T4 ....65.00N 176.00E
Anadyr, G. of Russian Fed. 59 U4 ....64.30N 177.50W
'Ānah Iraq 94 F5 ....34.29N 41.57E
Anambas Is. Indonesia 104 D4 ....3.00N 106.10E
Anamur Turkey 57 K2 ....36.06N 32.49E
Anápolis Brazil 77 F3 ....16.19S 48.58W
Anatahan i. N. Mariana Is. 105 L7 ....16.22N 145.38E
Anatolia f. Turkey 57 J3 ....38.30N 32.00E
Anchorage U.S.A. 62 D4 ....61.10N 150.00W
Ancona Italy 50 E5 ....43.37N 13.33E
Åndalsnes Norway 43 A3 ....62.33N 7.43E
Andaman Is. India 97 I3 ....12.00N 93.00E
Andaman Sea Indian Oc. 97 J3 ....11.00N 96.00E
Anderlecht Belgium 42 D2 ....50.51N 4.18E
Anderson U.S.A. 62 D4 ....64.25N 149.10W
Anderson r. Canada 62 F4 ....69.45N 129.00W
Andes mts. S. America 74 B5 ....15.00S 74.00W
Andfjorden est. Norway 43 D5 ....69.10N 16.20E
Andhra Pradesh d. India 97 F4 ....17.00N 79.00E
Andkhvoy Afghan. 95 K6 ....36.56N 65.05E
Andorra Europe 46 F5 ....42.30N 1.32E
Andorra La Vella Andorra 46 F5 ....42.30N 1.31E
Andover England 10 D2 ....51.13N 1.29W
Andoya i. Norway 43 D5 ....69.00N 15.30E
Andreas I.o.M. 14 C3 ....54.22N 4.26W
Andreas, C. Cyprus 57 L1 ....35.40N 34.35E
Andros i. Greece 56 G2 ....37.50N 24.50E
Andros i. The Bahamas 71 I5 ....24.30N 78.00W
Andújar Spain 46 C3 ....38.02N 4.03W
Anegada i. B.V.Is. 71 L4 ....18.46N 64.24W
Aneto, Pico de mtn. Spain 46 F5 ....42.40N 0.19E
Angara r. Russian Fed. 59 L3 ....58.00N 93.00E
Angarsk Russian Fed. 103 I8 ....52.31N 103.55E
Ånge Sweden 43 C3 ....62.31N 15.40E
Angel de la Guarda i. Mexico 70 B6 ....29.10N 113.20W
Ångelholm Sweden 43 C2 ....56.15N 12.50E
Angers France 44 C5 ....47.29N 0.32W
Angola Africa 86 A3 ....12.00S 18.00E
Angola Basin f. Atlantic Oc. 117 J5
............45.40N 0.10E
Angoulême France 44 D4 ....45.40N 0.10E
Angren Uzbekistan 102 C6 ....41.01N 70.10E
Anguilla i. Leeward Is. 71 L4 ....18.14N 63.05W
Angus d. Scotland 8 D5 ....56.45N 3.00W
Anhui d. China 103 L4 ....31.30N 116.45E
Ankara Turkey 57 K3 ....39.55N 32.50E
Anlaby England 15 G2 ....53.45N 0.27W
Annaba Algeria 84 E5 ....36.55N 7.47E
An Nafud des. Saudi Arabia 94 F4 ....28.40N 41.30E
An Najaf Iraq 94 F5 ....31.59N 44.19E
Annalee r. Rep. of Ire. 20 D4 ....54.08N 7.25W
Annalong N. Ireland 16 D2 ....54.06N 5.55W
Annan Scotland 17 F2 ....54.59N 3.16W
Annan r. Scotland 17 F3 ....54.58N 3.16W
Annapurna mtn. Nepal 97 G6 ....28.34N 83.50E
Ann Arbor U.S.A. 65 J5 ....42.18N 83.45W
An Nasiriyah Iraq 95 G5 ....31.04N 46.16E
An Nawfaliyah Libya 53 G3 ....30.47N 17.50E
Annecy France 44 G4 ....45.54N 6.07E
Ansbach Germany 48 E3 ....49.18N 10.36E
Anshan China 103 M6 ....41.05N 122.58E
Anshun China 103 J3 ....26.15N 105.51E
Anstruther Scotland 17 G4 ....56.14N 2.42W
Antakya Turkey 57 M2 ....36.12N 36.10E
Antalya Turkey 57 J2 ....36.53N 30.42E
Antalya, G. of Turkey 57 J2 ....36.38N 31.00E
Antananarivo Madagascar 86 D3 ....18.52S 47.30E
Antarctica 112
Antarctic Pen. f. Antarctica 116 F2 ....65.00S 64.00W
An Teallach mtn. Scotland 18 D2 ....57.48N 5.16W
Antequera Spain 46 C2 ....37.01N 4.34W
Antibes France 44 G3 ....43.35N 7.07E
Anticosti, Île d' Canada 63 L2 ....49.20N 63.00W
Antigua i. Leeward Is. 71 L4 ....17.09N 61.49W
Antigua and Barbuda Leeward Is. 71 L4
............17.30N 61.49W
Antikythira i. Greece 56 F1 ....35.52N 23.18E
Antipodes Is. Pacific Oc. 108 M2 ....49.42S 178.50E
Antofagasta Chile 75 B2 ....23.40S 70.23W
Antrim N. Ireland 16 C2 ....54.43N 6.14W
Antrim d. N. Ireland 20 E4 ....54.45N 6.15W
Antrim Hills N. Ireland 16 C2 ....55.00N 6.10W
Antsirañana Madagascar 86 D3 ....12.19S 49.17E
Antwerpen Belgium 42 D3 ....51.13N 4.25E
Antwerpen d. Belgium 42 D3 ....51.16N 4.45E
Anxi China 102 H6 ....40.32N 95.57E
Anyang China 103 K5 ....36.04N 114.20E
Anzhero-Sudzhensk Russian Fed. 58 K3
............56.10N 86.10E
Aomori Japan 106 D4 ....40.50N 140.43E
Aoraki mtn. New Zealand 111 G1 ....43.36S 170.09E
Aosta Italy 50 B6 ....45.43N 7.19E
Apa r. Brazil/Paraguay 77 E2 ....22.08S 57.55W
Apalachee B. U.S.A. 65 J2 ....29.30N 84.00W
Apaporis r. Colombia 76 D4 ....1.40S 69.20W
Aparri Phil. 105 G7 ....18.22N 121.40E
Apatity Russian Fed. 43 H4 ....67.32N 33.21E
Apeldoorn Neth. 42 E4 ....52.13N 5.57E
Apennines mts. Italy 50 D6 ....43.00N 11.00E
Apia Samoa 108 N5 ....13.50S 171.44E
Aporé r. Brazil 77 E3 ....19.30S 50.58S
Appalachian Mts. U.S.A. 65 K4 ....39.30N 78.00W
Appleby-in-Westmorland England 15 E3
............54.35N 2.29W
Appledore England 13 C3 ....51.03N 4.12W
Appleton U.S.A. 65 I5 ....44.16N 88.25W
Apucarana Brazil 77 E2 ....23.34S 51.28W
Apurímac r. Peru 76 C3 ....10.43S 73.55W
Aqaba Jordan 94 E4 ....29.32N 35.00E
Aqaba, G. of Saudi Arabia 94 E4 ....28.45N 34.45E
Ära Ärba Ethiopia 87 C4 ....5.50N 41.30E
Arabian Pen. Asia 90 G5 ....24.00N 45.00E
Arabian Sea Asia 96 C4 ....19.00N 65.00E
Aracaju Brazil 77 G3 ....10.54S 37.07W

Brazil S. America **77 E3** 10.00S 52.00W
Brazil Basin *f.* Atlantic Oc. **116 H5**
Brazilian Highlands Brazil **77 F3** 17.00S 48.00W
Brazos *r.* U.S.A. **64 G2** 28.55N 95.20W
Brazzaville Congo **86 F1** 4.14S 15.14E
Brechin Scotland **19 G1** 56.44N 2.40W
Breckland *f.* England **11 F3** 52.28N 0.40E
Břeclav Czech Rep. **54 E3** 48.46N 16.53E
Brecon Wales **12 D3** 51.57N 3.23W
Brecon Beacons *mts.* Wales **12 D3** 51.53N 3.27W
Breda Neth. **42 D3** 51.35N 4.46E
Bredhafjördhur *est.* Iceland **43 X2** 65.15N 23.00W
Breidhdalsvik Iceland **43 Z2** 64.48N 14.00W
Bremen Germany **48 D5** 53.05N 8.48E
Bremerhaven Germany **48 D5** 53.33N 8.35E
Brenner Pass Italy/Austria **50 D7** 47.00N 11.30E
Brentwood England **11 F2** 51.38N 0.18E
Brescia Italy **50 D6** 45.33N 10.12E
Bressay *i.* Scotland **19 Y9** 60.08N 1.05W
Bressuire France **44 C5** 46.50N 0.28W
Brest Belarus **55 H5** 52.08N 23.40E
Brest France **44 A6** 48.23N 4.30W
Brest-Nantes Canal France **44 B5** 47.52N 2.20W
Bretton Wales **12 E5** 53.10N 3.00W
Bria C.A.R. **85 G2** 6.32N 21.59E
Briançon France **44 G4** 44.53N 6.39E
Bride *r.* Rep. of Ire. **20 D2** 52.06N 7.50W
Bridgend Wales **13 D3** 51.30N 3.35W
Bridgend *d.* Wales **9 D2** 51.33N 3.35W
Bridgeport U.S.A. **65 L5** 41.12N 73.12W
Bridgetown Barbados **71 M3** 13.06N 59.37W
Bridgnorth England **10 C3** 52.33N 2.25W
Bridgwater England **13 D3** 51.08N 3.00W
Bridgwater B. England **13 D3** 51.15N 3.10W
Bridlington England **15 G3** 54.06N 0.11W
Bridlington B. England **15 G3** 54.03N 0.10W
Bridport England **10 C1** 50.43N 2.45W
Brig Switz. **44 G5** 46.19N 8.00E
Brigg England **15 G2** 53.33N 0.30W
Brighstone England **10 D1** 50.38N 1.24W
Brightlingsea England **11 G2** 51.49N 1.01E
Brighton England **11 E1** 50.50N 0.09W
Brighton and Hove *d.* England **9 E2** 50.50N 0.09W
Brindisi Italy **50 G4** 40.38N 17.57E
Brisbane Australia **110 E3** 27.30S 153.00E
Bristol England **10 C2** 51.26N 2.35W
Bristol *d.* England **9 D2** 51.26N 2.35W
Bristol B. U.S.A. **62 C3** 58.00N 158.50W
Bristol Channel England/Wales **13 D3** 51.17N 3.20W
British Columbia *d.* Canada **62 F3** 55.00N 125.00W
British Isles Europe **34 C3** 54.00N 5.00W
Briton Ferry *town* Wales **12 D3** 51.37N 3.50W
Brittany *f.* France **44 B6** 48.00N 3.00W
Brive-la-Gaillarde France **44 D4** 45.09N 1.32E
Brixham England **13 D2** 50.24N 3.31W
Brno Czech Rep. **54 E3** 49.11N 16.39E
Broad B. Scotland **18 C3** 58.15N 6.15W
Broad Law *mtn.* Scotland **17 F3** 55.30N 3.21W
Broadstairs England **11 G2** 51.22N 1.27E
Broadview Canada **64 F7** 50.20N 102.30W
Broadway England **10 D3** 52.02N 1.50W
Broadwey England **10 C1** 50.39N 2.29W
Broadwindsor England **10 C1** 50.49N 2.48W
Brockenhurst England **10 D1** 50.49N 1.34W
Brock I. Canada **62 G5** 78.00N 114.30W
Brodeur Pen. Canada **63 J5** 73.00N 88.00W
Brodick Scotland **16 D3** 55.34N 5.09W
Broken Hill *town* Australia **110 D2** 31.57S 141.30E
Bromley England **11 F2** 51.24N 0.02E
Bromsgrove England **10 C3** 52.20N 2.03W
Bromyard England **10 C3** 52.12N 2.30W
Brønderslev Denmark **43 B2** 57.16N 9.58E
Brønnøysund Norway **43 C4** 65.38N 12.15E
Brooke England **11 G3** 52.32N 1.25E
Brooke's Point *town* Phil. **104 F5** 8.50N 117.52E
Brooks Range *mts.* U.S.A. **62 C4** 68.50N 152.00W
Broome Australia **110 B4** 17.58S 122.15E
Broom, Loch Scotland **18 D2** 57.55N 5.15W
Brora Scotland **19 F3** 58.01N 3.52W
Brora *r.* Scotland **19 F3** 57.59N 3.51W
Brosna *r.* Rep. of Ire. **20 D3** 53.12N 7.59W
Brotton England **15 G3** 54.34N 0.55W
Brough Cumbria England **15 E3** 54.32N 2.19W
Brough E.Yorks. England **15 G2** 53.42N 0.34W
Brough Head Scotland **19 F4** 59.09N 3.19W
Brough Ness *c.* Scotland **19 G3** 58.44N 2.57W
Broughshane N. Ireland **16 C2** 54.54N 6.12W
Brownhills England **10 D3** 52.38N 1.57W
Broxburn Scotland **17 F3** 55.57N 3.29W
Bruay-en-Artois France **42 B2** 50.29N 2.36E
Brue *r.* England **13 E3** 51.13N 3.00W
Brugge Belgium **42 C3** 51.13N 3.14E
Brunei Asia **104 E4** 4.56N 114.58E
Brunflo Sweden **43 C3** 63.04N 14.50E
Brunswick U.S.A. **65 J3** 31.09N 81.21W
Brussels Belgium **42 D2** 50.50N 4.23E
Bruton England **13 E3** 51.06N 2.28W
Bryansk Russian Fed. **55 N5** 53.15N 34.09E
Bryher *i.* England **13 A1** 49.57N 6.21W
Brynamman Wales **12 D3** 51.49N 3.52W
Brynmawr Wales **12 D3** 51.48N 3.10W
Buca Turkey **56 H3** 38.22N 27.10E
Bucaramanga Colombia **71 J2** 7.08N 73.01W
Bucharest Romania **56 H6** 44.25N 26.06E
Buckhaven Scotland **17 F4** 56.11N 3.03W
Buckie Scotland **19 G2** 57.40N 2.58W
Buckingham England **10 E2** 52.00N 0.59W
Buckinghamshire *d.* England **9 E2** 51.50N 0.48W
Buckley Wales **12 D5** 53.11N 3.04W
Budapest Hungary **54 F2** 47.30N 19.03E
Buddon Ness *c.* Scotland **17 G4** 56.29N 2.42W
Bude England **13 C2** 50.49N 4.33W
Bude B. England **13 C2** 50.45N 4.40W
Buenaventura Colombia **74 B7** 3.54N 77.02W
Buenos Aires Argentina **75 D3** 34.40S 58.30W
Buffalo N.Y. U.S.A. **65 K5** 42.52N 78.55W
Buffalo Wyo. U.S.A. **64 E5** 44.21N 106.40W
Bug *r.* Poland **54 G5** 52.29N 21.11E
Bug *r.* Ukraine **55 L2** 46.55N 31.59E
Buhayrat al Asad *l.* Syria **57 N2** 36.10N 38.20E
Builth Wells Wales **12 D4** 52.09N 3.24W
Buir Nur *l.* Mongolia **103 L7** 47.50N 117.40E
Bujumbura Burundi **86 B4** 3.22S 29.21E
Bukavu Dem. Rep. of Congo **86 B4** 2.30S 28.49E
Bukittinggi Indonesia **104 C3** 0.18S 100.20E
Bukoba Tanzania **86 C4** 1.20S 31.49E
Bula Indonesia **105 I3** 3.07S 130.27E
Bulawayo Zimbabwe **86 B2** 20.10S 28.43E
Bulgan Mongolia **103 I7** 48.34N 103.12E
Bulgaria Europe **56 G5** 42.30N 25.00E
Bulukumba Indonesia **104 G2** 5.35S 120.13E
Bulun Russian Fed. **59 O5** 70.50N 127.20E
Buna Kenya **87 B3** 2.49N 39.27E
Bunbury Australia **110 A2** 33.20S 115.34E
Bunclody Rep. of Ire. **20 E2** 52.39N 6.39W
Buncrana Rep. of Ire. **20 D5** 55.08N 7.28W
Bunda Tanzania **87 A2** 2.00S 33.57E
Bundaberg Australia **110 E3** 24.50S 152.21E
Bundoran Rep. of Ire. **20 C4** 54.28N 8.20W
Bungay England **11 G3** 52.27N 1.26E
Bungoma Kenya **87 A3** 0.33N 34.33E
Buôn Mê Thuôt Vietnam **104 D6** 12.41N 108.02E
Bura Kenya **87 B2** 1.09S 39.55E
Buraydah Saudi Arabia **94 F4** 26.18N 43.58E
Burbage England **10 D2** 51.22N 1.40W
Burdur Turkey **57 J2** 37.44N 30.17E
Bure *r.* England **11 G3** 52.36N 1.44E
Burford England **10 D2** 51.48N 1.38W
Burgas Bulgaria **56 H5** 42.30N 27.29E
Burgess Hill *town* England **11 E1** 50.57N 0.07W
Burghead Scotland **19 F2** 57.42N 3.30W
Burgh le Marsh England **15 H2** 53.10N 0.15E
Burgos Spain **46 D5** 42.21N 3.41W
Burhanpur India **96 F5** 21.18N 76.08E
Burkina Africa **84 D3** 12.15N 1.30W
Burley U.S.A. **64 D5** 42.32N 113.48W
Burlington U.S.A. **65 L5** 44.25N 73.14W
Burnham England **10 E2** 51.35N 0.39W
Burnham-on-Crouch England **11 F2** 51.37N 0.50E
Burnham-on-Sea England **13 D3** 51.15N 3.00W
Burnie Australia **110 D1** 41.03S 145.55E
Burniston England **15 G3** 54.19N 0.27W
Burnley England **15 E2** 53.47N 2.15W
Burns Lake *town* Canada **62 F3** 54.14N 125.45W
Burntisland Scotland **17 F4** 56.03N 3.15W
Burntwood Green England **10 D3** 52.42N 1.54W
Burray *i.* Scotland **19 G3** 58.51N 2.54W
Burrow Head Scotland **16 E2** 54.41N 4.24W
Burry Inlet Wales **12 C3** 51.40N 4.11W
Burry Port Wales **12 C3** 51.41N 4.17W
Bursa Turkey **57 I4** 40.11N 29.04E
Bûr Safâga Egypt **94 D4** 26.45N 33.55E
Burscough Bridge England **14 E2** 53.37N 2.51W
Burton-in-Kendal England **14 E3** 54.12N 2.43W
Burton Latimer England **10 E3** 52.23N 0.41W
Burton upon Trent England **10 D3** 52.48N 1.39W
Buru *i.* Indonesia **105 H3** 3.30S 126.30E
Burundi Africa **86 B4** 3.30S 30.00E
Burwash Landing Canada **62 E4** 61.21N 139.01W
Burwell England **11 F3** 52.17N 0.20E
Burwick Scotland **19 G3** 58.44N 2.57W
Bury England **15 E2** 53.36N 2.19W
Bury St. Edmunds England **11 F3** 52.15N 0.42E
Bush *r.* N. Ireland **16 C3** 55.13N 6.33W
Bushbush *r.* Somalia **87 C2** 1.08S 41.52E
Bushehr Iran **95 H4** 28.57N 50.52E
Bushmills N. Ireland **16 C3** 55.12N 6.32W
Buta Dem. Rep. of Congo **86 B5** 2.49N 24.50E
Bute *i.* Scotland **16 D3** 55.51N 5.07W
Bute, Sd. of Scotland **16 D3** 55.44N 5.10W
Buton *i.* Indonesia **105 G2** 5.00S 122.50E
Butte U.S.A. **64 D6** 46.00N 112.31W
Butterworth Malaysia **104 C5** 5.24N 100.22E
Buttevant Rep. of Ire. **20 C2** 52.13N 8.40W
Butt of Lewis *c.* Scotland **18 C3** 58.31N 6.15W
Butuan Phil. **105 H5** 8.56N 125.31E
Buur Gaabo Somalia **87 C2** 1.10S 41.50E
Buvuma I. Uganda **87 A3** 0.12N 33.17E
Buxton England **15 F2** 53.16N 1.54W
Buzău Romania **56 H6** 45.10N 26.49E
Byarezina *r.* Belarus **55 L5** 52.30N 30.20E
Bydgoszcz Poland **54 E5** 53.16N 18.00E
Byfield England **10 D3** 52.10N 1.15W
Bylot I. Canada **63 K5** 73.00N 78.30W
Byrranga Mts. Russian Fed. **59 M5** 74.50N 101.00E
Bytom Poland **54 F4** 50.22N 18.54E

C

Caacupé Paraguay **77 E2** 25.23S 57.05W
Cabanatuan Phil. **104 G7** 15.30N 120.58E
Caban Coch Resr. Wales **12 D4** 52.17N 3.34W
Cabimas Venezuela **71 J3** 10.26N 71.27W
Cabinda Angola **84 F1** 5.34S 12.12E
Cabonga, Resr. Canada **65 K6** 47.20N 76.35W
Cabot Str. Canada **63 M2** 47.00N 59.00W
Cabrera, Sierra *mts.* Spain **46 B5** 42.10N 6.30W
Cabriel *r.* Spain **46 E3** 39.13N 1.07W
Cáceres Brazil **77 E3** 16.05S 57.40W
Cáceres Spain **46 B3** 39.29N 6.23W
Cachimbo, Serra do *mts.* Brazil **77 E4** 8.30S 55.00W
Cachoeiro de Itapemirim Brazil **77 F2** 20.51S 41.07W
Cadera, C. Venezuela **71 K3** 10.40N 66.05W
Cadillac U.S.A. **65 I5** 44.15N 85.23W
Cadiz Phil. **105 G6** 10.57N 123.18E
Cádiz Spain **46 B2** 36.32N 6.18W
Cádiz, G. of Spain **46 B2** 37.00N 7.10W
Caen France **44 C6** 49.11N 0.22W
Caerleon Wales **12 E3** 51.36N 2.57W
Caernarfon Wales **12 C5** 53.08N 4.17W
Caernarfon B. Wales **12 C5** 53.05N 4.25W
Caerphilly Wales **12 D3** 51.34N 3.13W
Caerphilly *d.* Wales **9 D2** 51.34N 3.13W
Cagayan de Oro Phil. **105 G5** 8.29N 124.40E
Cagliari Italy **50 C3** 39.14N 9.07E
Cagliari, G. of Med. Sea **50 C3** 39.07N 9.15E
Caha Mts. Rep. of Ire. **20 B1** 51.44N 9.45W
Cahirciveen Rep. of Ire. **20 A1** 51.51N 10.14W
Cahora Bassa, Lago de *l.* Mozambique **86 C3** 15.33S 32.42E
Cahore Pt. Rep. of Ire. **20 E2** 52.33N 6.11W
Cahors France **44 D4** 44.28N 0.26E
Cahul Moldova **55 K1** 45.58N 28.10E
Caiabis, Serra dos *mts.* Brazil **77 E3** 12.00S 56.00W
Caiapó, Serra do *mts.* Brazil **77 E3** 17.10S 52.00W
Caicos Is. Turks & Caicos Is. **71 J5** 21.30N 72.00W
Cairn Gorm *mtn.* Scotland **19 F2** 57.06N 3.39W
Cairngorm Mts. Scotland **19 F2** 57.04N 3.30W
Cairnryan Scotland **16 D2** 54.58N 5.02W
Cairns Australia **110 D4** 16.51S 145.43E
Cairn Toul *mtn.* Scotland **19 F2** 57.04N 3.44W
Cairo Egypt **94 D4** 30.03N 31.15E
Caister-on-Sea England **11 G3** 52.38N 1.43E
Caistor England **15 G2** 53.29N 0.20W
Caithness *f.* Scotland **19 F3** 58.25N 3.25W
Calabar Nigeria **84 E2** 4.56N 8.22E
Calais France **44 D7** 50.57N 1.50E
Calama Chile **76 D2** 22.30S 68.55W
Calamian Group *is.* Phil. **104 G6** 12.00N 120.05E
Calamocha Spain **46 E4** 40.54N 1.18W
Calanscio Sand Sea *f.* Libya **53 H2** 27.00N 23.00E
Calapan Phil. **104 G6** 13.23N 121.10E
Călăraşi Romania **56 H6** 44.11N 27.21E
Calatayud Spain **46 E4** 41.21N 1.39W
Calbayog Phil. **105 G6** 12.04N 124.58E
Calcanhar, Punta do *c.* Brazil **77 G4** 5.06S 35.30W
Calcutta *see* **Kolkata** India **97**
Caldas da Rainha Portugal **46 A3** 39.24N 9.08W
Caldey Island Wales **12 C3** 51.38N 4.43W
Caldicot Wales **12 E3** 51.36N 2.45W
Calf of Man *i.* I.o.M. **14 C3** 54.03N 4.49W
Calgary Canada **62 G3** 51.05N 114.05W
Cali Colombia **74 B7** 3.24N 76.30W
Caliente U.S.A. **64 D4** 37.36N 114.31W
California *d.* U.S.A. **64 B4** 37.00N 120.00W
California, G. of Mexico **70 B6** 28.30N 112.30W
Callan Rep. of Ire. **20 D2** 52.33N 7.23W
Callander Scotland **16 E4** 56.15N 4.13W
Callanish Scotland **18 C3** 58.12N 6.45W
Callao Peru **76 C3** 12.05S 77.08W
Callington England **13 C2** 50.30N 4.19W
Calne England **10 D2** 51.26N 2.00W
Caltanissetta Italy **50 F2** 37.30N 14.05E
Calvi France **44 H3** 42.34N 8.44E
Cam *r.* England **11 F3** 52.34N 0.21E
Camaçari Brazil **77 G3** 12.44S 38.16W
Camagüey Cuba **71 I5** 21.25N 77.55W
Camagüey, Archipelago de Cuba **71 I5** 22.30N 78.00W
Cambay, G. of India **96 E5** 20.30N 72.00E
Camberley England **10 E2** 51.21N 0.45W
Cambodia Asia **104 C6** 12.00N 105.00E
Camborne England **13 B2** 50.12N 5.19W
Cambrai France **42 C2** 50.10N 3.14E
Cambrian Mts. Wales **12 D4** 52.33N 3.33W
Cambridge England **11 F3** 52.13N 0.08E
Cambridge Bay *town* Canada **62 H4** 69.09N 105.00W
Cambridgeshire *d.* England **9 E3** 52.15N 0.05W
Camelford England **13 C2** 50.37N 4.41W
Cameroon Africa **84 F2** 6.00N 12.30E
Cameroun, Mt. Cameroon **84 E2** 4.20N 9.05E
Cametá Brazil **77 F4** 2.12S 49.30W
Campbell I. Pacific Oc. **108 L1** 52.30S 169.02E
Campbell River *town* Canada **62 F3** 50.00N 125.18W
Campbellton Canada **65 M6** 48.00N 66.40W
Campbeltown Scotland **16 D3** 55.25N 5.36W
Campeche Mexico **70 F4** 19.50N 90.30W
Campeche *d.* Mexico **70 F4** 19.00N 90.00W
Campeche B. Mexico **70 F4** 19.30N 94.00W
Campina Grande Brazil **77 G4** 7.15S 35.53W
Campinas Brazil **77 F2** 22.54S 47.06W
Campobasso Italy **50 F4** 41.34N 14.39E
Campo Grande Brazil **77 E2** 20.24S 54.35W
Campos Brazil **77 F2** 21.46S 41.21W
Cam Ranh Vietnam **104 D6** 11.54N 109.14E
Camrose Wales **12 B3** 51.50N 5.01W
Canada N. America **62 H3** 60.00N 105.00W
Canadian *r.* U.S.A. **64 G4** 35.20N 95.40W
Canadian Shield *f.* N. America **60 K7** 50.00N 80.00W
Çanakkale Turkey **56 H4** 40.09N 26.26E
Canal du Midi France **44 D3** 43.18N 2.00E
Canary Is. Atlantic Oc. **46 X2** 29.00N 15.00W
Canaveral, C. U.S.A. **65 J2** 28.28N 80.28W
Canberra Australia **110 D2** 35.18S 149.08E
Cancún Mexico **70 G5** 21.26N 86.51W
Caniapiscau *r.* Canada **63 L3** 57.40N 69.30W
Caniapiscau, Résr. Canada **63 K3** 55.05N 72.40W
Canindé *r.* Brazil **77 F4** 6.14S 42.51W
Canisp *mtn.* Scotland **18 D3** 58.07N 5.03W
Çankaya Turkey **57 K3** 39.52N 32.52E
Çankırı Turkey **57 K4** 40.35N 33.37E
Canna *i.* Scotland **18 C2** 57.03N 6.30W
Cannes France **44 G3** 43.33N 7.00E
Cannock England **10 C3** 52.42N 2.02W
Canôas Brazil **77 E2** 29.55S 51.10W
Canon City U.S.A. **64 E4** 38.27N 105.14W
Cantabrian Mts. Spain **46 B5** 43.00N 6.00W
Canterbury England **11 G2** 51.17N 1.05E
Cân Tho Vietnam **104 D6** 10.03N 105.46E
Canvey Island *town* England **11 F2** 51.32N 0.35E
Cao Bang Vietnam **104 D8** 22.40N 106.16E
Capbreton France **44 C3** 43.38N 1.15W
Cape Breton I. Canada **63 L2** 46.00N 61.00W
Capel St. Mary England **11 G2** 51.59N 1.02E
Cape Town R.S.A. **86 A1** 33.56S 18.28E
Cape Verde Atlantic Oc. **84 B3** 16.00N 24.00W
Cape York Pen. Australia **110 D4** 12.40S 142.20E
Cap Haïtien *town* Haiti **71 J4** 19.47N 72.17W
Capim *r.* Brazil **77 F4** 1.40S 47.47W
Capitol Hill N. Mariana Is. **105 L7** 15.12N 145.45E
Capraia *i.* Italy **50 C5** 43.03N 9.50E
Caprera *i.* Italy **50 C4** 41.48N 9.27E
Capri *i.* Italy **50 F4** 40.33N 14.13E
Capricorn Channel *str.* Australia **110 E3** 23.00S 152.00E
Caprivi Strip *f.* Namibia **86 B3** 17.50S 23.10E
Caquetá *r.* Colombia **76 C4** 1.20S 70.50W
Carabay, Cordillera de *mts.* Peru **76 C3** 13.50S 71.00W
Caracal Romania **56 G6** 44.08N 24.18E
Caracas Venezuela **71 K3** 10.35N 66.56W
Carajás, Serra dos *mts.* Brazil **77 E4** 5.00S 51.00W
Caratasca Lagoon Honduras **71 H4** 15.10N 84.00W
Caratinga Brazil **77 F3** 19.50S 42.06W
Caravaca de la Cruz Spain **46 E3** 38.06N 1.51W
Carbonara, C. Italy **50 C3** 39.06N 9.32E
Carcassonne France **44 E3** 43.13N 2.21E
Cardiff Wales **13 D3** 51.28N 3.11W
Cardiff *d.* Wales **9 D2** 51.30N 3.12W
Cardigan Wales **12 C4** 52.06N 4.41W
Cardigan B. Wales **12 C4** 52.30N 4.30W
Carei Romania **54 H2** 47.42N 22.28E
Carey, L. Australia **110 B3** 29.05S 122.15E
Cariacica Brazil **77 F2** 20.15S 40.23W
Caribbean Sea C. America **71 I4** 15.00N 75.00W
Caribou Mts. Canada **62 G3** 58.30N 115.00W
Cark Mtn. Rep. of Ire. **16 B2** 54.53N 7.53W
Carletonville R.S.A. **86 B2** 26.21S 27.23E
Carlingford Lough Rep. of Ire./N. Ireland **16 C2** 54.03N 6.09W
Carlisle England **14 E3** 54.54N 2.55W
Carlow Rep. of Ire. **20 E2** 52.50N 6.54W
Carlow *d.* Rep. of Ire. **20 E2** 52.43N 6.50W
Carluke Scotland **17 E3** 55.44N 3.51W
Carmacks Canada **62 E4** 62.04N 136.21W
Carmarthen Wales **12 C3** 51.52N 4.20W
Carmarthen B. Wales **12 C3** 51.40N 4.30W
Carmarthenshire *d.* Wales **9 C2** 52.00N 4.17W
Carmel Head Wales **12 C5** 53.24N 4.35W
Carndonagh Rep. of Ire. **16 B3** 55.15N 7.15W
Carnedd Llywelyn *mtn.* Wales **12 D5** 53.10N 3.58W
Carnedd y Filiast *mtn.* Wales **12 D4** 52.56N 3.40W
Carnegie, L. Australia **110 B3** 26.15S 123.00E
Carn Eighe *mtn.* Scotland **18 D2** 57.17N 5.07W
Carnforth England **14 E3** 54.08N 2.47W
Carnic Alps *mts.* Italy/Austria **50 E7** 46.40N 12.48E
Car Nicobar *i.* India **104 A5** 9.06N 92.57E
Carnlough N. Ireland **16 D2** 54.58N 6.00W
Carn nan Gabhar *mtn.* Scotland **19 F1** 56.49N 3.44W
Carnot C.A.R. **84 F2** 4.59N 15.56E
Carnot, C. Australia **110 C2** 34.57S 135.38E
Carnoustie Scotland **17 G4** 56.30N 2.44W
Carnsore Pt. Rep. of Ire. **20 E2** 52.10N 6.21W
Caroline I. *see* **Millennium I.** Kiribati **109**
Caroline Is. Pacific Oc. **108 J7** 5.00N 150.00E
Carpathian Mts. Europe **34 F2** 48.45N 23.45E
Carpentaria, G. of Australia **110 C4** 14.00S 140.00E
Carpentras France **44 F4** 44.03N 5.03E
Carra, Lough Rep. of Ire. **20 B3** 53.40N 9.15W
Carrara Italy **50 D6** 44.04N 10.06E
Carrauntoohill *mtn.* Rep. of Ire. **20 B2** 52.00N 9.45W
Carrickfergus N. Ireland **16 D2** 54.43N 5.49W
Carrickmacross Rep. of Ire. **20 E3** 53.59N 6.44W
Carrick-on-Shannon Rep. of Ire. **20 C3** 53.57N 8.06W
Carrick-on-Suir Rep. of Ire. **20 D2** 52.21N 7.26W
Carron *r.* Falkirk Scotland **17 F4** 56.01N 3.44W
Carron *r.* High. Scotland **19 E2** 57.53N 4.22W
Carrowmore Lake Rep. of Ire. **20 B4** 54.11N 9.48W
Carson City U.S.A. **64 C4** 39.10N 119.46W
Carsphairn Scotland **16 E3** 55.13N 4.15W
Cartagena Colombia **71 I3** 10.24N 75.33W
Cartagena Spain **46 E2** 37.36N 0.59W
Carter Bar *pass* Scotland/England **17 G3** 55.21N 2.27W
Carterton England **10 D2** 51.46N 1.35W
Cartmel England **14 E3** 54.12N 2.57W
Caruarú Brazil **77 G4** 8.15S 35.55W
Carvin France **42 B2** 50.30N 2.58E
Casablanca Morocco **84 D5** 33.39N 7.35W
Cascade Range *mts.* U.S.A. **64 B5** 44.00N 121.30W
Cascavel Brazil **77 E2** 24.59S 53.29W
Caserta Italy **50 F4** 41.06N 14.21E
Caseyr, C. Somalia **85 J3** 12.00N 51.30E
Cashel Rep. of Ire. **20 D2** 52.31N 7.54W
Casper U.S.A. **64 E5** 42.50N 106.20W
Caspian Depression *f.* Russian Fed./Kazakhstan **58 G2** 47.00N 48.00E
Caspian Sea Asia **90 H7** 42.00N 51.00E
Cassiar Mts. Canada **62 F3** 60.00N 131.00W
Cassley *r.* Scotland **18 E3** 57.58N 4.35W
Castanhal Brazil **77 F4** 1.16S 47.51W
Castelló de la Plana Spain **46 E3** 39.59N 0.03W
Castlebar Rep. of Ire. **20 B3** 53.52N 9.19W
Castleblayney Rep. of Ire. **20 E4** 54.08N 6.46W
Castle Cary England **13 E3** 51.06N 2.31W
Castlecomer Rep. of Ire. **20 D2** 52.48N 7.12W
Castleconnell Rep. of Ire. **20 C2** 52.43N 8.30W
Castlederg N. Ireland **16 B2** 54.43N 7.37W
Castle Donnington England **10 D3** 52.51N 1.19W
Castle Douglas Scotland **17 F2** 54.56N 3.56W
Castleford England **15 F2** 53.43N 1.21W
Castleisland *town* Rep. of Ire. **20 B2** 52.14N 9.29W
Castletown I.o.M. **14 C3** 54.04N 4.38W
Castres France **44 E3** 43.36N 2.14E
Castries St. Lucia **71 L3** 14.01N 60.59W
Catamarca Argentina **76 D2** 28.28S 65.46W
Catanduanes *i.* Phil. **105 G6** 13.45N 124.20E
Catania Italy **50 F2** 37.31N 15.05E
Catanzaro Italy **50 G3** 38.55N 16.35E
Catarman Phil. **105 G6** 12.28N 124.50E
Caterham England **11 E2** 51.17N 0.04W
Cat I. The Bahamas **71 I5** 24.30N 75.30W
Catoche, C. Mexico **70 G5** 21.38N 87.08W
Catterick England **15 F3** 54.23N 1.38W
Cauca *r.* Colombia **71 J2** 8.57N 74.30W
Caucaia Brazil **77 G4** 3.45S 38.45W
Caucasus *mts.* Europe **58 G2** 43.00N 44.00E
Caudry France **42 C2** 50.07N 3.25E
Cavan Rep. of Ire. **20 D3** 53.59N 7.22W
Cavan *d.* Rep. of Ire. **20 D3** 54.00N 7.15W
Cawston England **11 G3** 52.46N 1.10E
Caxias Brazil **77 F4** 4.53S 43.20W
Caxias do Sul Brazil **77 E2** 29.14S 51.10W
Cayenne French Guiana **74 D7** 4.55N 52.18W
Cayman Brac *i.* Cayman Is. **71 I4** 19.44N 79.48W
Cayman Is. C. America **71 H4** 19.00N 81.00W
Cayos Miskito *is.* Nicaragua **71 H3** 14.30N 82.40W
Ceará *d.* Brazil **77 G4** 4.50S 39.00W
Cebu Phil. **105 G6** 10.17N 123.56E
Cebu *i.* Phil. **105 G6** 10.15N 123.45E
Cedar *r.* U.S.A. **65 H5** 41.15N 91.20W
Cedar City U.S.A. **64 D4** 37.40N 113.04W
Cedar Rapids *town* U.S.A. **65 H5** 41.59N 91.31W
Cedros *i.* Mexico **64 C2** 28.15N 115.15W
Cefalù Italy **50 F3** 38.01N 14.03E
Cegléd Hungary **54 F2** 47.10N 19.48E
Celaya Mexico **70 D5** 20.32N 100.48W
Celebes Sea Indonesia **105 G4** 3.00N 122.00E
Celje Slovenia **54 D2** 46.15N 15.16E
Celle Germany **48 E5** 52.37N 10.05E
Cenderawasih G. Indonesia **105 J3** 2.30S 135.20E
Central *d.* Kenya **87 B2** 1.00S 37.00E
Central African Republic Africa **85 F2** 6.30N 20.00E
Central, Cordillera *mts.* Bolivia **76 D3** 20.00S 65.00W
Central, Cordillera *mts.* Colombia **74 B7** 5.00N 75.20W
Central, Cordillera *mts.* Peru **76 C4** 7.00S 79.00W
Central Range *mts.* P.N.G. **105 K3** 5.00S 142.00E
Central Russian Uplands *f.* Russian Fed. **55 O5** 53.00N 37.00E
Central Siberian Plateau *f.* Russian Fed. **59 M4** 66.00N 108.00E
Ceredigion *d.* Wales **9 C3** 52.15N 4.00W
Cernavodă Romania **57 I6** 44.20N 28.02E

Cerralvo *i.* Mexico **70 C5** 24.17N 109.52W
Cerro de Pasco Peru **76 C3** 10.43S 76.15W
Cervo Spain **46 B5** 43.40N 7.25W
České Budějovice Czech Rep. **54 D3** 49.00N 14.30E
Ceuta Spain **46 C1** 35.53N 5.19W
Cévennes *mts.* France **44 E4** 44.25N 3.30E
Ceyhan *r.* Turkey **57 L2** 36.54N 34.58E
Chad Africa **85 F3** 13.00N 19.00E
Chad, L. Africa **84 F3** 13.30N 14.00E
Chadan Russian Fed. **102 G8** 51.20N 91.39E
Chagai Hills Pakistan **95 J4** 29.10N 63.35E
Chaghcharan Afghan. **95 K5** 34.32N 65.15E
Chagos Archipelago *is.* Indian Oc. **90 J2** 7.00S 72.00E
Chah Bahar Iran **95 J4** 25.17N 60.41E
Chake Chake Tanzania **87 B1** 5.13S 39.46E
Chalbi Desert Kenya **87 B3** 3.00N 37.20E
Chale England **10 D1** 50.36N 1.19W
Chalkida Greece **56 F3** 38.27N 23.36E
Challenger Deep Pacific Oc. **91 Q4** 11.19N 142.15E
Châlons-en-Champagne France **44 F6** 48.58N 4.22E
Chalon-sur-Saône France **44 F5** 46.47N 4.51E
Chalus Iran **95 H6** 36.40N 51.25E
Chaman Pakistan **96 D7** 30.55N 66.27E
Chambéry France **44 F4** 45.34N 5.55E
Ch'amo Hayk' *l.* Ethiopia **87 B4** 5.49N 37.35E
Chamonix France **44 G4** 45.55N 6.52E
Champaqui *mtn.* Argentina **76 D1** 31.59S 64.59W
Champlain, L. U.S.A. **65 L5** 44.45N 73.20W
Chañaral Chile **76 C2** 26.21S 70.37W
Chandalar *r.* U.S.A. **62 D4** 66.40N 146.00W
Chandeleur Is. U.S.A. **65 I2** 29.50N 88.50W
Chandigarh India **96 F7** 30.44N 76.54E
Chandrapur India **97 F4** 19.58N 79.21E
Changchun China **103 N6** 43.50N 125.20E
Changde China **103 K3** 29.03N 111.35E
Changgi Gap *b.* S. Korea **106 A3** 36.00N 129.30E
Chang Jiang *r.* China **103 M4** 31.40N 121.15E
Changsha China **103 K3** 28.10N 113.00E
Changzhi China **103 K5** 36.09N 113.12E
Changzhou China **103 L4** 31.45N 119.57E
Chania Greece **56 G1** 35.30N 24.02E
Channel Is. U.K. **13 Z9** 49.28N 2.13W
Channel-Port aux Basques *town* Canada **63 M2** 47.35N 59.10W
Chanthaburi Thailand **104 C6** 12.38N 102.12E
Chantilly France **42 B1** 49.12N 2.28E
Chao Phraya *r.* Thailand **104 C6** 13.35N 100.37E
Chapada de Maracás *f.* Brazil **77 F3** 13.20S 40.00W
Chapada Diamantina *f.* Brazil **77 F3** 13.30S 42.30W
Chapala, Lago de *l.* Mexico **70 D5** 20.00N 103.00W
Chapecó Brazil **77 E2** 27.14S 52.41W
Chapel-en-le-Firth England **15 F2** 53.19N 1.54W
Chapeltown England **15 F2** 53.28N 1.27W
Chapleau Canada **63 J2** 47.50N 83.24W
Chaplynka Ukraine **55 M2** 46.23N 33.32E
Chard England **13 E2** 50.52N 2.59W
Chari *r.* Chad **82 E6** 13.00N 14.30E
Charīkar Afghan. **95 K6** 35.02N 69.13E
Charlbury England **10 D2** 51.53N 1.29W
Charleroi Belgium **42 D2** 50.25N 4.27E
Charleston S.C. U.S.A. **65 K3** 32.48N 79.58W
Charleston W.Va. U.S.A. **65 J4** 38.23N 81.20W
Charlestown Rep. of Ire. **20 C3** 53.57N 8.50W
Charleville-Mézières France **44 F6** 49.46N 4.43E
Charlotte U.S.A. **65 J4** 35.05N 80.50W
Charlottesville U.S.A. **65 K4** 38.02N 78.29W
Charlottetown Canada **63 L2** 46.14N 63.09W
Chartres France **44 D6** 48.27N 1.30E
Châteaubriant France **44 C5** 47.43N 1.22W
Châteaudun France **44 D6** 48.04N 1.20E
Châteauroux France **44 D5** 46.49N 1.41E
Château-Thierry France **44 E6** 49.03N 3.24E
Châtellerault France **44 D5** 46.49N 0.33E
Chatham England **11 F2** 51.23N 0.32E
Chatham Is. Pacific Oc. **111 H1** 44.00S 176.35W
Chattahoochee *r.* U.S.A. **65 J2** 30.52N 84.57W
Chattanooga U.S.A. **65 I4** 35.01N 85.18W
Chatteris England **11 F3** 52.27N 0.03E
Chaumont France **44 F6** 48.07N 5.08E
Chauny France **42 C1** 49.37N 3.13E
Cheadle England **10 D3** 52.59N 1.59W
Cheb Czech Rep. **54 C4** 50.04N 12.20E
Cheboksary Russian Fed. **58 G3** 56.08N 47.12E
Cheboygan U.S.A. **65 J6** 45.40N 84.28W
Cheddar England **13 E3** 51.16N 2.47W
Cheju do *i.* S. Korea **103 N4** 33.20N 126.30E
Chekhov Russian Fed. **55 O6** 55.21N 37.31E
Cheleken Turkmenistan **95 H6** 39.26N 53.11E
Chełm Poland **55 H4** 51.10N 23.28E
Chelmer *r.* England **11 F2** 51.43N 0.42E
Chelmsford England **11 F2** 51.44N 0.28E
Cheltenham England **10 C2** 51.53N 2.07W
Chelyabinsk Russian Fed. **58 I3** 55.10N 61.25E
Chelyuskin, C. Russian Fed. **59 M5** 77.20N 106.00E
Chemnitz Germany **48 F4** 50.50N 12.55E
Ch'ench'a Ethiopia **87 B4** 6.18N 37.37E
Chengde China **103 L6** 40.48N 118.06E
Chengdu China **103 I4** 30.37N 104.06E
Chennai India **97 G3** 13.05N 80.18E
Chenzhou China **103 K3** 25.45N 113.00E
Chepstow Wales **12 E3** 51.38N 2.40W
Cher *r.* France **44 D5** 47.21N 0.29E
Cherbourg France **44 C6** 49.38N 1.37W
Cherepovets Russian Fed. **58 F3** 59.05N 37.55E
Cherkasy Ukraine **55 M3** 49.27N 32.04E
Chernihiv Ukraine **55 L4** 51.30N 31.18E
Chernivtsi Ukraine **55 I3** 48.19N 25.52E
Chernyakhovsk Russian Fed. **54 G6** 54.36N 21.48E
Cherskogo Range *mts.* Russian Fed. **59 Q4** 65.50N 143.00E
Chervonohrad Ukraine **55 I4** 50.25N 24.10E
Cherwell *r.* England **10 D2** 51.45N 1.15W
Chesapeake B. U.S.A. **65 K4** 38.00N 76.00W
Chesham England **11 E2** 51.43N 0.38W
Cheshire *d.* England **9 D3** 53.14N 2.30W
Chëshskaya Bay Russian Fed. **35 H4** 67.20N 46.30E
Cheshunt England **11 E2** 51.43N 0.02W
Chesil Beach *f.* England **10 C1** 50.37N 2.33W
Chester England **14 E2** 53.12N 2.53W
Chesterfield England **15 F2** 53.14N 1.26W
Chester-le-Street England **15 F3** 54.53N 1.34W
Cheviot Hills U.K. **17 G3** 55.22N 2.24W
Che'w Bahir *l.* Ethiopia **87 B3** 4.40N 36.50E
Che'w Bahir Wildlife Res. Ethiopia **87 B4** 5.00N 36.50E
Chew Magna England **10 C2** 51.21N 2.37W
Chew Valley L. England **10 C2** 51.20N 2.37W
Cheyenne U.S.A. **64 F5** 41.08N 104.50W
Chiang Mai Thailand **104 B7** 18.48N 98.59E
Chiang Rai Thailand **104 B7** 19.56N 99.51E
Chiapas *d.* Mexico **70 F4** 16.30N 93.00W
Chiba Japan **106 D3** 35.38N 140.07E
Chibougamau Canada **63 K2** 49.56N 74.24W
Chicago U.S.A. **65 I5** 41.50N 87.45W
Chichester England **10 E1** 50.50N 0.47W
Chiclayo Peru **76 C4** 6.47S 79.47W
Chico U.S.A. **64 B4** 39.46N 121.50W
Chicoutimi Canada **63 K2** 48.26N 71.06W
Chidley, C. Canada **63 L4** 60.30N 65.00W
Chiemsee *l.* Germany **48 F2** 47.55N 12.30E
Chieti Italy **50 F5** 42.22N 14.12E
Chihli, G. of China **103 L5** 38.30N 119.30E
Chihuahua Mexico **70 C6** 28.40N 106.06W
Chihuahua *d.* Mexico **70 C6** 28.40N 105.00W
Chile S. America **75 B3** 33.00S 71.00W
Chilham England **11 F2** 51.15N 0.57E
Chillán Chile **75 B3** 36.37S 72.10W
Chiloé, Isla de Chile **75 B2** 43.00S 73.00W
Chilpancingo Mexico **70 E4** 17.33N 99.30W
Chiltern Hills England **10 E2** 51.40N 0.53W
Chimborazo *mtn.* Ecuador **76 C4** 1.10S 78.50W
Chimbote Peru **74 B6** 8.58S 78.34W
Chimoio Mozambique **86 C3** 19.04S 33.29E
China Asia **103 H4** 33.00N 103.00E
Chindwin *r.* Myanmar **97 J5** 21.30N 95.12E
Chingola Zambia **86 B3** 12.31S 27.53E
Chinhoyi Zimbabwe **86 C3** 17.22S 30.10E
Chios Greece **56 H3** 38.22N 26.08E
Chios *i.* Greece **56 G3** 38.23N 26.04E
Chipata Zambia **86 C3** 13.37S 32.40E
Chippenham England **10 C2** 51.27N 2.07W
Chipping Campden England **10 D3** 52.03N 1.46W
Chipping Norton England **10 D2** 51.56N 1.32W
Chipping Ongar England **11 F2** 51.43N 0.15E
Chipping Sodbury England **10 C2** 51.31N 2.23W
Chiriquí, G. of Panama **71 H2** 8.00N 82.20W
Chirk Wales **12 D4** 52.56N 3.03W
Chirnside Scotland **17 G3** 55.48N 2.12W
Chirripó *mtn.* Costa Rica **71 H2** 9.31N 83.30W
Chisasibi *see* **Fort George** Canada **63**
Chişinau Moldova **55 K2** 47.00N 28.50E
Chita Russian Fed. **59 N3** 52.03N 113.35E
Chitradurga India **96 F3** 14.16N 76.23E
Chitral Pakistan **96 E8** 35.52N 71.58E
Chittagong Bangla. **97 I5** 22.20N 91.48E
Chittoor India **97 F3** 13.13N 79.06E
Choiseul *i.* Solomon Is. **111 E5** 7.00S 157.00E
Chojnice Poland **54 E5** 53.42N 17.32E
Cholet France **44 C5** 47.04N 0.53W
Chon Buri Thailand **104 C6** 13.24N 100.59E
Chongjin N. Korea **103 N6** 41.55N 129.50E
Chongqing China **103 J3** 29.31N 106.35E
Chonju S. Korea **103 N5** 35.50N 127.05E
Chorley England **14 E2** 53.39N 2.39W
Chornobyl' Ukraine **55 L4** 51.17N 30.15E
Chortkiv Ukraine **55 I3** 49.01N 25.42E
Choshi Japan **106 D3** 34.53N 140.51E
Chott ech Chergui *f.* Algeria **52 D3** 34.00N 0.30E
Chott ech Hodna *f.* Algeria **52 D4** 35.40N 5.00E
Chott el Jerid *f.* Tunisia **84 E5** 33.30N 8.30E
Chott Melrhir *f.* Algeria **84 E5** 34.15N 7.00E
Choybalsan Mongolia **103 K7** 48.02N 114.32E
Christchurch England **10 D1** 50.44N 1.47W
Christchurch New Zealand **111 G1** 43.32S 172.37E
Christmas I. Indian Oc. **104 D1** 10.30S 105.40E
Chuckchi Pen. Russian Fed. **59 U4** 66.00N 174.30W
Chudovo Russian Fed. **54 G2** 59.10N 31.41E
Chugoku-sanchi *mts.* Japan **106 B3** 35.30N 133.00E
Chuhuyiv Ukraine **55 O3** 49.51N 36.44E
Chukchi Sea Arctic Oc. **59 U4** 69.30N 172.00W
Chulmleigh England **13 D2** 50.55N 3.52W
Chumba Ethiopia **87 B3** 4.30N 38.15E
Chumphon Thailand **104 B6** 10.35N 99.14E
Chunchon S. Korea **103 N5** 37.53N 127.45E
Chur Switz. **44 H5** 46.52N 9.32E
Churchill Canada **63 I3** 58.45N 94.00W
Churchill *r.* Man. Canada **63 I3** 58.20N 94.15W
Churchill *r.* Nfld. Canada **63 L3** 53.20N 60.00W
Churchill, C. Canada **63 I3** 58.50N 93.00W
Church Stretton England **10 C3** 52.32N 2.49W
Chuuk *i.* Fed. States of Micronesia **108 K7** 7.23N 151.46E
Chuxiong China **103 I3** 25.03N 101.33E
Chyulu Range *mts.* Kenya **87 B2** 2.40S 37.53E
Ciego de Avila Cuba **71 I5** 21.51N 78.47W
Cienfuegos Cuba **71 H5** 22.10N 80.27W
Cigüela *r.* Spain **46 D3** 39.08N 3.44W
Cihanbeyli Turkey **57 K3** 38.40N 32.55E
Cijara L. Spain **46 C3** 39.20N 4.50W
Cilacap Indonesia **104 D2** 7.44S 109.00E
Cinca *r.* Spain **46 F4** 41.22N 0.20E
Cincinnati U.S.A. **65 J4** 39.10N 84.30W
Cinderford England **10 C2** 51.49N 2.30W
Ciney Belgium **42 E2** 50.17N 5.06E
Cinto, Monte *mtn.* France **44 H3** 42.23N 8.57E
Cirebon Indonesia **104 D2** 6.46S 108.33E
Cirencester England **10 D2** 51.43N 1.59W
City of Edinburgh *d.* Scotland **8 D4** 55.57N 3.13W
Ciudad Bolívar Venezuela **71 L2** 8.06N 63.36W
Ciudad Camargo Mexico **70 C6** 27.41N 105.10W
Ciudad Delicias Mexico **70 C6** 28.10N 105.30W
Ciudad de Valles Mexico **70 E5** 22.00N 99.00W
Ciudad Guayana Venezuela **71 L2** 8.22N 62.40W
Ciudad Ixtepec Mexico **70 E4** 16.32N 95.10W
Ciudad Juárez Mexico **70 C7** 31.42N 106.29W
Ciudad Madero Mexico **70 E5** 22.19N 97.50W
Ciudad Obregón Mexico **70 C6** 27.28N 109.55W
Ciudad Real Spain **46 D3** 38.59N 3.55W
Ciudad-Rodrigo Spain **46 B4** 40.36N 6.33W
Ciudad Victoria Mexico **70 E5** 23.43N 99.10W
Ciutadella de Menorca Spain **46 G4** 40.00N 3.50E
Civitavecchia Italy **50 D5** 42.06N 11.48E
Clackmannanshire *d.* Scotland **8 D5** 56.10N 3.45W
Clacton-on-Sea England **11 G2** 51.47N 1.10E
Claerwen Resr. Wales **12 D4** 52.17N 3.40W
Claonaig Scotland **16 D3** 55.43N 5.24W
Clara Rep. of Ire. **20 D3** 53.21N 7.37W
Clare *r.* Rep. of Ire. **20 B3** 53.20N 9.03W
Clare *d.* Rep. of Ire. **20 C2** 52.52N 8.55W
Clare I. Rep. of Ire. **20 A3** 53.50N 10.00W
Claremorris Rep. of Ire. **20 C3** 53.44N 9.00W
Clarksville U.S.A. **65 I4** 36.31N 87.21W
Claro *r.* Brazil **77 E3** 19.05S 50.40W
Clay Cross England **15 F2** 53.11N 1.26W
Claydon England **11 G3** 52.06N 1.07E
Clay Head I.o.M. **14 C3** 54.12N 4.23W
Clayton U.S.A. **64 F4** 36.27N 103.12W
Clear, C. Rep. of Ire. **20 B1** 51.25N 9.31W
Clear I. Rep. of Ire. **20 B1** 51.26N 9.30W
Cleator Moor *town* England **14 D3** 54.30N 3.32W
Cleethorpes England **15 G2** 53.33N 0.02W
Cleobury Mortimer England **10 C3** 52.23N 2.28W
Clermont France **42 B1** 49.23N 2.24E
Clermont-Ferrand France **44 E4** 45.47N 3.05E
Clevedon England **10 C2** 51.26N 2.52W
Cleveland U.S.A. **65 J5** 41.30N 81.41W
Cleveland Hills England **15 F3** 54.25N 1.10W
Cleveleys England **14 D2** 53.52N 3.01W
Clew B. Rep. of Ire. **20 B3** 53.50N 9.47W
Cliffe England **11 F2** 51.28N 0.30E
Clipperton I. Pacific Oc. **109 U8** 10.17N 109.13W
Clitheroe England **15 E2** 53.52N 2.23W
Clogher Head Rep. of Ire. **20 E3** 53.48N 6.13W
Clonakilty Rep. of Ire. **20 C1** 51.37N 8.55W
Clones Rep. of Ire. **20 D4** 54.11N 7.15W
Clonmel Rep. of Ire. **20 D2** 52.21N 7.44W
Cloud Peak *mtn.* U.S.A. **64 E5** 44.23N 107.11W
Clovis U.S.A. **64 F3** 34.14N 103.13W
Cluanie, Loch Scotland **18 D2** 57.08N 5.05W
Cluj-Napoca Romania **55 H2** 46.47N 23.37E
Clun England **10 B3** 52.26N 3.02W
Clydach Wales **12 D3** 51.42N 3.53W
Clyde *r.* Scotland **16 E3** 55.58N 4.53W
Clydebank Scotland **16 E3** 55.53N 4.23W
Clyde River *town* Canada **63 L5** 70.30N 68.30W
Coahuila *d.* Mexico **70 D6** 27.00N 103.00W
Coalville England **10 D3** 52.43N 1.21W
Coari Brazil **76 D4** 4.08S 63.07W
Coari *r.* Brazil **76 D4** 4.08S 63.07W
Coast *d.* Kenya **87 B2** 3.00S 40.00E
Coast Mts. Canada **62 F3** 55.30N 128.00W
Coast Range *mts.* U.S.A. **64 B5** 40.00N 123.00W
Coatbridge Scotland **17 E3** 55.52N 4.02W
Coats I. Canada **63 J4** 62.30N 83.00W
Coatzacoalcos Mexico **70 F4** 18.10N 94.25W
Cobh Rep. of Ire. **20 C1** 51.50N 8.18W
Cobija Bolivia **76 D3** 11.01S 68.45W
Coburg Germany **48 E4** 50.15N 10.58E
Cochabamba Bolivia **76 D3** 17.26S 66.10W
Cochin India **96 F2** 9.56N 76.15E
Cochrane Canada **65 J6** 49.00N 81.00W
Cochrane Chile **75 B2** 47.20S 72.30W
Cockburnspath Scotland **17 G3** 55.56N 2.22W
Cockburn Town Turks & Caicos Is. **71 J5** 21.30N 71.30W
Cockermouth England **14 D3** 54.40N 3.22W
Coco *r.* Honduras **71 H3** 14.58N 83.15W
Coco, Isla del *i.* Pacific Oc. **60 K2** 5.32N 87.04W
Cod, C. U.S.A. **65 L5** 42.08N 70.10W
Coddington England **15 G2** 53.04N 0.45W
Codó Brazil **77 F4** 4.28S 43.51W
Códoba, Sierras de *mts.* Argentina **76 D1** 30.30S 64.40W
Codsall England **10 C3** 52.37N 2.11W
Coffs Harbour Australia **110 E2** 30.19S 153.05E
Coggeshall England **11 F2** 51.53N 0.41E
Coiba, I. Panama **71 H2** 7.23N 81.45W
Coihaique Chile **75 B2** 45.35S 72.08W
Coimbatore India **96 F3** 11.00N 76.57E
Coimbra Portugal **46 A4** 40.12N 8.25W
Colatina Brazil **77 F3** 19.35S 40.37W
Colchester England **11 F2** 51.54N 0.55E
Cold Bay *town* U.S.A. **62 B3** 55.10N 162.47W
Coldstream Scotland **17 G3** 55.39N 2.15W
Coleford England **10 C2** 51.46N 2.38W
Coleraine N. Ireland **16 C3** 55.08N 6.41W
Colima Mexico **70 D4** 19.14N 103.41W
Colima *mtn.* Mexico **70 D4** 19.32N 103.36W
Colima *d.* Mexico **70 D4** 19.05N 104.00W
Coll *i.* Scotland **16 C4** 56.38N 6.34W
Collier B. Australia **110 B4** 16.10S 124.15E
Colmar France **44 G6** 48.05N 7.21E
Colne England **15 E2** 53.51N 2.11W
Colne *r.* England **11 F2** 51.50N 0.59E
Cologne Germany **48 C4** 50.56N 6.57E
Colombia S. America **74 B7** 5.00N 75.00W
Colombo Sri Lanka **97 F2** 6.55N 79.52E
Colón Panama **71 I2** 9.21N 79.54W
Colonsay *i.* Scotland **16 C4** 56.04N 6.13W
Colorado *r.* Argentina **75 C3** 39.50S 62.02W
Colorado *r.* Tex. U.S.A. **64 G2** 28.30N 96.00W
Colorado *r.* U.S.A./Mexico **64 D3** 31.45N 114.40W
Colorado *d.* U.S.A. **64 E4** 39.00N 106.00W
Colorado Plateau *f.* U.S.A. **64 D4** 36.00N 111.00W
Colorado Springs *town* U.S.A. **64 F4** 38.50N 104.40W
Coltishall England **11 G3** 52.44N 1.22E
Columbia U.S.A. **65 J3** 34.00N 81.00W
Columbia *r.* U.S.A. **64 B6** 46.10N 123.30W
Columbia, Mt. Canada **62 G3** 52.09N 117.25W
Columbus Ga. U.S.A. **65 J3** 32.28N 84.59W
Columbus Ohio U.S.A. **65 J4** 39.59N 83.03W
Colville *r.* U.S.A. **62 C5** 70.06N 151.30W
Colwyn Bay *town* Wales **12 D5** 53.18N 3.43W
Combe Martin England **13 C3** 51.12N 4.02W
Comber N. Ireland **16 D2** 54.33N 5.45W
Comeragh Mts. Rep. of Ire. **20 D2** 52.15N 7.35W
Como Italy **50 C6** 45.48N 9.04E
Como, L. Italy **50 C7** 46.05N 9.17E
Comodoro Rivadavia Argentina **75 C2** 45.50S 67.30W
Comorin, C. India **90 J3** 8.04N 77.35E
Comoros Africa **86 D3** 12.15S 44.00E
Compiègne France **42 B1** 49.25N 2.50E
Comrie Scotland **17 F4** 56.23N 4.00W
Conakry Guinea **84 C2** 9.30N 13.43W
Concarneau France **44 B5** 47.53N 3.55W
Concepción Chile **75 B3** 36.50S 73.03W
Conception, Pt. U.S.A. **64 B3** 34.27N 120.26W
Conchos *r.* Chihuahua Mexico **70 D6** 29.34N 104.30W
Conchos *r.* Tamaulipas Mexico **70 E6** 25.00N 97.30W
Concord U.S.A. **65 L5** 43.13N 71.34W
Concordia Argentina **77 E1** 31.25S 58.00W
Condor, Cordillera del *mts.* Ecuador/Peru **76 C4** 4.00S 78.30W
Congleton England **15 E2** 53.10N 2.12W
Congo Africa **84 F1** 1.00S 16.00E
Congo *r.* Africa **84 F1** 6.00S 12.30E
Congo Basin *f.* Africa **82 E4** 1.00S 20.00E
Congo, Dem. Rep. of Africa **85 G1** 1.00S 21.00E
Coningsby England **15 G2** 53.07N 0.09W
Coniston England **14 D3** 54.22N 3.06W
Coniston Water *l.* England **14 D3** 54.20N 3.05W
Connah's Quay *town* Wales **12 D5** 53.13N 3.03W
Connecticut *d.* U.S.A. **65 L5** 41.30N 73.00W
Connemara *f.* Rep. of Ire. **20 B3** 53.30N 9.50W
Conn, Lough Rep. of Ire. **20 B4** 54.01N 9.15W
Conon Bridge Scotland **19 E2** 57.33N 4.26W
Consett England **15 F3** 54.52N 1.50W
Con Son *is.* Vietnam **104 D5** 8.30N 106.30E
Constanța Romania **57 I6** 44.10N 28.31E
Constantine Algeria **84 E5** 36.22N 6.38E
Conwy Wales **12 D5** 53.17N 3.50W
Conwy *r.* Wales **12 D5** 53.17N 3.49W
Conwy *d.* Wales **9 D3** 53.10N 3.45W
Conwy B. Wales **12 D5** 53.19N 3.55W
Cook Is. Pacific Oc. **108 O5** 15.00S 160.00W
Cook, Mt. *see* **Aoraki** New Zealand **111**
Cookstown N. Ireland **16 C2** 54.39N 6.46W
Cook Str. New Zealand **111 G1** 41.15S 174.30E
Cooktown Australia **110 D4** 15.29S 145.15E
Coolangatta Australia **110 E3** 28.10S 153.26E
Cooper Creek *r.* Australia **110 C3** 28.33S 137.46E
Copenhagen Denmark **43 C1** 55.43N 12.34E
Copiapo Chile **76 C2** 27.20S 70.23W
Copinsay *i.* Scotland **19 G3** 58.54N 2.41W
Coppermine *see* **Kugluktuk** Canada **62**
Coquimbo Chile **76 C1** 30.00S 71.25W
Coral Harbour *town* Canada **63 J4** 64.10N 83.15W
Coral Sea Pacific Oc. **110 E4** 13.00S 150.00E
Coral Sea Islands Territory Austa. **110 E4** 15.00S 153.00E
Corbie France **42 B1** 49.55N 2.31E
Corbridge England **15 E3** 54.58N 2.01W
Corby England **10 E3** 52.29N 0.41W
Córdoba Argentina **76 D1** 31.25S 64.11W
Córdoba Mexico **70 E4** 18.55N 96.55W
Córdoba Spain **46 C2** 37.53N 4.46W
Corfe Castle *town* England **10 C1** 50.38N 2.04W
Corfu Greece **56 D3** 39.37N 19.50E
Corfu *i.* Greece **56 D3** 39.35N 19.50E
Corigliano Calabro Italy **50 G3** 39.36N 16.31E
Corinth Greece **56 F2** 37.56N 22.55E
Corinth, G. of Greece **56 F3** 38.15N 22.30E
Corixa Grande *r.* Brazil/Bolivia **77 E3** 17.30S 57.55W
Cork Rep. of Ire. **20 C1** 51.54N 8.28W
Cork *d.* Rep. of Ire. **20 C1** 52.00N 8.40W
Çorlu Turkey **57 H4** 41.11N 27.48E
Corner Brook *town* Canada **63 M2** 48.58N 57.58W
Corno, Monte *mtn.* Italy **50 E5** 42.29N 13.33E
Cornwall *d.* England **9 C2** 50.26N 4.40W
Cornwall, C. England **13 B2** 50.07N 5.44W
Cornwallis I. Canada **63 I5** 75.00N 95.00W
Coro Venezuela **71 K3** 11.27N 69.41W
Coronada B. Costa Rica **71 H2** 9.00N 83.50W
Coronation G. Canada **62 G4** 68.00N 112.00W
Coronel Oviedo Paraguay **77 E2** 25.24S 56.30W
Coropuna *mtn.* Peru **76 C3** 15.31S 72.45W
Corpus Christi U.S.A. **64 G2** 27.47N 97.26W
Corralejo Canary Is. **46 Z2** 28.43N 13.53W
Corrib, Lough Rep. of Ire. **20 B3** 53.26N 9.14W
Corrientes Argentina **77 E2** 27.30S 58.48W
Corrientes *r.* Argentina **77 E2** 29.55S 59.32W
Corrientes, C. Mexico **70 C5** 20.25N 105.42W
Corse, Cap *c.* France **44 H3** 43.00N 9.21E
Corserine *mtn.* Scotland **16 E3** 55.09N 4.22W
Corsham England **10 C2** 51.25N 2.11W
Corsica *i.* France **44 H3** 42.00N 9.10E
Corte France **44 H3** 42.18N 9.08E
Cortegana Spain **46 B2** 37.55N 6.49W
Corton England **11 G3** 52.32N 1.44E
Çorum Turkey **57 L4** 40.31N 34.57E
Corumbá Brazil **77 E3** 19.00S 57.25W
Corwen Wales **12 D4** 52.59N 3.23W
Cosenza Italy **50 G3** 39.17N 16.14E
Cosmoledo Is. Indian Oc. **86 D4** 9.30S 49.00E
Cosne France **44 E5** 47.25N 2.55E
Costa Blanca *f.* Spain **46 F3** 38.30N 0.05E
Costa Brava *f.* Spain **46 G4** 41.30N 3.00E
Costa del Sol *f.* Spain **46 C2** 36.30N 4.00W
Costa Rica C. America **71 H3** 10.00N 84.00W
Cotabato Phil. **105 G5** 7.14N 124.15E
Côte d'Azur *f.* France **44 G3** 43.20N 6.45E
Côte d'Ivoire Africa **84 D2** 7.00N 5.30W
Cothi *r.* Wales **12 C3** 51.51N 4.10W
Cotonou Benin **84 E2** 6.24N 2.31E
Cotopaxi *mtn.* Ecuador **76 C4** 0.40S 78.30W
Cotswold Hills England **10 C2** 51.50N 2.00W
Cottbus Germany **48 G4** 51.43N 14.21E
Cottenham England **11 F3** 52.18N 0.08E
Cottesmore England **10 E3** 52.43N 0.39W
Coulogne France **42 A2** 50.55N 1.54E
Council Bluffs U.S.A. **64 G5** 41.14N 95.54W
Coupar Angus Scotland **17 F4** 56.33N 3.17W
Courland Lagoon Russian Fed. **54 G6** 55.00N 21.00E
Coutances France **44 C6** 49.03N 1.29W
Coventry England **10 D3** 52.25N 1.31W
Covilhã Portugal **46 B4** 40.17N 7.30W
Cowan, L. Australia **110 B2** 32.00S 122.00E
Cowbridge Wales **13 D3** 51.28N 3.28W
Cowdenbeath Scotland **17 F4** 56.07N 3.21W
Cowes England **10 D1** 50.45N 1.18W
Cowfold England **11 E1** 50.59N 0.17W
Cow Green Resr. England **15 E3** 54.40N 2.19W
Cox's Bazar Bangla. **97 I5** 21.25N 91.59E
Cozumel I. Mexico **70 G5** 20.30N 87.00W
Craigavon N. Ireland **16 C2** 54.28N 6.25W
Craig Goch Resr. Wales **12 D4** 52.20N 3.35W
Craignure Scotland **16 D4** 56.28N 5.42W
Crail Scotland **17 G4** 56.16N 2.38W
Craiova Romania **56 F6** 44.18N 23.46E
Cramlington England **15 F4** 55.06N 1.33W
Cranbrook Canada **62 G2** 49.29N 115.48W
Cranleigh England **11 E2** 51.08N 0.29W
Crawley England **11 E2** 51.07N 0.10W
Creag Meagaidh *mtn.* Scotland **19 E1** 56.57N 4.38W
Credenhill England **10 C3** 52.06N 2.49W
Crediton England **13 D2** 50.47N 3.39W
Cree L. Canada **62 H3** 57.20N 108.30W
Creil France **42 B1** 49.16N 2.29E
Cremona Italy **50 D6** 45.08N 10.03E
Crepy-en-Valois France **42 B1** 49.14N 2.54E
Cres *i.* Croatia **56 B6** 44.50N 14.20E
Crescent City U.S.A. **64 B5** 41.46N 124.13W
Creston U.S.A. **65 H5** 41.04N 94.20W
Creswell England **15 F2** 53.16N 1.12W

Dunstable England 11 E2 51.53N 0.32W
Dunvegan, Loch Scotland 18 C2 57.30N 6.40W
Durance *r.* France 44 F3 43.55N 4.48E
Durango Mexico 70 D5 24.01N 104.00W
Durango *d.* Mexico 70 D5 24.00N 104.00W
Durban R.S.A. 86 C2 29.53S 31.00E
Düren Germany 42 F2 50.48N 6.30E
Durham England 15 F3 54.47N 1.34W
Durham U.S.A. 65 K4 36.00N 78.54W
Durham *d.* England 9 D4 54.42N 1.45W
Durness Scotland 18 E3 58.33N 4.45W
Durrës Albania 56 D4 41.19N 19.27E
Dursey Head Rep. of Ire. 20 A1 51.35N 10.15W
Dursley England 10 C2 51.41N 2.21W
Dushanbe Tajikistan 115 M7 38.38N 68.51E
Düsseldorf Germany 48 C4 51.13N 6.47E
Duyun China 103 J3 26.16N 107.29E
Düzce Turkey 57 J4 40.51N 31.09E
Dyat'kovo Russian Fed. 55 N5 53.35N 34.22E
Dyce Scotland 19 G2 57.12N 2.11W
Dyer, C. Canada 63 L4 67.45N 61.45W
Dyfi *r.* Wales 12 D4 52.33N 3.56W
Dymchurch England 11 G2 51.02N 1.00E
Dymock England 10 C2 51.59N 2.26W
Dzerzhinsk Russian Fed. 58 G3 56.15N 43.30E
Dzhankoy Ukraine 55 N1 45.42N 34.23E
Dzhigudzhak Russian Fed. 59 R4 64.33N 157.19E
Dzhugdzhur Range *mts.* Russian Fed. 59 P3 57.30N 138.00E
Dzisna *r.* Belarus 55 K6 55.30N 28.20E
Dzungarian Basin *f.* Asia 102 F7 45.20N 86.30E
Dzyarzhynsk Belarus 55 J5 53.40N 27.01E

E

Eagle Pass *town* U.S.A. 64 F2 28.44N 100.31W
Ealing England 11 E2 51.31N 0.20W
Earn *r.* Scotland 17 F4 56.21N 3.18W
Earn, Loch Scotland 16 E4 56.23N 4.12W
Easington England 15 F3 54.47N 1.21W
Easingwold England 15 F3 54.08N 1.11W
East Anglian Heights *hills* England 11 F3 52.03N 0.15E
East Ayrshire *d.* Scotland 8 C4 55.25N 4.20W
Eastbourne England 11 F1 50.46N 0.18E
East C. New Zealand 111 G2 37.45S 178.30E
East China Sea Asia 103 N3 29.00N 125.00E
East Dereham England 11 F3 52.40N 0.57E
East Dunbartonshire *d.* Scotland 8 C4 56.00N 4.15W
Easter I. Pacific Oc. 109 U4 27.08S 109.23W
Eastern *d.* Kenya 87 B3 1.00N 38.40E
Eastern Desert Egypt 94 D4 28.15N 31.55E
Eastern Ghāts *mts.* India 97 G4 16.30N 80.30E
Eastern Sayan *mts.* Russian Fed. 59 L3 53.30N 98.00E
Easter Ross *f.* Scotland 19 E2 57.46N 4.25W
East Falkland *i.* Falkland Is. 75 D1 51.45S 58.50W
East Fen *f.* England 15 H2 53.06N 0.02E
Eastfield England 15 G3 54.15N 0.25W
East Frisian Is. Germany 48 C5 53.45N 7.00E
East Grinstead England 11 E2 51.08N 0.01W
East Kilbride Scotland 16 E3 55.46N 4.09W
Eastleigh England 10 D1 50.58N 1.21W
East Linton Scotland 17 G3 55.59N 2.39W
East Loch Tarbert Scotland 18 C2 57.52N 6.43W
East London R.S.A. 86 B1 33.00S 27.54E
East Looe England 13 C2 50.21N 4.26W
East Lothian *d.* Scotland 8 D4 55.55N 2.40W
Eastmain Canada 63 K3 52.10N 78.30W
Eastmain *r.* Canada 63 K3 52.10N 78.30W
Easton England 10 C1 50.32N 2.26W
East Renfrewshire *d.* Scotland 8 C3 55.45N 4.20W
East Riding of Yorkshire *d.* England 9 E3 53.48N 0.35W
Eastry England 11 G2 51.15N 1.18E
East Siberian Sea Russian Fed. 59 S5 73.00N 160.00E
East Sussex *d.* England 9 F2 50.56N 0.12E
East Timor Asia 105 H2 8.33S 126.10E
Eastwood England 15 F2 53.02N 1.17W
Eaton Socon England 11 E3 52.13N 0.18W
Eau Claire, Lac á l' Canada 63 K3 56.10N 74.30W
Eauripik Atoll *i.* Fed. States of Micronesia 105 K5 6.42N 143.04E
Ebbw Vale *town* Wales 12 D3 51.47N 3.12W
Eberswalde Germany 48 F5 52.50N 13.50E
Ebinur Hu *l.* China 102 E6 45.00N 83.00E
Ebro *r.* Spain 46 F4 40.43N 0.54E
Ecclefechan Scotland 17 F3 55.03N 3.18W
Eccleshall England 10 C3 52.52N 2.14W
Ech Chélif Algeria 84 E5 36.20N 1.30E
Écija Spain 46 C2 37.33N 5.04W
Eckington England 10 C3 52.04N 2.07W
Eck, Loch Scotland 16 D4 56.05N 5.00W
Ecuador S. America 76 C4 2.00S 78.00W
Eday *i.* Scotland 19 G4 59.11N 2.47W
Ed Da'ein Sudan 94 C1 11.26N 26.09E
Ed Damazin Sudan 85 H3 11.52N 34.23E
Ed Damer Sudan 94 D2 17.37N 33.59E
Ed Debba Sudan 94 D2 18.03N 30.57E
Ed Dueim Sudan 94 D1 14.00N 32.19E
Eden *r.* England 14 E3 54.57N 3.02W
Edessa Greece 56 F4 40.47N 22.03E
Edinburgh Scotland 17 F3 55.57N 3.13W
Edirne Turkey 56 H4 41.40N 26.35E
Edmonton Canada 62 G3 53.34N 113.25W
Edmundston Canada 63 L2 47.22N 68.20W
Edremit Turkey 56 H3 39.35N 27.02E
Edson Canada 62 G3 53.36N 116.28W
Edward, L. Uganda/Dem. Rep. of Congo 86 B4 0.30S 29.30E
Edwards Plateau *f.* U.S.A. 64 F3 30.30N 100.30W
Éfaté *i.* Vanuatu 111 F4 17.40S 168.25E
Egilsay *i.* Scotland 19 G4 59.09N 2.56W
Egilsstadhir Iceland 43 Z2 65.16N 14.25W
Eğirdir Turkey 57 J2 37.52N 30.51E
Eğirdir L. Turkey 57 J3 38.04N 30.55E
Eglinton N. Ireland 16 B3 55.02N 7.12W
Egmont, Mt. *see* **Taranaki, Mt.** New Zealand 111
Egremont England 14 D3 54.28N 3.33W
Egypt Africa 94 C4 26.30N 29.30E
Eifel *f.* Germany 48 C4 50.10N 6.45E
Eigg *i.* Scotland 18 C1 56.53N 6.09W
Eighty Mile Beach *f.* Australia 110 B4 19.00S 121.00E
Eilean Shona *i.* Scotland 18 D1 56.50N 5.48W
Eindhoven Neth. 42 E3 51.26N 5.30E
Eisenhüttenstadt Germany 48 G5 52.09N 14.41E
Eishort, Loch Scotland 18 D2 57.09N 5.58W
Eisleben Germany 48 E4 51.32N 11.33E
Eivissa *town* Spain 46 F3 38.55N 1.30E
Ekibastuz Kazakhstan 102 D8 51.45N 75.22E
Ekostrovskaya, Ozero *l.* Russian Fed. 43 H4 67.30N 33.00E
Eksjö Sweden 43 C2 57.40N 15.00E
Ekwan *r.* Canada 63 J3 53.30N 84.00W
Elat Israel 94 D4 29.33N 34.56E
Elazığ Turkey 57 N3 38.41N 39.14E
Elba *i.* Italy 50 D5 42.47N 10.17E
El'ban Russian Fed. 103 P8 50.05N 136.35E
Elbasan Albania 56 E4 41.07N 20.04E
El Bayadh Algeria 52 D3 33.40N 1.00E
Elbe *r.* Germany 48 E5 53.33N 10.00E
Elbistan Turkey 57 M3 38.14N 37.11E
Elbląg Poland 54 F6 54.10N 19.25E
Elbrus *mtn.* Russian Fed. 58 G2 43.21N 42.29E
Elburz Mts. Iran 95 H6 36.00N 52.30E
Elche Spain 46 E3 38.16N 0.41W
Elda Spain 46 E3 38.29N 0.47W
El Dera Kenya 87 B3 0.39N 38.43E
El Djouf *des.* Africa 82 C7 21.00N 8.00W
Eldoret Kenya 87 B3 0.31N 35.17E
El Eulma Algeria 52 E4 36.09N 5.41E
Eleuthera *i.* The Bahamas 71 I6 25.00N 76.00W
El Faiyûm Egypt 94 D4 29.19N 30.50E
El Fasher Sudan 85 G3 13.37N 25.22E
El Geneina Sudan 85 G3 13.27N 22.30E
Elgin Scotland 19 F2 57.39N 3.20W
Elgin U.S.A. 65 I5 42.03N 88.19W
El Gîza Egypt 94 D5 30.01N 31.12E
El Goléa Algeria 84 E5 30.35N 2.51E
Elgon, Mt. Kenya/Uganda 87 A3 1.07N 34.35E
El Hierro *i.* Canary Is. 46 X1 27.45N 18.00W
Elista Russian Fed. 58 G2 46.18N 44.14E
El Jadida Morocco 84 D5 33.16N 8.30W
Ełk Poland 54 H5 53.50N 22.22E
Ēl Kerē Ethiopia 87 C4 5.48N 42.10E
El Khârga Egypt 94 D4 25.27N 30.32E
Elkhovo Bulgaria 56 H5 42.09N 26.35E
Elko U.S.A. 64 C5 40.50N 115.46W
Ellef Ringnes I. Canada 62 H5 78.30N 102.00W
Ēl Lēh Ethiopia 87 B3 3.47N 39.09E
Ellensburg U.S.A. 64 B6 47.00N 120.34W
Ellesmere England 10 C3 52.55N 2.53W
Ellesmere I. Canada 63 J5 78.00N 82.00W
Ellesmere Port England 14 E2 53.17N 2.55W
Ellice *r.* Canada 62 H4 67.30N 104.00W
Ellon Scotland 19 G2 57.22N 2.05W
Ellsworth Land Antarctica 112 77.00S 100.00W
Elmalı Turkey 57 I2 36.43N 29.56E
El Mansûra Egypt 94 D5 31.03N 31.23E
El Meghaïer Algeria 52 E3 33.58N 5.56E
El Minya Egypt 94 D4 28.06N 30.45E
El Muglad Sudan 85 G3 11.01N 27.50E
El Obeid Sudan 85 H3 13.11N 30.10E
El Oued Algeria 52 E3 33.20N 6.53E
El Paso U.S.A. 64 E3 31.45N 106.30W
El Puerto de Santa Maria Spain 46 B2 36.36N 6.14W
El Salvador C. America 70 G3 13.30N 89.00W
Eluru India 97 G4 16.45N 81.10E
Elvas Portugal 46 B3 38.53N 7.10W
Elverum Norway 43 B3 60.54N 11.33E
El Wak Kenya 87 C3 2.45N 40.52E
Ely England 11 F3 52.24N 0.16E
Emamrud Iran 95 I6 36.25N 55.00E
Embu Kenya 87 B2 0.32S 37.27E
Emden Germany 48 C5 53.23N 7.13E
Emerson Canada 64 G6 49.00N 97.12W
Emi Koussi *mtn.* Chad 85 F3 19.58N 18.30E
Emmeloord Neth. 42 E4 52.43N 5.46E
Emmen Neth. 42 F4 52.48N 6.55E
Emory Peak *mtn.* U.S.A. 64 F2 29.15N 103.19W
Emporia U.S.A. 64 G4 38.24N 96.10W
Ems *r.* Germany 48 C5 53.14N 7.25E
Encantada, Cerro de la Mexico 70 A7 31.00N 115.23W
Encarnación Paraguay 77 E2 27.20S 55.50W
Endeh Indonesia 105 G2 8.51S 121.40E
Enderby Land *f.* Antarctica 112 67.00S 53.00E
Endicott Mts. U.S.A. 62 C4 68.00N 152.00W
Enfield England 11 E2 51.40N 0.05W
Engels Russian Fed. 58 G3 51.30N 46.07E
Enggano *i.* Indonesia 104 C2 5.20S 102.15E
England U.K. 10-15
English Channel France/U.K. 44 C7 50.15N 1.00W
Enköping Sweden 43 D2 59.38N 17.07E
En Nahud Sudan 94 C1 12.41N 28.28E
Ennerdale Water *l.* England 14 D3 54.31N 3.21W
Ennis Rep. of Ire. 20 B2 52.51N 9.00W
Enniscorthy Rep. of Ire. 20 E2 52.30N 6.35W
Enniskillen N. Ireland 16 B2 54.20N 7.39W
Ennistymon Rep. of Ire. 20 B2 52.56N 9.20W
Enns *r.* Austria 54 D3 48.14N 14.22E
Enschede Neth. 42 F4 52.13N 6.54E
Ensenada Mexico 70 A7 31.53N 116.35W
Enshi China 103 J4 30.18N 109.29E
Enugu Nigeria 84 E2 6.20N 7.29E
Envira *r.* Brazil 76 C4 7.29S 70.00W
Épinal France 44 G6 48.10N 6.28E
Epping England 11 F2 51.42N 0.07E
Epsom England 11 E2 51.20N 0.16W
Epworth England 15 G2 53.30N 0.50W
Eqlīd Iran 95 H5 30.55N 52.40E
Equatorial Guinea Africa 84 E2 1.30N 10.30E
Erciyas, Mt. Turkey 57 L3 38.33N 35.25E
Erdi *f.* Chad 94 B2 19.00N 23.00E
Ereğli Turkey 57 L2 37.30N 34.02E
Erfurt Germany 48 E4 50.58N 11.02E
Ergani Turkey 57 N3 38.17N 39.44E
Ergene *r.* Turkey 56 H4 40.52N 26.12E
Eriboll, Loch Scotland 18 E3 58.28N 4.41W
Ericht, Loch Scotland 19 E1 56.52N 4.20W
Erie U.S.A. 65 J5 42.07N 80.05W
Erie, L. Canada/U.S.A. 65 J5 42.15N 81.00W
Eriskay *i.* Scotland 18 B2 57.04N 7.17W
Eritrea Africa 85 H3 15.20N 38.50E
Erlangen Germany 48 E3 49.36N 11.02E
Erode India 97 F3 11.21N 77.43E
Er Rachidia Morocco 84 D5 31.58N 4.25W
Er Rahad Sudan 85 H3 12.42N 30.33E
Errigal *mtn.* Rep. of Ire. 20 C5 55.02N 8.08W
Erris Head Rep. of Ire. 20 A4 54.19N 10.00W
Errochty, Loch Scotland 19 E1 56.45N 4.08W
Erromango *i.* Vanuatu 111 F4 18.45S 169.00E
Erzincan Turkey 57 N3 39.44N 39.30E
Erzurum Turkey 94 F6 39.57N 41.17E
Esbjerg Denmark 43 B1 55.28N 8.28E
Escanaba U.S.A. 65 I6 45.47N 87.04W
Esch-sur-Alzette Lux. 42 F1 49.31N 6.01E
Eschweiler Germany 42 F2 50.49N 6.16E
Esfahan Iran 95 H5 32.42N 51.40E
Esha Ness *c.* Scotland 19 Y9 60.29N 1.37W
Esk *r.* D. and G. Scotland 17 F3 54.58N 3.08W
Esk *r.* Midlothian Scotland 17 F3 55.56N 3.03W
Eske, L. Rep. of Ire. 16 A2 54.42N 8.03W
Eskilstuna Sweden 43 D2 59.22N 16.31E
Eskişehir Turkey 57 J3 39.46N 30.30E
Esla *r.* Spain 46 B4 41.29N 6.03W
Eslamabade Gharb Iran 95 G5 34.08N 46.35E
Esmeraldas Ecuador 76 C5 0.56N 79.40W
Esperance Australia 110 B2 33.49S 121.52E
Espigüete *mtn.* Spain 46 C5 42.56N 4.48W
Espinhaç, Serra da *mts.* Brazil 77 F3 17.15S 43.10W
Espinosa Brazil 77 F3 14.58S 42.49W
Espírito Santo *d.* Brazil 77 F3 20.00S 40.30W
Espiritu Santo *i.* Mexico 70 B5 24.30N 110.20W
Espíritu Santo *i.* Vanuatu 111 F4 15.50S 166.50E
Espoo Finland 43 F3 60.13N 24.40E
Esquel Argentina 75 B2 42.55S 71.20W
Essen Germany 48 C4 51.27N 6.57E
Essequibo *r.* Guyana 74 D7 6.48N 58.23W
Essex *d.* England 9 F2 51.46N 0.30E
Estats, Pic d' *mtn.* Spain 46 F5 42.40N 1.23E
Estevan Canada 62 H2 49.09N 103.00W
Estonia Europe 43 F2 58.45N 25.30E
Estrela, Serra da *mts.* Portugal 46 B4 40.20N 7.40W
Estrondo, Serra Brazil 77 F4 9.00S 49.00W
Étaples France 44 D7 50.31N 1.39E
Ethiopia Africa 85 H3 10.00N 39.00E
Ethiopian Highlands Ethiopia 82 G6 10.00N 37.00E
Etive, Loch Scotland 16 D4 56.27N 5.15W
Etna, Mt. Italy 50 F2 37.43N 14.59E
Etosha Pan *f.* Namibia 86 A3 18.50S 16.30E
Ettelbruck Lux. 42 F1 49.51N 6.06E
Ettrick Forest *f.* Scotland 17 F3 55.30N 3.00W
Ettrick Water *r.* Scotland 17 G3 55.36N 2.49W
Etwall England 10 D3 52.54N 1.35W
Eugene U.S.A. 64 B5 44.03N 123.07W
Eugenia, Punta *c.* Mexico 64 C2 27.50N 115.50W
Euphrates *r.* Asia 95 G5 31.00N 47.27E
Eureka U.S.A. 64 B5 40.49N 124.10W
Europa, Picos de *mts.* Spain 46 C5 43.10N 4.40W
Europe 34-35
Evans, L. Canada 63 K3 50.50N 77.00W
Evansville U.S.A. 65 I4 38.02N 87.24W
Evaton R.S.A. 86 B2 26.32S 27.51E
Everest, Mt. Asia 97 H6 27.59N 86.56E
Evesham England 10 D3 52.06N 1.57W
Évora Portugal 46 B3 38.34N 7.54W
Évreux France 44 D6 49.03N 1.11E
Evvoia *i.* Greece 56 G3 38.30N 23.50E
Ewaso Ngiro Eastern *r.* Kenya 87 B3 1.06N 39.26E
Ewaso Ngiro Rift Valley *r.* Kenya 87 B2 2.08S 36.03E
Ewe, Loch Scotland 18 D2 57.52N 5.40W
Exe *r.* England 13 D2 50.40N 3.28W
Exeter England 13 D2 50.43N 3.31W
Exminster England 13 D2 50.41N 3.29W
Exmoor *hills* England 13 D3 51.08N 3.45W
Exmouth England 13 D2 50.37N 3.24W
Exuma Is. The Bahamas 71 I5 24.00N 76.00W
Exuma Sd. The Bahamas 71 I5 24.10N 76.00W
Eyasi, L. Tanzania 86 C4 3.40S 35.00E
Eye Cambs. England 11 E3 52.36N 0.11W
Eye Suffolk England 11 G3 52.19N 1.09E
Eyemouth Scotland 17 G3 55.52N 2.05W
Eye Pen. Scotland 18 C3 58.13N 6.12W
Eyre (North), L. Australia 110 C3 28.30S 137.25E
Eyre (South), L. Australia 110 C3 29.30S 137.25E
Ezine Turkey 56 H3 39.48N 26.12E

F

Faafxadhuun Somalia 87 C3 2.14N 41.38E
Faeroes *is.* Europe 34 C4 62.00N 7.00W
Fagatogo American Samoa 108 N5 14.17S 170.41W
Fagersta Sweden 43 C3 59.59N 15.49E
Fairbanks U.S.A. 62 D4 64.50N 147.50W
Fair Head N. Ireland 16 C3 55.14N 6.10W
Fair Isle Scotland 19 Y8 59.32N 1.38W
Fairweather, Mt. U.S.A. 62 E3 59.00N 137.30W
Fais *i.* Fed. States of Micronesia 105 K5 9.45N 140.31E
Faisalabad Pakistan 96 E7 31.25N 73.09E
Fakaofo Pacific Oc. 108 O6 9.30S 171.15W
Fakenham England 11 F3 52.50N 0.51E
Fakfak Indonesia 105 I3 2.55S 132.17E
Fal *r.* England 13 C2 50.14N 4.58W
Falcon Lake U.S.A./Mexico 64 G2 26.37N 99.11W
Falkenberg Sweden 43 C2 56.55N 12.30E
Falkirk Scotland 17 F3 56.00N 3.48W
Falkirk *d.* Scotland 8 D5 56.00N 3.48W
Falkland Is. S. America 75 C1 52.00S 60.00W
Falköping Sweden 43 C2 58.10N 13.32E
Falmouth England 13 B2 50.09N 5.05W
Falmouth B. England 13 B2 50.06N 5.05W
Falster *i.* Denmark 43 C1 54.30N 12.00E
Falun Sweden 43 C3 60.37N 15.40E
Famagusta Cyprus 57 K1 35.07N 33.57E
Famatina, Sierra de *mts.* Argentina 76 D2 28.30S 67.50W
Fanad Head Rep. of Ire. 16 B3 55.17N 7.38W
Fane *r.* Rep. of Ire. 20 E3 53.56N 6.22W
Fangzheng China 106 A5 45.50N 128.50E
Fannich, Loch Scotland 18 E2 57.38N 4.58W
Farafra Oasis Egypt 94 C4 27.00N 28.20E
Farah Afghan. 95 J5 32.23N 62.07E
Farah *r.* Afghan. 95 J5 31.25N 61.30E
Farallon de Pajaros *i.* N. Mariana Is. 105 K8 20.33N 144.59E
Fareham England 10 D1 50.52N 1.11W
Farewell, C. Greenland 63 N3 60.00N 44.20W
Farewell, C. New Zealand 111 G1 40.30S 172.35E
Fargo U.S.A. 64 G6 46.52N 96.59W
Farnborough England 10 E2 51.17N 0.46W
Farndon England 14 E2 53.06N 2.53W
Farne Is. England 15 F4 55.38N 1.36W
Farnham England 10 E2 51.13N 0.49W
Faro Portugal 46 B2 37.01N 7.56W
Fårö *i.* Sweden 43 D2 57.55N 19.10E
Farsala Greece 56 F3 39.17N 22.22E
Fasa Iran 95 H4 28.55N 53.38E
Fastiv Ukraine 55 K4 50.08N 29.59E
Fatehgarh India 97 F6 27.22N 79.38E
Faversham England 11 F2 51.18N 0.54E
Faxaflói *b.* Iceland 43 X2 64.30N 22.50W
Fayetteville U.S.A. 65 K4 35.03N 78.53W
Fdérik Mauritania 84 C4 22.30N 12.30W
Feale *r.* Rep. of Ire. 20 B2 52.28N 9.38W
Fear, C. U.S.A. 65 K3 33.51N 77.59W
Federated States of Micronesia Pacific Oc. 108 K7 7.50N 155.00E
Fehmarn *i.* Germany 48 E6 54.30N 11.05E
Feijó Brazil 76 C4 8.09S 70.21W
Feira de Santana Brazil 77 G3 12.17S 38.53W
Felixstowe England 11 G2 51.58N 1.20E
Femunden *l.* Norway 43 B3 62.05N 11.55E
Feniton England 13 D2 50.47N 3.17W
Feodosiya Ukraine 55 N1 45.03N 35.23E
Ferdows Iran 95 I5 34.00N 58.10E
Fergus Falls *town* U.S.A. 64 G6 46.18N 96.00W
Ferkessédougou Côte d'Ivoire 84 D2 9.30N 5.10W
Fermanagh *d.* N. Ireland 20 D4 54.15N 7.45W
Fermoy Rep. of Ire. 20 C2 52.08N 8.17W
Fernando de Noronha *i.* Atlantic Oc. 72 G6 3.50S 32.25W
Ferndown England 10 D1 50.48N 1.55W
Ferrara Italy 50 D6 44.49N 11.38E
Ferrol Spain 46 A5 43.29N 8.14W
Ferryhill England 15 F3 54.41N 1.33W
Fethaland, Pt. of Scotland 19 Y9 60.38N 1.18W
Fethard Rep. of Ire. 20 D2 52.28N 7.41W
Fethiye Turkey 57 I2 36.37N 29.06E
Fetlar *i.* Scotland 19 Z9 60.37N 0.52W
Feuilles *r.* Canada 63 K3 58.47N 70.06W
Feyzabad Afghan. 95 L6 37.05N 70.40E
Fez Morocco 84 D5 34.05N 5.00W
Ffestiniog Wales 12 D4 52.58N 3.56W
Fianarantsoa Madagascar 86 D2 21.27S 47.05E
Fife *d.* Scotland 8 D5 56.10N 3.10W
Fife Ness *c.* Scotland 17 G4 56.17N 2.36W
Figeac France 44 E4 44.32N 2.01E
Figueira da Foz Portugal 46 A4 40.09N 8.51W
Figueres Spain 46 G5 42.16N 2.57E
Figuig Morocco 52 C3 32.10N 1.15W
Fiji Pacific Oc. 111 G4 18.00S 178.00E
Filadélfia Paraguay 77 D2 22.17S 60.03W
Filey England 15 G3 54.13N 0.18W
Filton England 10 C2 51.28N 2.35W
Fīltu Ethiopia 87 C4 5.05N 40.42E
Findhorn *r.* Scotland 19 F2 57.37N 3.40W
Finisterre, C. Spain 46 A5 42.54N 9.16W
Finland Europe 43 F3 64.30N 27.00E
Finland, G. of Finland/Estonia 43 F2 60.00N 26.50E
Finlay *r.* Canada 62 F3 56.30N 124.40W
Finn *r.* Rep. of Ire. 20 D4 54.50N 7.30W
Finspång Sweden 43 C2 58.42N 15.45E
Fintona N. Ireland 16 B2 54.29N 7.19W
Fionn Loch Scotland 18 D2 57.45N 5.27W
Fionnphort Scotland 16 C4 56.19N 6.23W
Firth of Clyde *est.* Scotland 16 E3 55.35N 5.00W
Firth of Forth *est.* Scotland 17 G4 56.05N 3.00W
Firth of Lorn *est.* Scotland 16 D4 56.20N 5.40W
Firth of Tay *est.* Scotland 17 F4 56.24N 3.08W
Firuzabad Iran 95 H4 28.50N 52.35E
Fisher Str. Canada 63 J4 63.00N 84.00W
Fishguard Wales 12 C3 51.59N 4.59W
Fishguard B. Wales 12 C4 52.06N 4.54W
Fismes France 42 C1 49.18N 3.41E
Flagstaff U.S.A. 64 D4 35.12N 111.38W
Flamborough England 15 G3 54.07N 0.07W
Flamborough Head England 15 G3 54.06N 0.05W
Flannan Isles *is.* Scotland 18 B3 58.16N 7.40W
Flåsjön *l.* Sweden 43 C4 64.05N 15.50E
Flathead L. U.S.A. 64 D6 47.50N 114.05W
Flat Holm *i.* England 13 D3 51.23N 3.08W
Flattery, C. U.S.A. 64 B6 48.23N 124.43W
Fleet England 10 F2 51.16N 0.50W
Fleetwood England 14 D2 53.55N 3.01W
Flensburg Germany 48 D6 54.47N 9.27E
Flevoland *d.* Neth. 42 E4 52.30N 5.40E
Flinders *r.* Australia 110 D4 15.12S 141.40E
Flinders I. Australia 110 D2 40.00S 148.00E
Flinders Range *mts.* Australia 110 C2 31.00S 138.30E
Flin Flon Canada 62 H3 54.47N 101.51W
Flint U.S.A. 65 J5 43.03N 83.40W
Flint Wales 12 D5 53.15N 3.07W
Flint I. Kiribati 108 P5 11.26S 151.48W
Flintshire *d.* Wales 9 D3 53.12N 3.10W
Florence Italy 50 D5 43.46N 11.15E
Florence U.S.A. 65 K3 34.12N 79.44W
Florencia Colombia 74 B7 1.37N 75.37W
Flores *i.* Indonesia 105 G2 8.40S 121.20E
Flores Sea Indonesia 104 G2 7.00S 121.00E
Floresta Brazil 77 G4 8.33S 38.35W
Florianópolis Brazil 77 F2 27.35S 48.31W
Florida *d.* U.S.A. 65 J2 29.00N 82.00W
Florida, Straits of U.S.A. 71 H5 24.00N 81.00W
Florida Keys U.S.A. 71 H5 25.00N 81.00W
Florina Greece 56 E4 40.48N 21.25E
Florø Norway 43 A3 61.45N 4.55E
Flotta *i.* Scotland 19 F3 58.49N 3.07W
Fly *r.* P.N.G. 105 K2 8.22S 142.23E
Fochabers Scotland 19 F2 57.37N 3.07W
Focşani Romania 55 J1 45.40N 27.12E
Foggia Italy 50 F4 41.28N 15.33E
Fogo *i.* Cape Verde 84 B3 15.00N 24.28W
Foinaven *mtn.* Scotland 18 E3 58.24N 4.53W
Foix France 44 D3 42.57N 1.35E
Foligno Italy 50 E5 42.56N 12.43E
Folkestone England 11 G2 51.05N 1.11E
Fonseca, G. of Honduras 70 G3 13.10N 87.30W
Fontainebleau France 44 E6 48.24N 2.42E
Fontur *c.* Iceland 43 Z2 66.30N 14.30W
Førde Norway 43 A3 61.28N 5.51E
Fordingbridge England 10 D1 50.56N 1.48W
Foreland *c.* England 10 D1 50.42N 1.06W
Foreland Pt. England 13 D3 51.15N 3.47W
Forest of Atholl *f.* Scotland 19 F1 56.53N 3.55W
Forest of Bowland *hills* England 14 E2 53.57N 2.30W
Forest of Dean *f.* England 10 C2 51.48N 2.32W
Forfar Scotland 19 G1 56.38N 2.54W
Forlì Italy 50 E6 44.13N 12.02E

Formby England **14 D2** 53.34N 3.04W
Formentera *i.* Spain **46 F3** 38.41N 1.30E
Formosa Argentina **77 E2** 26.06S 58.14W
Formosa, Serra *mts.* Brazil **77 E3** 12.00S 55.20W
Forres Scotland **19 F2** 57.37N 3.38W
Forssa Finland **43 E3** 60.49N 23.40E
Forst Germany **48 G4** 51.46N 14.39E
Fort Albany Canada **63 J3** 52.15N 81.35W
Fortaleza Brazil **77 G4** 3.45S 38.45W
Fort Augustus Scotland **18 E2** 57.09N 4.41W
Fort Chipewyan Canada **62 G3** 58.46N 111.09W
Fort Collins U.S.A. **64 E5** 40.35N 105.05W
Fort-de-France Martinique **71 L3** 14.36N 61.05W
Fortescue *r.* Australia **110 A3** 21.00S 116.06E
Fort Frances Canada **63 I2** 48.37N 93.23W
Fort George Canada **63 K3** 53.50N 79.01W
Fort Good Hope Canada **62 F4** 66.16N 128.37W
Forth *r.* Scotland **17 F4** 56.06N 3.48W
Fort Liard Canada **62 F4** 60.14N 123.28W
Fort McMurray Canada **62 G3** 56.45N 111.27W
Fort McPherson Canada **62 E4** 67.29N 134.50W
Fort Nelson Canada **62 F3** 58.48N 122.44W
Fort Norman *see* **Tulít'a** Canada **62**
Fort Peck Resr. U.S.A. **64 E6** 47.55N 107.00W
Fortrose Scotland **19 E2** 57.34N 4.07W
Fort Rupert Canada **63 K3** 51.30N 79.45W
Fort St. John Canada **62 F3** 56.14N 120.55W
Fort Scott U.S.A. **65 H4** 37.52N 94.43W
Fort Severn Canada **63 J3** 56.00N 87.40W
Fort-Shevchenko Kazakhstan **58 H2** 44.31N 50.15E
Fort Simpson Canada **62 F4** 61.46N 121.15W
Fort Smith Canada **62 G4** 60.00N 111.51W
Fort Smith U.S.A. **65 H4** 35.22N 94.27W
Fortuneswell England **10 C1** 50.33N 2.27W
Fort Wayne U.S.A. **65 I5** 41.05N 85.08W
Fort William Scotland **18 D1** 56.49N 5.07W
Fort Worth U.S.A. **64 G3** 32.45N 97.20W
Fort Yukon U.S.A. **62 D4** 66.35N 145.20W
Foshan China **103 K2** 23.03N 113.08E
Fougères France **44 C6** 48.21N 1.12W
Foula *i.* Scotland **19 X9** 60.08N 2.05W
Foulness Pt. England **11 F2** 51.37N 0.57E
Fouta Djallon *f.* Guinea **84 C3** 11.30N 12.30W
Fowey *r.* England **13 C2** 50.22N 4.40W
Foxe Basin *b.* Canada **63 K4** 67.30N 79.00W
Foxe Channel Canada **63 J4** 65.00N 80.00W
Foxe Pen. Canada **63 K4** 65.00N 76.00W
Foxford Rep. of Ire. **20 B3** 53.59N 9.07W
Foyle *r.* N. Ireland **16 B2** 55.00N 7.20W
Foyle, Lough Rep. of Ire./N. Ireland **16 B3** 55.07N 7.06W
Foz do Iguaçú Brazil **77 E2** 25.33S 54.31W
Framlingham England **11 G3** 52.14N 1.20E
Franca Brazil **77 F2** 20.33S 47.27W
France Europe **44 D5** 47.00N 2.00E
Franceville Gabon **84 F1** 1.40S 13.31E
Francistown Botswana **86 B2** 21.11S 27.32E
Frankfort U.S.A. **65 J4** 38.11N 84.53W
Frankfurt Germany **48 G5** 52.20N 14.32E
Frankfurt am Main Germany **48 D4** 50.06N 8.41E
Franklin D. Roosevelt L. U.S.A. **64 C6** 47.55N 118.20W
Franz Josef Land *is.* Russian Fed. **58 H6** 81.00N 54.00E
Fraser *r.* Canada **62 F2** 49.05N 123.00W
Fraserburgh Scotland **19 H2** 57.42N 2.00W
Fraser I. Australia **110 E3** 25.15S 153.10E
Freckleton England **14 E2** 53.45N 2.50W
Fredericia Denmark **43 B1** 55.34N 9.47E
Fredericksburg U.S.A. **65 K4** 38.18N 77.30W
Fredericton Canada **63 L2** 45.57N 66.40W
Frederikshavn Denmark **43 B2** 57.26N 10.32E
Fredrikstad Norway **43 B2** 59.15N 10.55E
Freeport City The Bahamas **71 I6** 26.40N 78.30W
Freetown Sierra Leone **84 C2** 8.30N 13.17W
Freiberg Germany **48 F4** 50.54N 13.20E
Freiburg im Breisgau Germany **48 C2** 48.00N 7.52E
Fréjus France **44 G3** 43.26N 6.44E
Fremantle Australia **110 A2** 32.07S 115.44E
French Guiana S. America **74 D7** 3.40N 53.00W
French Polynesia Pacific Oc. **109 Q5** 20.00S 140.00W
Freshwater England **10 D1** 50.40N 1.30W
Fresno U.S.A. **64 C4** 36.41N 119.57W
Fria Guinea **84 C3** 10.13N 13.48W
Friedrichshafen Germany **48 D2** 47.39N 9.29E
Friesland *d.* Neth. **42 E5** 53.05N 5.45E
Frinton-on-Sea England **11 G2** 51.50N 1.16E
Frio, Cabo *c.* Brazil **75 E4** 22.50S 42.10W
Frisa, Loch Scotland **16 C4** 56.33N 6.05W
Frisian Is. Europe **34 D3** 54.00N 7.00E
Frizington England **14 D3** 54.30N 3.30W
Frobisher B. Canada **63 L4** 63.00N 66.45W
Frodsham England **14 E2** 53.17N 2.45W
Frogmore England **10 E2** 51.20N 0.49W
Frohavet *est.* Norway **43 B3** 63.55N 9.05E
Frome England **13 E3** 51.16N 2.17W
Frome *r.* England **10 C1** 50.41N 2.05W
Frome, L. Australia **110 C2** 30.45S 139.45E
Frontera Canary Is. **46 W1** 27.46N 18.01W
Frosinone Italy **50 E4** 41.36N 13.21E
Frøya *i.* Norway **43 B3** 63.45N 8.30E
Fuenlabrada Spain **46 D4** 40.16N 3.49W
Fuerteventura *i.* Canary Is. **46 Y2** 28.20N 14.10W
Fujairah U.A.E. **95 I4** 25.10N 56.20E
Fujian *d.* China **103 L3** 26.30N 118.00E
Fuji-san *mtn.* Japan **106 C3** 35.23N 138.42E
Fukui Japan **106 C3** 36.04N 136.12E
Fukuoka Japan **106 B2** 33.39N 130.21E
Fukushima Japan **106 B3** 37.44N 140.28E
Fulda Germany **48 D4** 50.35N 9.45E
Fulford England **15 F2** 53.56N 1.04W
Fulham England **11 E2** 51.30N 0.14W
Fumay France **42 D1** 49.59N 4.42E
Funabashi Japan **106 C3** 35.42N 139.59E
Funchal Madeira Is. **84 C5** 32.38N 16.54W
Fundy, B. of N. America **65 M5** 44.30N 66.30W
Fürth Germany **48 E3** 49.28N 11.00E
Fushun China **103 M6** 41.51N 123.53E
Fuxin China **103 M6** 42.08N 121.39E
Fuzhou Fujian China **103 L3** 26.01N 119.20E
Fuzhou Jiangxi China **103 L3** 28.03N 116.15E
Fyn *i.* Denmark **43 B1** 55.10N 10.30E
Fyne, Loch Scotland **16 D3** 55.55N 5.23W

G

Gabès Tunisia **84 F5** 33.52N 10.06E
Gabès, G. of Tunisia **84 F5** 34.00N 11.00E
Gabon Africa **84 F1** 0.00 12.00E
Gaborone Botswana **86 B2** 24.45S 25.55E
Gadsden U.S.A. **65 I3** 34.00N 86.00W
Gaer Wales **12 D3** 51.54N 3.11W
Gaeta Italy **50 E4** 41.13N 13.35E
Gaeta, G. of Med. Sea **50 E4** 41.05N 13.30E
Gafsa Tunisia **84 E5** 34.28N 8.43E
Gagarin Russian Fed. **55 N6** 55.38N 35.00E
Gagnon Canada **63 L3** 51.56N 68.16W
Gagra Georgia **57 O5** 43.21N 40.16E
Gainesville Fla. U.S.A. **65 J2** 29.37N 82.31W
Gainesville Tex. U.S.A. **64 G3** 33.37N 97.08W
Gainsborough England **15 G2** 53.23N 0.46W
Gairdner, L. Australia **110 C2** 31.30S 136.00E
Gair Loch Scotland **18 D2** 57.43N 5.43W
Gairloch *town* Scotland **18 D2** 57.43N 5.41W
Galana *r.* Kenya **87 B2** 3.10S 40.10E
Galapagos Is. Pacific Oc. **76 A4** 0.30S 90.30W
Galashiels Scotland **17 G3** 55.37N 2.49W
Galati Romania **55 J1** 45.27N 27.59E
Gala Water *r.* Scotland **17 F3** 55.36N 2.48W
Gáldar Canary Is. **46 Y2** 28.09N 15.40W
Galdhøpiggen *mtn.* Norway **43 B3** 61.38N 8.19E
Gallabat Sudan **94 E1** 12.58N 36.09E
Galle Sri Lanka **97 G2** 6.01N 80.13E
Galley Head Rep. of Ire. **20 C1** 51.31N 8.57W
Gallinas Pt. Colombia **71 J3** 12.20N 71.30W
Gallipoli Italy **50 H4** 40.02N 18.01E
Gallipoli Turkey **56 H4** 40.25N 26.31E
Gällivare Sweden **43 E4** 67.10N 20.40E
Gallup U.S.A. **64 E4** 35.32N 108.46W
Galole Kenya **87 B2** 1.34S 40.01E
Galston Scotland **16 E3** 55.36N 4.23W
Galtee Mts. Rep. of Ire. **20 C2** 52.20N 8.10W
Galveston U.S.A. **65 H2** 29.17N 94.48W
Galveston B. U.S.A. **65 H2** 29.40N 94.40W
Galway Rep. of Ire. **20 B3** 53.17N 9.04W
Galway *d.* Rep. of Ire. **20 B3** 53.25N 9.00W
Galway B. Rep. of Ire. **20 B3** 53.12N 9.07W
Gambia *r.* The Gambia **84 C3** 13.28N 15.55W
Gambier Is. Pacific Oc. **109 R4** 23.10S 135.00W
Gamud *mtn.* Ethiopia **87 B3** 4.08N 38.04E
Gäncä Azerbaijan **58 G2** 40.39N 46.20E
Gandadiwata, Bukit *mtn.* Indonesia **104 F3** 2.45S 119.25E
Gander Canada **63 M2** 48.58N 54.34W
Gandhidham India **96 E5** 23.07N 70.10E
Gandhinagar India **96 E5** 23.15N 72.45E
Gandía Spain **46 E3** 38.59N 0.11W
Gand-i-Zureh *des.* Afghan. **95 J4** 30.00N 62.00E
Ganges *r.* India **97 I5** 23.30N 90.25E
Ganges, Mouths of the India/Bangla. **97 H5** 22.00N 89.35E
Gannett Peak *mtn.* U.S.A. **64 E5** 43.10N 109.38W
Gansu *d.* China **103 I5** 36.00N 103.00E
Gantamaa Somalia **87 C3** 2.25N 41.49E
Ganzhou China **103 K3** 25.52N 114.51E
Gao Mali **84 E3** 16.19N 0.09W
Gaoxiong Taiwan **103 M2** 22.36N 120.17E
Gap France **44 G4** 44.33N 6.05E
Gar China **102 E4** 32.10N 80.00E
Gara, Lough Rep. of Ire. **20 C3** 53.56N 8.28W
Garanhuns Brazil **77 G4** 8.53S 36.28W
Garbahaarey Somalia **87 C3** 3.20N 42.11E
Garba Tula Kenya **87 B3** 0.31N 38.30E
Gard *r.* France **44 F3** 43.52N 4.40E
Garda, L. Italy **50 D6** 45.40N 10.40E
Gardēz Afghan. **95 K5** 33.37N 69.07E
Garelochhead Scotland **16 E4** 56.05N 4.49W
Garforth England **15 F2** 53.48N 1.22W
Gargzdai Lithuania **43 E1** 55.42N 21.21E
Garissa Kenya **87 B2** 0.27S 39.39E
Garmisch-Partenkirchen Germany **48 E2** 47.30N 11.05E
Garonne *r.* France **44 C4** 45.00N 0.37W
Garoowe Somalia **85 I2** 8.17N 48.20E
Garoua Cameroon **84 F2** 9.17N 13.22E
Garron Pt. N. Ireland **16 D3** 55.03N 5.58W
Garry *r.* Scotland **18 E2** 57.05N 4.49W
Garry, Loch Scotland **19 E1** 56.47N 4.13W
Garsen Kenya **87 A2** 2.18S 40.08E
Garstang England **14 E2** 53.53N 2.47W
Garvagh N. Ireland **16 C2** 54.59N 6.42W
Gary U.S.A. **65 I5** 41.34N 87.20W
Gascony, G. of France **44 B3** 44.00N 2.40W
Gascoyne *r.* Australia **110 A3** 25.00S 113.40E
Gashua Nigeria **84 F3** 12.53N 11.05E
Gaspé Canada **63 L2** 48.50N 64.30W
Gaspé Pen. Canada **63 L2** 48.30N 65.00W
Gastonia U.S.A. **65 J4** 35.14N 81.12W
Gata, Cabo de *c.* Spain **46 D2** 36.45N 2.11W
Gatehouse of Fleet Scotland **16 E2** 54.53N 4.12W
Gateshead England **15 F3** 54.57N 1.35W
Gävle Sweden **43 D3** 60.41N 17.10E
Gaya India **97 H5** 24.48N 85.00E
Gaya Niger **84 E3** 11.52N 3.28E
Gaza Asia **94 D5** 31.20N 34.20E
Gaza *town* Gaza **94 D5** 31.30N 34.28E
Gaziantep Turkey **57 M2** 37.04N 37.21E
Gdańsk Poland **54 F6** 54.22N 18.38E
Gdańsk, G. of Poland **54 F6** 54.45N 19.15E
Gdynia Poland **54 F6** 54.31N 18.30E
Geal Charn *mtn.* Scotland **19 F2** 57.10N 3.31W
Gebze Turkey **57 I3** 40.48N 29.26E
Gedaref Sudan **85 H3** 14.01N 35.24E
Gediz Turkey **57 I3** 39.04N 29.25E
Gediz *r.* Turkey **56 H3** 38.37N 26.47E
Gedser Odde *c.* Denmark **54 C6** 54.35N 11.57E
Geel Belgium **42 E3** 51.10N 5.00E
Geelong Australia **110 D2** 38.10S 144.26E
Gejiu China **103 I2** 23.25N 103.05E
Gela Italy **50 F2** 37.03N 14.15E
Gelderland *d.* Neth. **42 E4** 52.05N 6.00E
Gelligaer Wales **12 D3** 51.40N 3.18W
Gelsenkirchen Germany **48 C4** 51.30N 7.05E
Gemlik Turkey **57 I4** 40.26N 29.10E
Genale Wenz *r.* Ethiopia **87 C3** 4.15N 42.10E
General Santos Phil. **105 H5** 6.05N 125.15E
Geneva Switz. **44 G5** 46.13N 6.09E
Geneva, L. Switz. **44 G5** 46.30N 6.30E
Genil *r.* Spain **46 C2** 37.42N 5.20W
Genk Belgium **42 E2** 50.58N 5.34E
Genoa Italy **50 C6** 44.24N 8.54E
Genoa, G. of Italy **50 C5** 43.50N 8.55E
Gent Belgium **42 C3** 51.02N 3.42E
George *r.* Canada **63 L3** 58.30N 66.00W
Georgetown Guyana **74 D7** 6.48N 58.08W
George Town Malaysia **104 C5** 5.30N 100.16E
Georgia Asia **58 G2** 42.00N 43.30E
Georgia *d.* U.S.A. **65 J3** 33.00N 83.00W
Georgian B. Canada **63 J2** 45.15N 80.45W
Georgina *r.* Australia **110 C3** 23.12S 139.33E
Georgiyevka Kazakhstan **102 E7** 49.21N 81.35E
Gera Germany **48 F4** 50.51N 12.11E
Geral de Goiás, Serra *mts.* Brazil **77 F3** 13.00S 45.40W
Geraldton Australia **110 A3** 28.49S 114.36E
Gereshk Afghan. **95 J5** 31.48N 64.34E
Germany Europe **48 D4** 51.00N 10.00E
Gevgelija Macedonia **56 F4** 41.09N 22.30E
Gexto Spain **46 D5** 43.21N 3.01W
Geyik Dag *mtn.* Turkey **57 K2** 36.53N 32.12E
Geyve Turkey **57 J4** 40.32N 30.18E
Gezira *f.* Sudan **82 G6** 14.30N 33.00E
Ghadamis Libya **84 E5** 30.10N 9.30E
Gham Shahr Iran **95 H6** 36.28N 52.53E
Ghaghara *r.* India **97 G6** 25.45N 84.50E
Ghana Africa **84 D2** 8.00N 1.00W
Ghardaïa Algeria **84 E5** 32.20N 3.40E
Gharyan Libya **52 F3** 32.10N 13.01E
Ghazaouet Algeria **52 C4** 35.08N 1.50W
Ghaziabad India **97 F6** 28.40N 77.26E
Ghaznī Afghan. **95 K5** 33.33N 68.28E
Giant's Causeway *f.* N. Ireland **16 C3** 55.14N 6.31W
Gibraltar Europe **46 C2** 36.07N 5.22W
Gibraltar, Str. of Africa/Europe **46 C1** 36.00N 5.25W
Gibson Desert Australia **110 B3** 23.10S 125.35E
Gīdolē Ethiopia **87 B4** 5.38N 37.28E
Gien France **44 E5** 47.42N 2.38E
Giessen Germany **48 D4** 50.35N 8.42E
Gifu Japan **106 C3** 35.27N 136.50E
Gigha *i.* Scotland **16 D3** 55.41N 5.44W
Gijón Spain **46 C5** 43.32N 5.40W
Gila *r.* U.S.A. **64 D3** 32.45N 114.30W
Gilbert Is. Kiribati **108 M7** 1.20N 173.00E
Gilf Kebir Plateau *f.* Egypt **94 C3** 23.30N 26.00E
Gilgil Kenya **87 B2** 0.30S 36.19E
Gilgit Jammu & Kashmir **96 E8** 35.54N 74.20E
Gillette U.S.A. **64 E5** 44.18N 105.30W
Gillingham Dorset England **10 C2** 51.02N 2.17W
Gillingham Kent England **11 F2** 51.24N 0.33E
Gill, Lough Rep. of Ire. **20 C4** 54.15N 8.25W
Giluwe, Mt. P.N.G. **110 D5** 6.06S 143.54E
Gilwern Wales **12 D3** 51.51N 3.06W
Gimbala, Jebel *mtn.* Sudan **85 G3** 13.00N 24.20E
Giresun Turkey **57 N4** 40.55N 38.25E
Girona Spain **46 G4** 41.59N 2.49E
Gironde *r.* France **44 C4** 45.35N 1.00W
Girvan Scotland **16 E3** 55.15N 4.51W
Gisborne New Zealand **111 G2** 38.41S 178.02E
Gisors France **42 A1** 49.17N 1.47E
Gizhiga Russian Fed. **59 S4** 62.00N 160.34E
Gizhiga G. Russian Fed. **59 R4** 61.00N 158.00E
Gjøvik Norway **43 B3** 60.47N 10.41E
Glace Bay *town* Canada **63 M2** 46.11N 60.00W
Glacier Peak *mtn.* U.S.A. **64 B6** 48.07N 121.06W
Gladstone Australia **110 E3** 23.52S 151.16E
Glanton England **15 F4** 55.25N 1.53W
Glasgow Scotland **16 E3** 55.52N 4.15W
Glasgow U.S.A. **64 E6** 48.12N 106.37W
Glasgow City *d.* Scotland **8 C4** 55.52N 4.15W
Glass, Loch Scotland **19 E2** 57.43N 4.30W
Glastonbury England **13 E3** 51.09N 2.42W
Glenarm N. Ireland **16 D2** 54.58N 5.58W
Glen Coe *f.* Scotland **16 E4** 56.40N 4.55W
Glendale U.S.A. **64 D3** 33.32N 112.11W
Glendive U.S.A. **64 F6** 47.08N 104.42W
Glengad Head Rep. of Ire. **20 D5** 55.20N 7.11W
Glen Garry *f.* Scotland **18 D2** 57.03N 5.05W
Glengormley N. Ireland **16 D2** 54.41N 5.59W
Glenluce Scotland **16 E2** 54.53N 4.48W
Glen More *f.* Scotland **19 E2** 57.15N 4.30W
Glen Moriston *f.* Scotland **18 E2** 57.10N 4.50W
Glennallen U.S.A. **62 D4** 62.08N 145.38W
Glenrothes Scotland **17 F4** 56.12N 3.10W
Glenshee *f.* Scotland **19 F1** 56.50N 3.28W
Glinton England **11 E3** 52.39N 0.17W
Gliwice Poland **54 F4** 50.17N 18.40E
Głogów Poland **54 E4** 51.40N 16.06E
Glomfjord *town* Norway **43 C4** 66.49N 14.00E
Glorieuses, Is. Indian Oc. **86 D3** 11.34S 47.19E
Glossop England **15 F2** 53.27N 1.56W
Gloucester England **10 C2** 51.52N 2.15W
Gloucestershire *d.* England **9 D2** 51.45N 2.00W
Glusburn England **15 F2** 53.54N 2.00W
Glyder Fawr *mtn.* Wales **12 C5** 53.06N 4.01W
Glynneath Wales **12 D3** 51.45N 3.37W
Gmünd Austria **54 D3** 48.47N 14.59E
Gniezno Poland **54 E5** 52.32N 17.32E
Goa *d.* India **96 E4** 15.30N 74.00E
Goat Fell *mtn.* Scotland **16 D3** 55.37N 5.12W
Gobabis Namibia **86 A2** 22.30S 18.58E
Gobi *des.* Asia **103 I6** 43.30N 103.30E
Gobowen England **10 B3** 52.54N 3.02W
Goch Germany **42 F3** 51.41N 6.10E
Godalming England **11 E2** 51.11N 0.37W
Godavari *r.* India **97 G4** 16.40N 82.15E
Godmanchester England **11 E3** 52.19N 0.11W
Godthåb *see* **Nuuk** Greenland **63**
Goes Neth. **42 C3** 51.30N 3.54E
Goiânia Brazil **77 F3** 16.43S 49.18W
Goiás *d.* Brazil **77 E3** 15.00S 48.00W
Gökçeada *i.* Turkey **56 G4** 40.10N 25.51E
Göksun Turkey **57 M3** 38.03N 36.30E
Gölcük Turkey **57 I4** 40.44N 29.50E
Gold Coast *town* Australia **110 E3** 28.00S 153.22E
Golmud China **102 G5** 36.23N 94.49E
Golpayegan Iran **95 H5** 33.23N 50.18E
Golspie Scotland **19 F2** 57.58N 3.58W
Gómez Palacio Mexico **70 D6** 25.39N 103.30W
Gonaïves Haiti **71 J4** 19.29N 72.42W
Gonbad-e Kavus Iran **95 I6** 37.15N 55.11E
Gondar Ethiopia **85 H3** 12.39N 37.29E
Gondia India **97 G5** 21.27N 80.12E
Gongga Shan *mtn.* China **103 I3** 29.57N 101.55E
Good Hope, C. of R.S.A. **86 A1** 34.20S 18.25E
Goodwick Wales **12 C4** 52.00N 5.00W
Goole England **15 G2** 53.42N 0.52W
Goose L. U.S.A. **64 B5** 41.55N 120.25W
Gorakhpur India **97 G6** 26.45N 83.23E
Gorebridge Scotland **17 F3** 55.51N 3.02W
Gorey Rep. of Ire. **20 E2** 52.40N 6.18W
Gorgan Iran **95 H6** 36.50N 54.29E
Goris Armenia **95 G6** 39.31N 46.22E
Gorizia Italy **50 E6** 45.58N 13.37E
Görlitz Germany **48 G4** 51.09N 15.00E
Gorno-Altaysk Russian Fed. **102 F8** 51.57N 85.58E
Gornyak Russian Fed. **102 E8** 50.59N 81.30E
Gorontalo Indonesia **105 G4** 0.33N 123.05E
Gort Rep. of Ire. **20 C3** 53.03N 8.50W
Gorzów Wielkopolski Poland **54 D5** 52.42N 15.12E
Gosberton England **11 E3** 52.52N 0.09W
Gosford Australia **110 E2** 33.25S 151.18E
Gosforth Cumbria England **14 D3** 54.26N 3.27W
Gosforth T. and W. England **15 F4** 55.02N 1.35W
Goshogawara Japan **106 D4** 40.48N 140.27E
Gospić Croatia **56 B6** 44.34N 15.23E
Gosport England **10 D1** 50.48N 1.08W
Göteborg Sweden **43 B2** 57.45N 12.00E
Gotha Germany **48 E4** 50.57N 10.43E
Gotland *i.* Sweden **43 D2** 57.30N 18.30E
Göttingen Germany **48 D4** 51.32N 9.57E
Gouda Neth. **42 D4** 52.01N 4.43E
Gouin, Résr. Canada **63 K2** 48.38N 74.54W
Goulburn Australia **110 D2** 34.47S 149.43E
Gourdon France **44 D4** 44.45N 1.22E
Governador Valadares Brazil **77 F3** 18.51S 42.00W
Gower *pen.* Wales **12 C3** 51.37N 4.10W
Gowna, L. Rep. of Ire. **20 D3** 53.50N 7.34W
Goya Argentina **77 E2** 29.10S 59.20W
Gozo *i.* Malta **50 F2** 36.03N 14.16E
Graciosa *i.* Canary Is. **46 Z2** 29.15N 13.31W
Gradaús, Serra dos *mts.* Brazil **77 E4** 8.00S 50.30W
Grafham Water *l.* England **11 E3** 52.19N 0.16W
Grafton Australia **110 E3** 29.40S 152.56E
Grafton U.S.A. **64 G6** 48.28N 97.25W
Graham Land *f.* Antarctica **112** 67.00S 60.00W
Grahamstown R.S.A. **86 B1** 33.19S 26.32E
Grain England **11 F2** 51.28N 0.43E
Grampian Mts. Scotland **19 E1** 56.55N 4.00W
Granada Spain **46 D2** 37.10N 3.35W
Gran Canaria *i.* Canary Is. **46 Y1** 28.00N 15.30W
Gran Chaco *f.* S. America **75 C4** 23.30S 60.00W
Grand Bahama *i.* The Bahamas **71 I6** 26.35N 78.00W
Grand Canyon *town* U.S.A. **64 D4** 36.04N 112.07W
Grand Canyon *f.* U.S.A. **64 D4** 36.15N 113.00W
Grand Cayman *i.* Cayman Is. **71 H4** 19.20N 81.30W
Grande *r.* Bahia Brazil **77 F3** 11.05S 43.09W
Grande *r.* Minas Gerais Brazil **77 E3** 20.00S 51.00W
Grande, Bahía *b.* Argentina **75 C1** 50.45S 68.00W
Grande Prairie *town* Canada **62 G3** 55.10N 118.52W
Grand Falls *town* Canada **63 M2** 48.57N 55.40W
Grand Forks U.S.A. **64 G6** 47.57N 97.05W
Grand Island *town* U.S.A. **64 G5** 40.56N 98.21W
Grand Junction U.S.A. **64 E4** 39.04N 108.33W
Grand Manan I. Canada **65 M5** 44.40N 66.50W
Grândola Portugal **46 A3** 38.10N 8.34W
Grand Rapids *town* U.S.A. **65 I5** 42.57N 85.40W
Grangemouth Scotland **17 F4** 56.01N 3.44W
Grange-over-Sands England **14 E3** 54.12N 2.55W
Gran Paradiso *mtn.* Italy **50 B6** 45.31N 7.15E
Grantham England **10 E3** 52.55N 0.39W
Grantown-on-Spey Scotland **19 F2** 57.19N 3.38W
Grants Pass *town* U.S.A. **64 B5** 42.26N 123.20W
Grasse France **44 G3** 43.40N 6.56E
Grassington England **15 F3** 54.04N 1.59W
Grave, Pointe de *c.* France **44 C4** 45.35N 1.04W
Gravesend England **11 F2** 51.27N 0.24E
Grays England **11 F2** 51.29N 0.20E
Graz Austria **54 D2** 47.05N 15.22E
Great Abaco *i.* The Bahamas **71 I6** 26.30N 77.00W
Great Australian Bight Australia **110 B2** 33.20S 130.00E
Great Baddow England **11 F2** 51.43N 0.29E
Great Barrier Reef *f.* Australia **110 D4** 16.30S 146.30E
Great Basin *f.* U.S.A. **64 C4** 39.00N 115.30W
Great Bear L. Canada **62 G4** 66.00N 120.00W
Great Bend *town* U.S.A. **64 G4** 38.22N 98.47W
Great Bernera *i.* Scotland **18 C3** 58.13N 6.50W
Great Blasket I. Rep. of Ire. **20 A2** 52.05N 10.32W
Great Clifton England **14 D3** 54.38N 3.30W
Great Cumbrae *i.* Scotland **16 E3** 55.45N 4.57W
Great Dividing Range *mts.* Australia **110 D2** 33.00S 151.00E
Great Driffield England **15 G3** 54.01N 0.26W
Great Dunmow England **11 F2** 51.53N 0.22E
Greater Antilles *is.* C. America **71 J4** 17.00N 70.00W
Greater London *d.* England **9 E2** 51.31N 0.06W
Greater Manchester *d.* England **9 D3** 53.30N 2.18W
Great Exuma *i.* The Bahamas **71 I5** 23.00N 76.00W
Great Falls *town* U.S.A. **64 D6** 47.30N 111.16W
Great Gonerby England **10 E3** 52.56N 0.40W
Greatham England **15 F3** 54.39N 1.14W
Great Harwood England **15 E2** 53.48N 2.24W
Great Inagua *i.* The Bahamas **71 J5** 21.00N 73.20W
Great Karoo *f.* R.S.A. **86 B1** 32.50S 22.30E
Great Linford England **10 E3** 52.03N 0.46W
Great Malvern England **10 C3** 52.07N 2.19W
Great Nicobar *i.* India **104 A5** 7.00N 93.50E
Great Ormes Head Wales **12 D5** 53.20N 3.52W
Great Ouse *r.* England **11 F3** 52.47N 0.23E
Great Plains *f.* N. America **60 I6** 45.00N 100.00W
Great Rhos *mtn.* Wales **12 D4** 52.16N 3.13W
Great Rift Valley *f.* Africa **82 G4** 7.00S 33.00E
Great St. Bernard Pass Italy/Switz. **44 G4** 45.52N 7.11E
Great Salt L. U.S.A. **64 D5** 41.10N 112.40W
Great Sand Sea *f.* Egypt/Libya **94 C4** 28.00N 25.00E
Great Sandy Desert Australia **110 B3** 21.00S 125.00E
Great Shelford England **11 F3** 52.09N 0.08E
Great Slave L. Canada **62 G4** 61.30N 114.20W
Great Stour *r.* England **11 G2** 51.19N 1.15E
Great Torrington England **13 C2** 50.57N 4.09W
Great Victoria Desert Australia **110 B3** 29.00S 127.30E
Great Whernside *mtn.* England **15 F3** 54.09N 1.59W
Great Yarmouth England **11 G3** 52.36N 1.45E
Gréboun, Mt. Niger **82 D6** 19.55N 8.35E
Greco, Monte *mtn.* Italy **50 E4** 41.48N 14.00E
Gredos, Sierra de *mts.* Spain **46 C4** 40.18N 5.20W
Greece Europe **56 E3** 39.00N 22.00E
Greeley U.S.A. **64 F5** 40.26N 104.43W
Green *r.* U.S.A. **64 E4** 38.20N 109.53W
Green Bay *town* U.S.A. **65 I5** 44.32N 88.00W

Greenland N. America **63 N4** 68.00N 45.00W
Greenlaw Scotland **17 G3** 55.43N 2.28W
Greenock Scotland **16 E3** 55.57N 4.45W
Greensboro U.S.A. **65 K4** 36.03N 79.50W
Greenstone Pt. Scotland **18 D2** 57.55N 5.37W
Greenville Miss. U.S.A. **65 H3** 33.23N 91.03W
Greenwich England **11 F2** 51.29N 0.00
Greenville S.C. U.S.A. **65 J3** 34.52N 82.25W
Greifswald Germany **48 F6** 54.06N 13.24E
Grenå Denmark **43 B2** 56.25N 10.53E
Grenada C. America **71 L3** 12.15N 61.45W
Grenade France **44 D3** 43.47N 1.10E
Grenoble France **44 F4** 45.11N 5.43E
Greta *r.* England **15 F3** 54.31N 1.52W
Gretna Scotland **17 F2** 55.00N 3.04W
Grey Range *mts.* Australia **110 D3** 28.30S 142.15E
Grimsby England **15 G2** 53.35N 0.05W
Grímsey *i.* Iceland **43 Y2** 66.33N 18.00W
Grímsvötn *mtn.* Iceland **43 Y2** 64.30N 17.10W
Grodno Belarus **55 H5** 53.40N 23.50E
Grodzisk Wielkopolski Poland **54 E5** ..52.14N 16.22E
Groningen Neth. **42 F5** 53.13N 6.35E
Groningen *d.* Neth. **42 F5** 53.15N 6.45E
Groote Eylandt *i.* Australia **110 C4** 14.00S 136.30E
Grosseto Italy **50 D5** 42.46N 11.08E
Gross Glockner *mtn.* Austria **54 C2** 47.05N 12.50E
Groundhog *r.* Canada **63 J3** 49.40N 82.06W
Groznyy Russian Fed. **58 G2** 43.21N 45.42E
Grudziądz Poland **54 F5** 53.29N 18.45E
Gruinard B. Scotland **18 D2** 57.52N 5.26W
Guadalajara Mexico **70 D5** 20.30N 103.20W
Guadalajara Spain **46 D4** 40.37N 3.10W
Guadalcanal *i.* Solomon Is. **111 E5** 9.30S 160.00E
Guadalete *r.* Spain **46 B2** 36.37N 6.15W
Guadalope *r.* Spain **46 E4** 41.15N 0.03W
Guadalquivir *r.* Spain **46 B2** 36.50N 6.20W
Guadalupe *i.* Mexico **64 C2** 29.00N 118.25W
Guadalupe, Sierra de *mts.* Spain **46 C3** ..39.30N 5.25W
Guadarrama, Sierra de *mts.* Spain **46 D4** 41.00N 3.50W
Guadeloupe C. America **71 L4** 16.20N 61.40W
Guadiana *r.* Portugal **46 B2** 37.10N 7.36W
Guadix Spain **46 D2** 37.19N 3.08W
Guajira Pen. Colombia **71 J3** 12.00N 72.00W
Guam *i.* Pacific Oc. **105 K6** 13.30N 144.40E
Guanajuato Mexico **70 D5** 21.00N 101.16W
Guanajuato *d.* Mexico **70 D5** 21.00N 101.00W
Guanare Venezuela **71 K2** 9.04N 69.45W
Guangdong *d.* China **103 K2** 23.00N 113.00E
Guangxi Zhuangzu Zizhiqu *d.* China **103 J2** 23.50N 109.00E
Guangyuan China **103 J4** 32.29N 105.55E
Guangzhou China **103 K2** 23.20N 113.30E
Guanipa *r.* Venezuela **71 L2** 10.00N 62.20W
Guantánamo Cuba **71 I5** 20.09N 75.14W
Guaporé *r.* Brazil **76 D3** 12.00S 65.15W
Guarapuava Brazil **77 E2** 25.22S 51.28W
Guara, Sierra de *mts.* Spain **46 E5** 42.20N 0.00
Guarda Portugal **46 B4** 40.32N 7.17W
Guatemala C. America **70 F4** 15.40N 90.00W
Guatemala City Guatemala **70 F3** 14.38N 90.22W
Guaviare *r.* Colombia **74 C7** 4.00N 67.35W
Guayaquil Ecuador **76 C4** 2.13S 79.54W
Guayaquil, Golfo de *g.* Ecuador **76 B4** ..2.30S 80.00W
Guaymas Mexico **70 B6** 27.59N 110.54W
Guba Ethiopia **85 H3** 11.17N 35.20E
Gubkin Russian Fed. **55 O4** 51.18N 37.32E
Gudbrandsdalen *f.* Norway **43 B3** 62.00N 9.10E
Guelma Algeria **52 E4** 36.28N 7.26E
Guelmine Morocco **84 D4** 28.56N 10.04W
Guéret France **44 D5** 46.10N 1.52E
Guernsey *i.* Channel Is. **13 Y9** 49.27N 2.35W
Guerrero *d.* Mexico **70 D4** 18.00N 100.00W
Guge *mtn.* Ethiopia **87 B4** 6.16N 37.25E
Guiana Highlands S. America **74 D7** 4.00N 59.00W
Guildford England **11 E2** 51.14N 0.35W
Guilin China **103 K3** 25.21N 110.11E
Guinea Africa **84 C3** 10.30N 10.30W
Guinea, G. of Africa **82 D5** 3.00N 3.00E
Guinea-Bissau Africa **84 C3** 12.00N 15.30W
Guînes France **42 A2** 50.52N 1.52E
Güiria Venezuela **71 L3** 10.37N 62.21W
Guisborough England **15 F3** 54.32N 1.02W
Guise France **42 C1** 49.54N 3.39E
Guiyang China **103 J3** 26.35N 106.40E
Guizhou *d.* China **103 J3** 27.00N 106.30E
Gujarat *d.* India **96 E5** 22.45N 71.30E
Gujranwala Pakistan **96 E7** 32.06N 74.11E
Gulbarga India **96 F4** 17.22N 76.47E
Gullane Scotland **17 G4** 56.02N 2.49W
Gulu Uganda **86 C5** 2.46N 32.21E
Gumdag Turkmenistan **95 H6** 39.14N 54.33E
Gümüshane Turkey **57 N4** 40.26N 39.26E
Guna India **96 F5** 24.39N 77.19E
Gunnbjørn Fjeld *mtn.* Greenland **63 P4** 68.54N 29.48W
Guntur India **97 G4** 16.20N 80.27E
Gurgueia *r.* Brazil **77 F4** 6.45S 43.35W
Gurupi *r.* Brazil **77 F4** 1.13S 46.06W
Gushgy Turkmenistan **95 J6** 35.14N 62.15E
Guwahati India **97 I6** 26.05N 91.55E
Guyana S. America **74 D7** 5.00N 59.00W
Gwadar Pakistan **96 C6** 25.09N 62.21E
Gwalior India **97 F6** 26.12N 78.09E
Gweebarra B. Rep. of Ire. **20 C4** 54.52N 8.30W
Gweru Zimbabwe **86 B3** 19.25S 29.50E
Gwynedd *d.* Wales **9 D3** 53.00N 4.00W
Gydanskiy Pen. Russian Fed. **58 J5** 70.00N 78.30E
Gyöngyös Hungary **54 F2** 47.47N 19.56E
Györ Hungary **54 E2** 47.41N 17.40E
Gypsumville Canada **62 I3** 51.47N 98.38W
Gyzylarbat Turkmenistan **95 I6** 39.00N 56.23E

H

Haapajärvi Finland **43 F3** 63.45N 25.20E
Haapsalu Estonia **43 E2** 58.58N 23.32E
Haarlem Neth. **42 D4** 52.22N 4.38E
Habaswein Kenya **87 B3** 1.06N 39.26E
Habban Yemen **95 G1** 14.21N 47.04E
Hachijo-jima *i.* Japan **106 C2** 33.00N 139.50E
Hachinohe Japan **106 D4** 40.30N 141.30E
Haddington Scotland **17 G3** 55.57N 2.47W
Haderslev Denmark **43 B1** 55.15N 9.30E
Hadhramaut *f.* Yemen **95 G2** 16.30N 49.30E
Hadleigh England **11 F3** 52.03N 0.58E
Haēabja Iraq **95 G6** 35.11N 45.59E
Haeju N. Korea **103 N5** 38.04N 125.40E
Hagadera Kenya **87 C3** 0.01N 40.21E
Hagar Nish Plateau *f.* Eritrea **94 E2** 17.00N 38.00E
Hagåtña *see* **Agana** Guam **105**
Hagen Germany **42 G3** 51.22N 7.27E
Hags Head Rep. of Ire. **20 B2** 52.56N 9.29W
Hai Tanzania **87 B2** 3.19S 37.08E
Haifa Israel **94 D5** 32.49N 34.59E
Haikou China **103 K1** 20.05N 110.25E
Ha'il Saudi Arabia **94 F4** 27.31N 41.45E
Hailar China **103 L7** 49.15N 119.41E
Hailsham England **11 F1** 50.52N 0.17E
Hainan *i.* China **103 J1** 18.30N 109.40E
Hainaut *d.* Belgium **42 C2** 50.30N 3.45E
Haines U.S.A. **62 E3** 59.11N 135.23W
Hai Phong Vietnam **104 D8** 20.58N 106.41E
Haiti C. America **71 J4** 19.00N 73.00W
Haiya Sudan **94 E2** 18.17N 36.21E
Hajmah Oman **95 I2** 19.55N 56.15E
Hakodate Japan **106 D4** 41.46N 140.44E
Halden Norway **43 B2** 59.08N 11.13E
Halesowen England **10 C3** 52.27N 2.02W
Halesworth England **11 G3** 52.21N 1.30E
Halifax Canada **63 L2** 44.38N 63.35W
Halifax England **15 F2** 53.43N 1.51W
Halkirk Scotland **19 F3** 58.30N 3.30W
Halladale *r.* Scotland **19 F3** 58.32N 3.53W
Hall Beach *town* Canada **63 J4** 68.40N 81.30W
Halle Belgium **42 D2** 50.45N 4.14E
Halle Germany **48 E4** 51.28N 11.58E
Hall Is. Fed. States of Micronesia **108 K7** 8.37N 152.00E
Hall's Creek *town* Australia **110 B4** 18.13S 127.39E
Halmahera *i.* Indonesia **105 H4** 0.45N 128.00E
Halmstad Sweden **43 C2** 56.41N 12.55E
Halstead England **11 F2** 51.57N 0.39E
Haltwhistle England **15 E3** 54.58N 2.27W
Ham France **42 C1** 49.45N 3.04E
Hamadan Iran **95 G5** 34.47N 48.33E
Ḩamāh Syria **94 E6** 35.09N 36.44E
Hamamatsu Japan **106 C2** 34.42N 137.42E
Hamar Norway **43 B3** 60.47N 10.55E
Hambleton Hills England **15 F3** 54.15N 1.11W
Hamburg Germany **48 D5** 53.33N 10.00E
Hämeenlinna Finland **43 F3** 61.00N 24.25E
Hamersley Range *mts.* Australia **110 A3** 22.00S 118.00E
Hamhung N. Korea **103 N5** 39.54N 127.35E
Hami China **102 G6** 42.40N 93.30E
Hamilton Bermuda **71 L7** 32.18N 64.48W
Hamilton Canada **65 K5** 43.15N 79.50W
Hamilton New Zealand **111 G2** 37.47S 175.17E
Hamilton Scotland **16 E3** 55.46N 4.02W
Hamim, Wadi al *r.* Libya **53 H3** 32.06N 23.58E
Hamina Finland **43 F3** 60.33N 27.15E
Hamm Germany **48 C4** 51.40N 7.49E
Hammamet, G. of Tunisia **52 F4** 36.05N 10.40E
Hammerdal Sweden **43 C3** 63.35N 15.20E
Hammerfest Norway **43 E5** 70.40N 23.44E
Hampshire *d.* England **9 E2** 51.10N 1.20W
Hampshire Downs *hills* England **10 D2** .51.18N 1.25W
Hamstreet England **11 F2** 51.03N 0.52E
Hamun-e Jaz Murian *l.* Iran **95 I4** 27.00N 59.20E
Hanamaki Japan **106 D3** 39.23N 141.07E
Handa I. Scotland **18 D3** 58.23N 5.12W
Handan China **103 K5** 36.37N 114.26E
Handeni Tanzania **87 B2** 5.26S 38.02E
Hanggin Houqi China **103 J6** 40.52N 107.04E
Hangzhou China **103 M4** 30.10N 120.07E
Hanmni Mashkel *l.* Pakistan **95 J4** 28.15N 63.00E
Hannibal U.S.A. **65 H4** 39.41N 91.25W
Hannover Germany **48 D5** 52.23N 9.44E
Hanoi Vietnam **104 D8** 21.01N 105.52E
Hantsavichy Belarus **55 J5** 52.49N 26.29E
Hanzhong China **103 J4** 33.08N 107.04E
Haparanda Sweden **43 E4** 65.50N 24.05E
Happy Valley-Goose Bay *town* Canada **63 L3** 53.16N 60.14W
Harare Zimbabwe **86 C3** 17.43S 31.05E
Harbin China **103 N7** 45.45N 126.41E
Hardangervidda *f.* Norway **43 A3** 60.20N 8.00E
Harderwijk Neth. **42 E4** 52.21N 5.37E
Haren Germany **42 G4** 52.48N 7.15E
Hargele Ethiopia **87 C4** 5.19N 42.04E
Hargeysa Somalia **85 I2** 9.31N 44.02E
Har Hu *l.* China **102 H5** 38.20N 97.40E
Hari *r.* Afghan. **95 J6** 35.42N 61.12E
Haria Canary Is. **46 Z2** 29.09N 13.30W
Harlech Wales **12 C4** 52.52N 4.08W
Harleston England **11 G3** 52.25N 1.18E
Harlingen Neth. **42 E5** 53.10N 5.25E
Harlow England **11 F2** 51.47N 0.08E
Harney Basin *f.* U.S.A. **64 C5** 43.20N 119.00W
Härnösand Sweden **43 D3** 62.37N 17.55E
Har Nuur *l.* Mongolia **102 G7** 48.10N 93.30E
Harpenden England **11 E2** 51.49N 0.22W
Harray, Loch of Scotland **19 F4** 59.03N 3.15W
Harricana *r.* Canada **63 K3** 51.10N 79.45W
Harris *i.* Scotland **18 C2** 57.50N 6.55W
Harris, Sd. of Scotland **18 B2** 57.43N 7.05W
Harrisburg U.S.A. **65 K5** 40.35N 76.59W
Harrison, C. Canada **63 M3** 55.00N 58.00W
Harrogate England **15 F2** 53.59N 1.32W
Hârsova Romania **57 H6** 44.41N 27.56E
Harstad Norway **43 D5** 68.48N 16.30E
Harteigan *mtn.* Norway **43 A3** 60.11N 7.05E
Harter Fell *mtn.* England **14 E3** 54.27N 2.51W
Hart Fell *mtn.* Scotland **17 F3** 55.25N 3.25W
Hartford U.S.A. **65 L5** 41.40N 72.51W
Hartland England **13 C2** 50.59N 4.29W
Hartland Pt. England **13 C3** 51.01N 4.32W
Hartlepool England **15 F3** 54.42N 1.11W
Hartlepool *d.* England **9 E3** 54.42N 1.11W
Har Us Nuur *l.* Mongolia **102 G7** 48.10N 92.10E
Harwich England **11 G2** 51.56N 1.18E
Haryana *d.* India **96 F6** 29.15N 76.00E
Haslemere England **10 E2** 51.05N 0.41W
Hasselt Belgium **42 E2** 50.56N 5.20E
Hassi Messaoud Algeria **84 E5** 31.43N 6.03E
Hässleholm Sweden **43 C2** 56.09N 13.45E
Hastings England **11 F1** 50.51N 0.36E
Hatfield England **15 F2** 53.36N 0.59W
Ha Tinh Vietnam **104 D7** 18.21N 105.55E
Hatteras, C. U.S.A. **65 K4** 35.14N 75.31W
Hattiesburg U.S.A. **65 I3** 31.25N 89.19W
Haud *f.* Ethiopia **85 I2** 8.00N 46.00E
Haugesund Norway **43 A2** 59.25N 5.16E
Haukivesi *l.* Finland **43 G3** 62.10N 28.30E
Haut Folin *mtn.* France **44 E5** 47.00N 4.00E
Hauts Plateaux Algeria **52 C3** 34.00N 0.10W
Havana Cuba **71 H5** 23.07N 82.25W
Havant England **10 E1** 50.51N 0.59W
Havel *r.* Germany **48 F5** 52.51N 11.57E
Haverfordwest Wales **12 C3** 51.48N 4.59W
Haverhill England **11 F3** 52.06N 0.27E
Havre U.S.A. **64 E6** 48.34N 109.45W
Havre-St.-Pierre Canada **63 L3** 50.15N 63.36W
Hawaii *i.* Hawaiian Is. **108 P8** 19.30N 155.30W
Hawaiian Is. Pacific Oc. **108 O9** 21.00N 160.00W
Hawarden Wales **12 D5** 53.11N 3.02W
Hawes England **15 E3** 54.18N 2.12W
Haweswater Resr. England **14 E3** 54.30N 2.45W
Hawick Scotland **17 G3** 55.25N 2.47W
Hawke B. New Zealand **111 G2** 39.18S 177.15E
Hawkhurst England **11 F2** 51.02N 0.31E
Hawthorne U.S.A. **64 C4** 38.13N 118.37W
Haxby England **15 F3** 54.02N 1.06W
Hay Australia **110 D2** 34.21S 144.31E
Hay *r.* Canada **62 G3** 60.49N 115.52W
Haydarabad Iran **94 G6** 37.09N 45.27E
Haydon Bridge England **15 E3** 54.58N 2.14W
Hayle England **13 B2** 50.12N 5.25W
Hay-on-Wye Wales **12 D4** 52.04N 3.09W
Hay River *town* Canada **62 G4** 60.51N 115.42W
Haywards Heath *town* England **11 E1** 51.00N 0.05W
Hazarajat *f.* Afghan. **95 K5** 33.00N 66.00E
Hazebrouck France **42 B2** 50.43N 2.32E
Heacham England **11 F3** 52.55N 0.30E
Headcorn England **11 F2** 51.11N 0.37E
Heanor England **15 F2** 53.01N 1.20W
Heathfield England **11 F1** 50.58N 0.18E
Hebei *d.* China **103 L5** 39.20N 117.15E
Hebron Jordan **94 E5** 31.32N 35.06E
Hecate Str. Canada **62 E3** 53.00N 131.00W
Hechi China **103 J2** 24.42N 108.02E
Heckington England **15 G1** 52.59N 0.18W
Hede Sweden **43 C3** 62.27N 13.30E
Heerenveen Neth. **42 E4** 52.57N 5.55E
Heerlen Neth. **42 E2** 50.53N 5.59E
Hefei China **103 L4** 31.55N 117.18E
Hegang China **103 O7** 47.36N 130.30E
Heidelberg Germany **48 D3** 49.25N 8.42E
Heighington England **15 G2** 53.12N 0.28W
Heilbronn Germany **48 D3** 49.08N 9.14E
Heilongjiang *d.* China **103 N7** 47.00N 126.00E
Heinola Finland **43 F3** 61.13N 26.05E
Hekla *mtn.* Iceland **43 Y2** 64.00N 19.45W
Helena U.S.A. **64 D6** 46.35N 112.00W
Helensburgh Scotland **16 E4** 56.01N 4.44W
Heligoland B. Germany **48 D6** 54.00N 8.15E
Hellín Spain **46 E3** 38.31N 1.43W
Helmand *r.* Asia **95 J5** 31.10N 61.20E
Helmond Neth. **42 E3** 51.28N 5.40E
Helmsdale Scotland **19 F3** 58.08N 3.40W
Helmsdale *r.* Scotland **19 F3** 58.06N 3.40W
Helmsley England **15 F3** 54.14N 1.04W
Helong China **106 A4** 42.38N 128.58E
Helsingborg Sweden **43 C2** 56.05N 12.45E
Helsingør Denmark **43 C2** 56.03N 12.38E
Helsinki Finland **43 F3** 60.08N 25.00E
Helston England **13 B2** 50.07N 5.17W
Helvellyn *mtn.* England **14 D3** 54.31N 3.00W
Hemel Hempstead England **11 E2** 51.46N 0.28W
Henan *d.* China **103 K4** 33.45N 113.00E
Henares *r.* Spain **46 D4** 40.26N 3.35W
Henderson I. Pacific Oc. **109 S4** 24.20S 128.20W
Hendon England **11 E2** 51.35N 0.14W
Henfield England **11 E1** 50.56N 0.17W
Hengelo Neth. **42 F4** 52.16N 6.46E
Hengoed Wales **12 D3** 51.39N 3.14W
Hengyang China **103 K3** 26.58N 112.31E
Heniches'k Ukraine **55 N2** 46.10N 34.49E
Henley-on-Thames England **10 E2** 51.32N 0.53W
Hennef Germany **42 G2** 50.47N 7.17E
Henrietta Maria, C. Canada **63 J3** 55.00N 82.15W
Henzada Myanmar **97 J4** 17.38N 95.35E
Herat Afghan. **95 J5** 34.21N 62.10E
Hereford England **10 C3** 52.04N 2.43W
Herefordshire *d.* England **9 D3** 52.04N 2.43W
Herm *i.* Channel Is. **13 Y9** 49.28N 2.27W
Herma Ness *c.* Scotland **19 Z9** 60.50N 0.54W
Hermosillo Mexico **70 B6** 29.15N 110.59W
Herne Germany **42 G3** 51.32N 7.12E
Herne Bay *town* England **11 G2** 51.23N 1.10E
Herning Denmark **43 B2** 56.08N 9.00E
Hertford England **11 E2** 51.48N 0.05W
Hertfordshire *d.* England **9 E2** 51.51N 0.05W
Heswall England **14 D2** 53.20N 3.06W
Hetton England **15 E3** 54.01N 2.05W
Hexham England **15 E3** 54.58N 2.06W
Heysham England **14 E3** 54.03N 2.53W
Heywood England **15 E2** 53.36N 2.13W
Hidaka-sammyaku *mts.* Japan **106 D4** .42.50N 143.00E
Hidalgo *d.* Mexico **70 E5** 20.50N 98.30W
Hidalgo del Parral Mexico **70 C6** 26.58N 105.40W
Higashi-suido *str.* Japan **106 A2** 34.00N 129.30E
Higham Ferrers England **11 E3** 52.18N 0.36W
High Atlas *mts.* Morocco **84 D5** 32.00N 5.50W
Highbridge England **13 E3** 51.13N 2.59W
Highclere England **10 D2** 51.22N 1.22W
Highland *d.* Scotland **8 C5** 57.42N 5.00W
High Peak *mtn.* England **15 F2** 53.22N 1.48W
High Seat *hill* England **15 E3** 54.23N 2.18W
Highworth England **10 D2** 51.38N 1.42W
High Wycombe England **10 E2** 51.38N 0.46W
Hiiumaa *i.* Estonia **43 E2** 58.50N 22.30E
Hijaz *f.* Saudi Arabia **94 E4** 26.00N 37.30E
Hildesheim Germany **48 D5** 52.09N 9.58E
Hillerød Denmark **43 C1** 55.56N 12.18E
Hillside Scotland **19 G1** 56.45N 2.29W
Hilpsford Pt. England **14 D3** 54.02N 3.10W
Hilversum Neth. **42 E4** 52.14N 5.12E
Himachal Pradesh *d.* India **96 F7** 31.45N 77.30E
Himalaya *mts.* Asia **97 G6** 29.00N 84.00E
Hinckley England **10 D3** 52.33N 1.21W
Hinderwell England **15 G3** 54.32N 0.46W
Hindhead England **10 E2** 51.06N 0.42W
Hindley England **14 E2** 53.32N 2.35W
Hindu Kush *mts.* Asia **95 K6** 36.40N 70.00E
Hinnøya *i.* Norway **43 C5** 68.30N 16.00E
Hiraman *r.* Kenya **87 B2** 1.05S 39.55E
Hirosaki Japan **106 D4** 40.34N 140.28E
Hiroshima Japan **106 B2** 34.30N 132.27E
Hirson France **42 D1** 49.56N 4.05E
Hirwaun Wales **12 D3** 51.43N 3.30W
Hispaniola *i.* C. America **71 J5** 20.00N 71.00W
Hitachi Japan **106 D3** 36.35N 140.40E
Hitchin England **11 E2** 51.57N 0.16W
Hitra *i.* Norway **43 B3** 63.30N 8.50E
Hiva Oa *i.* Marquesas Is. **109 R5** 9.45S 139.00W
Hjälmaren *l.* Sweden **43 C2** 59.10N 15.45E
Hjørring Denmark **43 B2** 57.28N 9.59E
Hlybokaye Belarus **55 J6** 55.07N 27.42E
Hobart Australia **110 D1** 42.54S 147.18E
Hobro Denmark **43 B2** 56.38N 9.48E
Hô Chi Minh City Vietnam **104 D6** 10.46N 106.43E
Hoddesdon England **11 E2** 51.46N 0.01W
Hodeida Yemen **94 F1** 14.50N 42.58E
Hodnet England **10 C3** 52.51N 2.35W
Hoek van Holland Neth. **42 D3** 51.59N 4.08E
Hof Germany **48 E4** 50.19N 11.56E
Höfn Iceland **43 Z2** 64.16N 15.10W
Hofsjökull *mtn.* Iceland **43 Y2** 64.50N 19.00W
Hofuf Saudi Arabia **95 G4** 25.20N 49.34E
Hoggar *mts.* Algeria **84 E4** 24.00N 5.50E
Hohhot China **103 K6** 40.49N 111.37E
Hokkaido *i.* Japan **106 D4** 43.00N 144.00E
Holbæk Denmark **43 B1** 55.42N 11.41E
Holbeach England **11 F3** 52.48N 0.01E
Holbeach Marsh England **11 F3** 52.50N 0.05E
Holbrook U.S.A. **64 E3** 34.58N 110.00W
Holderness *f.* England **15 G2** 53.45N 0.05W
Holguín Cuba **71 I5** 20.54N 76.15W
Hollabrunn Austria **54 E3** 48.34N 16.05E
Holland Fen *f.* England **15 G2** 53.02N 0.12W
Hollesley B. England **11 G3** 52.02N 1.33E
Hollington England **11 F1** 50.51N 0.32E
Hollingworth England **15 F2** 53.28N 1.59W
Holme-on-Spalding-Moor England **15 G2** 53.50N 0.47W
Holmfirth England **15 F2** 53.34N 1.48W
Holon Israel **94 D5** 32.01N 34.46E
Holstebro Denmark **43 B2** 56.22N 8.38E
Holsworthy England **13 C2** 50.48N 4.21W
Holt England **11 G3** 52.55N 1.04E
Holyhead Wales **12 C5** 53.18N 4.38W
Holyhead B. Wales **12 C5** 53.22N 4.40W
Holy I. England **15 F4** 55.41N 1.47W
Holy I. Wales **12 C5** 53.15N 4.38W
Holywell Wales **12 D5** 53.17N 3.13W
Homa Bay *town* Kenya **87 A2** 0.32S 34.27E
Homayunshahr Iran **95 H5** 32.42N 51.28E
Homburg Germany **42 G1** 49.19N 7.20E
Home B. Canada **63 L4** 69.00N 66.00W
Homs Syria **94 E5** 34.44N 36.43E
Homyel Belarus **55 L5** 52.25N 31.00E
Hondo *r.* Mexico **70 G4** 18.33N 88.22W
Honduras C. America **70 G4** 15.00N 87.00W
Honduras, G. of Carib. Sea **60 K3** 16.20N 87.30W
Hønefoss Norway **43 B3** 60.10N 10.16E
Hong Kong China **103 K2** 22.30N 114.10E
Honiara Solomon Is. **111 E5** 9.27S 159.57E
Honiton England **13 D2** 50.48N 3.13W
Honley England **15 F2** 53.34N 1.46W
Honolulu Hawaiian Is. **108 P9** 21.19N 157.50W
Honshu *i.* Japan **106 C3** 36.00N 138.00E
Hood, Mt. U.S.A. **64 B6** 45.23N 121.41W
Hood Pt. Australia **110 A2** 34.23S 119.34E
Hoogeveen Neth. **42 F4** 52.44N 6.29E
Hook England **10 E2** 51.17N 0.55W
Hook Head Rep. of Ire. **20 E2** 52.07N 6.55W
Hooper Bay *town* U.S.A. **62 B4** 61.29N 166.10W
Hoorn Neth. **42 E4** 52.38N 5.03E
Hopedale Canada **63 L3** 55.30N 60.10W
Hope, Loch Scotland **19 E3** 58.27N 4.38W
Hope, Pt. U.S.A. **59 V4** 68.00N 167.00W
Horley England **11 E2** 51.11N 0.11W
Horlivka Ukraine **55 P3** 48.17N 38.05E
Hormuz, Str. of Asia **95 I4** 26.35N 56.20E
Horn *c.* Iceland **43 X2** 66.28N 22.27W
Horn, C. S. America **75 C1** 55.47S 67.00W
Hornavan *l.* Sweden **43 D4** 66.15N 17.40E
Horncastle England **15 G2** 53.13N 0.08W
Hornepayne Canada **65 J6** 49.14N 84.48W
Hornsea England **15 G2** 53.55N 0.10W
Horodnya Ukraine **55 L4** 51.54N 31.37E
Horodok Ukraine **55 H3** 49.48N 23.39E
Horrabridge England **13 C2** 50.30N 4.05W
Horsens Denmark **43 B1** 55.53N 9.53E
Horsham Australia **110 D2** 36.45S 142.15E
Horsham England **11 E2** 51.04N 0.20W
Horten Norway **43 B2** 59.25N 10.30E
Horwich England **14 E2** 53.37N 2.33W
Hospitalet de Llobregat Spain **46 G4** 41.20N 2.06E
Hotan China **102 E5** 37.07N 79.57E
Houffalize Belgium **42 E2** 50.08N 5.50E
Houghton-le-Spring England **15 F3** 54.51N 1.28W
Houghton Regis England **11 E2** 51.51N 0.30W
Hourn, Loch Scotland **18 D2** 57.05N 5.35W
Houston U.S.A. **65 G2** 29.45N 95.25W
Hovd Mongolia **102 G7** 48.00N 91.45E
Hove England **11 E1** 50.50N 0.10W
Hoveton England **11 G3** 52.45N 1.23E
Howden England **15 G2** 53.45N 0.52W
Howland I. Pacific Oc. **108 N7** 0.48N 176.38W
Hoy *i.* Scotland **19 F3** 58.51N 3.17W
Hoyanger Norway **43 A3** 61.13N 6.05E
Hoyerswerda Germany **48 G4** 51.26N 14.14E
Hövsgöl Nuur *l.* Mongolia **103 I8** 51.00N 100.30E
Hradec Králové Czech Rep. **54 D4** 50.13N 15.50E
Huacho Peru **76 C3** 11.05S 77.36W
Huaibei China **103 L4** 33.58N 116.50E
Huainan China **103 L4** 32.41N 117.06E
Huallaga *r.* Peru **76 C4** 5.05S 75.36W
Huambo Angola **86 A3** 12.47S 15.44E
Huancayo Peru **76 C3** 12.15S 75.12W
Huang He *r.* China **103 L5** 37.55N 118.46E
Huangshi China **103 L4** 30.13N 115.05E
Huascaran *mtn.* Peru **74 B6** 9.20S 77.36W
Hubei *d.* China **103 K4** 31.15N 112.15E
Hubli India **96 E4** 15.20N 75.14E
Hucknall England **15 F2** 53.03N 1.12W
Huddersfield England **15 F2** 53.38N 1.49W
Hudiksvall Sweden **43 D3** 61.45N 17.10E
Hudson *r.* U.S.A. **65 L5** 40.45N 74.00W
Hudson B. Canada **63 J3** 58.00N 86.00W

Kotzebue U.S.A. **62 B4** 66.51N 162.40W
Kouvola Finland **43 F3** 60.54N 26.45E
Kowloon China **103 K2** 22.30N 114.10E
Kovel' Ukraine **55 I4** 51.12N 24.48E
Kozan Turkey **57 L2** 37.27N 35.47E
Kozani Greece **56 E4** 40.18N 21.48E
Kozhikode India **96 F3** 11.15N 75.45E
Krabi Thailand **104 B5** 8.04N 98.52E
Krâcheh Cambodia **104 D6** 12.30N 106.03E
Kragujevac Yugo. **56 E6** 44.01N 20.55E
Kraków Poland **54 F4** 50.03N 19.55E
Kraljevo Yugo. **56 E5** 43.44N 20.41E
Kramators'k Ukraine **55 O3** 48.43N 37.33E
Kramfors Sweden **43 D3** 62.55N 17.50E
Kranj Slovenia **54 D2** 46.15N 14.21E
Krasnoarmiys'k Ukraine **55 O3** 48.17N 37.14E
Krasnodar Russian Fed. **58 F2** 45.02N 39.00E
Krasnodar Resr. Russian Fed. **57 N6** 45.00N 39.15E
Krasnohrad Ukraine **55 N3** 49.22N 35.28E
Krasnokamensk Russian Fed. **59 N3** 50.10N 118.00E
Krasnovodsk Turkmenistan **58 H2** 40.01N 53.00E
Krasnoyarsk Russian Fed. **59 L3** 56.05N 92.46E
Krefeld Germany **48 C4** 51.20N 6.32E
Kremenchuk Ukraine **55 M3** 49.03N 33.25E
Kremenchuk Resr. Ukraine **55 M3** 49.20N 32.30E
Krishna *r.* India **97 G4** 16.00N 81.00E
Kristiansand Norway **43 A2** 58.08N 7.59E
Kristianstad Sweden **43 C2** 56.02N 14.10E
Kristiansund Norway **43 A3** 63.15N 7.55E
Kristinehamn Sweden **43 C2** 59.17N 14.09E
Krk *i.* Croatia **56 B6** 45.04N 14.36E
Kropotkin Russian Fed. **53 L6** 45.25N 40.35E
Krosno Poland **54 G3** 49.42N 21.46E
Kruševac Yugo. **56 E5** 43.34N 21.20E
Krychaw Belarus **55 L5** 53.40N 31.44E
Krymsk Russian Fed. **57 M6** 44.56N 38.00E
Kryvyy Rih' Ukraine **55 M2** 47.55N 33.24E
Ksar El Boukhari Algeria **52 D4** 35.53N 2.45E
Ksar el Kebir Morocco **52 B3** 35.01N 5.54W
Kuala Lumpur Malaysia **104 C4** 3.08N 101.42E
Kuala Terengganu Malaysia **104 C5** 5.10N 103.10E
Kuantan Malaysia **104 C4** 3.50N 103.19E
Kuban *r.* Russian Fed. **57 M6** 45.20N 37.17E
Kuching Malaysia **104 E4** 1.32N 110.20E
Kugluktuk Canada **62 G4** 67.49N 115.12W
Kuh-e Baba Afghan. **95 K5** 34.40N 67.30E
Kuh-e Dinar *mtn.* Iran **95 H5** 30.45N 51.39E
Kuh-e Sahand *mtn.* Iran **95 G6** 37.37N 46.27E
Kuhmo Finland **43 G4** 64.04N 29.30E
Kuito Angola **86 A3** 12.25S 16.58E
Kūl *r.* Iran **95 I4** 27.00N 55.45E
Kulal, Mt. Kenya **87 B3** 2.44N 36.56E
Kuldiga Latvia **43 E2** 56.58N 21.59E
Külob Tajikistan **95 K6** 37.55N 69.47E
Kulunda Russian Fed. **102 D8** 52.34N 78.58E
Kumamoto Japan **106 B2** 32.50N 130.42E
Kumanovo Macedonia **56 E5** 42.08N 21.40E
Kumasi Ghana **84 D2** 6.45N 1.35W
Kumbakonam India **97 F3** 10.59N 79.24E
Kumi Uganda **87 A3** 1.26N 33.54E
Kumla Sweden **43 C2** 59.08N 15.09E
Kumluca Turkey **57 J2** 36.23N 30.17E
Kumo Nigeria **84 F3** 10.02N 11.50E
Kunashir *i.* Russian Fed. **106 E4** 44.25N 146.00E
Kungrad Uzbekistan **58 H2** 43.05N 58.23E
Kunlun Shan *mts.* China **102 E5** 36.40N 85.00E
Kunming China **103 I3** 25.04N 102.41E
Kuohijärvi *l.* Finland **43 F3** 61.20N 24.10E
Kuopio Finland **43 F3** 62.51N 27.30E
Kupa *r.* Croatia **56 C6** 45.30N 16.20E
Kupang Indonesia **105 G1** 10.13S 123.38E
Kup"yans'k Ukraine **55 O3** 49.41N 37.37E
Kuqa China **102 E6** 41.43N 82.58E
Kŭrdzhali Bulgaria **56 G4** 41.39N 25.22E
Kure Atoll Hawaiian Is. **108 N9** 28.25N 178.25W
Kuressaare Estonia **43 E2** 58.12N 22.30E
Kurgan Russian Fed. **58 I3** 55.20N 65.20E
Kuria Muria Is. Oman **95 I2** 17.30N 56.00E
Kurikka Finland **43 E3** 62.37N 22.25E
Kuril Is. Russian Fed. **59 R2** 46.00N 150.30E
Kuril Trench *f.* Pacific Oc. **117 Q8**
Kurmuk Sudan **85 H3** 10.33N 34.17E
Kurnool India **97 F4** 15.51N 78.01E
Kursk Russian Fed. **55 O4** 51.45N 36.14E
Kuršumlija Yugo. **56 E5** 43.09N 21.16E
Kurume Japan **106 B2** 33.20N 130.29E
Kurun *r.* Sudan **87 A4** 5.40N 33.50E
Kurunegala Sri Lanka **97 G2** 7.28N 80.23E
Kushiro Japan **106 D4** 42.58N 144.24E
Kuskokwim *r.* U.S.A. **62 B4** 60.50N 161.20W
Kuskokwim B. U.S.A. **62 B3** 59.45N 162.25W
Kuskokwim Mts. U.S.A. **62 C4** 62.50N 156.00W
Kustanay Kazakhstan **58 I3** 53.15N 63.40E
Kütahya Turkey **57 I3** 39.25N 29.56E
Kuujjuaq Canada **63 L3** 58.10N 68.15W
Kuusamo Finland **43 G4** 65.57N 29.15E
Kuwait Asia **95 G4** 29.20N 47.40E
Kuwait *town* Kuwait **95 G4** 29.20N 48.00E
Kuznetsk Russian Fed. **58 G3** 53.08N 46.36E
Kvaløya *i.* Norway **43 D5** 69.45N 18.20E
Kwale Kenya **87 B2** 4.10S 39.27E
KwaMashu R.S.A. **86 C2** 29.45S 30.56E
Kwa Mtoro Tanzania **87 B1** 5.14S 35.22E
Kwangju S. Korea **103 N5** 35.07N 126.52E
Kwania, L. Uganda **87 A3** 1.48N 32.45E
Kwanobuhle R.S.A. **86 B1** 33.30S 25.26E
Kwilu *r.* Dem. Rep. of Congo **86 A4** 3.18S 17.22E
Kwoka *mtn.* Indonesia **105 I3** 1.30S 132.30E
Kyakhta Russian Fed. **103 J8** 50.22N 106.30E
Kyle of Durness *est.* Scotland **18 E3** 58.32N 4.50W
Kyle of Lochalsh *town* Scotland **18 D2** 57.17N 5.43W
Kyle of Tongue *est.* Scotland **19 E3** 58.27N 4.26W
Kyllini *mtn.* Greece **56 F2** 37.56N 22.22E
Kyoga, L. Uganda **86 C5** 1.30N 33.00E
Kyoto Japan **106 C2** 35.04N 135.50E
Kyparissia Greece **56 E2** 37.15N 21.40E
Kyrgyzstan Asia **102 C6** 41.30N 75.00E
Kythira *i.* Greece **56 F2** 36.15N 23.00E
Kythnos *i.* Greece **56 G2** 37.25N 24.25E
Kyushu *i.* Japan **106 B2** 32.00N 131.00E
Kyustendil Bulgaria **56 F5** 42.18N 22.39E
Kyzyl Russian Fed. **59 L3** 51.42N 94.28E
Kzyl-Orda Kazakhstan **58 J2** 44.52N 65.28E

L

Laâyoune W. Sahara **84 C4** 27.10N 13.11W
La Banda Argentina **75 C5** 27.44S 64.14W
Labé Guinea **84 C3** 11.17N 12.11W
Labrador *f.* Canada **63 L3** 54.00N 61.30W
Labrador City Canada **63 L3** 52.54N 66.50W
Labrador Sea Canada/Greenland **63 M4** 60.00N 55.00W
Laccadive Is. Indian Oc. **96 E3** 11.00N 72.00E
Laceby England **15 G2** 53.33N 0.10W
La Chaux-de-Fonds Switz. **44 G5** 47.07N 6.51E
Lachlan *r.* Australia **110 D2** 34.21S 143.58E
La Crosse U.S.A. **65 H5** 43.48N 91.15W
La Demanda, Sierra de *mts.* Spain **46 D5** 42.10N 3.20W
Ladhar Bheinn *mtn.* Scotland **18 D2** 57.04N 5.35W
Ladiz Iran **95 J4** 28.57N 61.18E
Ladoga, L. Russian Fed. **58 F4** 61.00N 32.00E
Ladybank Scotland **17 F4** 56.17N 3.08W
Ladybower Resr. England **15 F2** 53.23N 1.42W
Lae P.N.G. **110 D5** 6.45S 146.30E
Læsø *i.* Denmark **43 B2** 57.16N 11.01E
Lafayette U.S.A. **65 H3** 30.12N 92.18W
Lagan *r.* N. Ireland **16 D2** 54.37N 5.54W
Laggan, Loch Scotland **19 E1** 56.57N 4.30W
Lagh Bogal *r.* Kenya **87 B3** 1.04N 40.10E
Lagh Bor *r.* Kenya/Somalia **87 C3** 0.30N 42.04E
Lagh Dima *r.* Kenya/Somalia **87 C3** 0.30N 41.09E
Lagh Kutulu *r.* Kenya **87 C3** 2.10N 40.53E
Laghouat Algeria **52 D3** 33.49N 2.55E
Lagh Walde Kenya **87 B3** 2.48N 38.54E
La Gomera *i.* Canary Is. **46 X2** 28.08N 17.14W
Lagos Nigeria **84 E2** 6.27N 3.28E
Lagos Portugal **46 A2** 37.05N 8.40W
La Grande 2, Résr de Canada **63 K3** 53.35N 77.10W
La Grande 3, Résr de Canada **63 K3** 53.35N 74.55W
La Grande 4, Résr de Canada **63 K3** 53.50N 73.30W
La Gran Sabana *f.* Venezuela **71 L2** 5.20N 61.30W
Lahad Datu Malaysia **104 F5** 5.05N 118.20E
Lahat Indonesia **104 C3** 3.46S 103.32E
Laḥij Yemen **94 F1** 13.04N 44.53E
Lahore Pakistan **96 E7** 31.34N 74.22E
Lahti Finland **43 F3** 61.00N 25.40E
Laidon, Loch Scotland **16 E4** 56.39N 4.38W
Laihia Finland **43 E3** 62.58N 22.01E
Lainio *r.* Sweden **43 E4** 67.26N 22.37E
Lairg Scotland **19 E3** 58.01N 4.25W
Laisamis Kenya **87 B3** 1.38N 37.47E
Lajes Brazil **77 E2** 27.48S 50.20W
La Junta U.S.A. **64 F4** 37.59N 103.34W
Lake City U.S.A. **65 J3** 30.05N 82.40W
Lake District *f.* England **14 D3** 54.30N 3.10W
Lakeland *town* U.S.A. **65 J2** 28.02N 81.59W
Lakeview U.S.A. **64 B5** 42.13N 120.21W
Lakki Pakistan **95 L5** 32.35N 70.58E
Lakonia, G. of Greece **56 F2** 36.35N 22.42E
Laksefjorden *est.* Norway **43 F5** 70.40N 26.50E
Lakshadweep *d.* India **96 E3** 11.00N 72.00E
La Línea de la Concepción Spain **46 C2** 36.10N 5.21W
La Louvière Belgium **42 D2** 50.29N 4.11E
La Mancha *f.* Spain **46 D3** 39.00N 2.30W
Lamar U.S.A. **64 F4** 38.04N 102.37W
Lamard Iran **95 H4** 27.22N 53.20E
Lambay I. Rep. of Ire. **20 E3** 53.30N 6.01W
Lambourn England **10 D2** 51.31N 1.31W
Lambourn Downs *hills* England **10 D2** 51.32N 1.36W
Lamego Portugal **46 B4** 41.05N 7.49W
Lamia Greece **56 F3** 38.53N 22.25E
Lamlash Scotland **16 D3** 55.32N 5.08W
Lammermuir Hills Scotland **17 G3** 55.51N 2.40W
Lampang Thailand **104 B7** 18.16N 99.30E
Lampeter Wales **12 C4** 52.06N 4.06W
Lamu Kenya **87 A2** 2.20S 40.54E
Lamu I. Kenya **87 A2** 2.17S 40.52E
La Nao, Cabo de *c.* Spain **46 F3** 38.42N 0.15E
Lanark Scotland **17 F3** 55.41N 3.47W
Lancashire *d.* England **9 D3** 53.53N 2.30W
Lancaster England **14 E3** 54.03N 2.48W
Lancaster Canal England **14 E2** 54.00N 2.48W
Lancaster Sd. Canada **63 J5** 74.00N 85.00W
Landes *f.* France **44 C4** 44.40N 0.40W
Landguard Pt. England **11 G2** 51.55N 1.18E
Land's End *c.* England **13 B2** 50.03N 5.45W
Landshut Germany **48 F3** 48.31N 12.10E
Lanesborough Rep. of Ire. **20 D3** 53.40N 8.00W
Langavat, Loch Scotland **18 C2** 57.48N 6.58W
Langavat, Loch Scotland **18 C3** 58.04N 6.45W
Langeland *i.* Denmark **43 B1** 54.50N 10.50E
Langholm Scotland **17 F3** 55.09N 3.00W
Langøya *i.* Norway **43 C5** 68.50N 15.00E
Langport England **13 E3** 51.02N 2.51W
Langres France **44 F5** 47.53N 5.20E
Langsa Indonesia **104 B4** 4.28N 97.59E
Langstrothdale Chase *hills* England **15 E3** 54.13N 2.15W
Languedoc *f.* France **44 E3** 43.30N 3.00E
Lanivet England **13 C2** 50.27N 4.45W
Länkäran Azerbaijan **95 G6** 38.45N 48.50E
Lannion France **44 B6** 48.44N 3.27W
Lansing U.S.A. **65 J5** 42.44N 84.34W
Lanzarote *i.* Canary Is. **46 Z2** 29.00N 13.55W
Lanzhou China **103 I5** 36.01N 103.45E
Laoag Phil. **104 G7** 18.14N 120.36E
Lao Cai Vietnam **104 C8** 22.30N 104.00E
Laois *d.* Rep. of Ire. **20 D2** 53.00N 7.20W
Laon France **42 C1** 49.34N 3.37E
Laos Asia **104 C7** 19.00N 104.00E
La Palma *i.* Canary Is. **46 X2** 28.50N 18.00W
La Palma del Condado Spain **46 B2** 37.23N 6.33W
La Paz Bolivia **76 D3** 16.30S 68.10W
La Paz Mexico **70 B5** 24.10N 110.17W
La Perouse Str. Russian Fed. **106 D5** 45.50N 142.30E
La Plata Argentina **75 D3** 34.52S 57.55W
La Plata, Río de *est.* S. America **75 D3** 35.15S 56.45W
Lappajärvi *l.* Finland **43 E3** 63.05N 23.30E
Lappeenranta Finland **43 G3** 61.04N 28.05E
Lappland *f.* Sweden/Finland **43 E5** 68.10N 24.00E
Laptev Sea Russian Fed. **59 O5** 74.30N 125.00E
Lapua Finland **43 E3** 62.57N 23.00E
L'Aquila Italy **50 E5** 42.22N 13.25E
Larache Morocco **46 B1** 35.12N 6.10W
Laramie U.S.A. **64 E5** 41.20N 105.38W
Laredo U.S.A. **64 G2** 27.32N 99.22W
Largs Scotland **16 E3** 55.48N 4.52W
La Rioja Argentina **76 D2** 29.26S 66.50W
Larisa Greece **56 F3** 39.36N 22.24E
Larkana Pakistan **96 D6** 27.32N 68.18E
Larnaca Cyprus **57 K1** 34.54N 33.39E
Larne N. Ireland **16 D2** 54.51N 5.50W
Larne Lough N. Ireland **16 D2** 54.50N 5.47W
La Rochelle France **44 C5** 46.10N 1.10W
La Roche-sur-Yon France **44 C5** 46.40N 1.25W
La Romana Dom. Rep. **71 K4** 18.27N 68.57W
La Ronge Canada **62 H3** 55.07N 105.18W
Larvik Norway **43 B2** 59.04N 10.02E
La Sagra *mtn.* Spain **46 D2** 37.58N 2.35W
Las Cruces U.S.A. **64 E3** 32.18N 106.47W
La Seine, Baie de France **44 C6** 49.40N 0.30W
La Selle *mtn.* Haiti **71 J4** 18.23N 71.59W
La Serena Chile **76 C2** 29.54S 71.16W
Las Marismas *f.* Spain **46 B2** 37.05N 6.20W
Las Nieves, Pico de *mtn.* Canary Is. **46 Y1** 27.56N 15.34W
Las Palmas de Gran Canaria Canary Is. **46 Y2** 28.08N 15.27W
La Spezia Italy **50 C6** 44.07N 9.49E
Las Vegas U.S.A. **64 C4** 36.10N 115.10W
Latacunga Ecuador **76 C4** 0.58S 78.36W
Latakia Syria **94 E6** 35.31N 35.47E
Latheron Scotland **19 F3** 58.17N 3.22W
Latina Italy **50 E4** 41.28N 12.52E
La Tortuga *i.* Venezuela **71 K3** 11.00N 65.20W
Latvia Europe **43 E2** 57.00N 25.00E
Launceston Australia **110 D1** 41.25S 147.07E
Launceston England **13 C2** 50.38N 4.21W
Laurencekirk Scotland **19 G1** 56.50N 2.30W
Lausanne Switz. **44 G5** 46.32N 6.39E
Laut *i.* Indonesia **104 F3** 3.45S 116.20E
Laval France **44 C6** 48.04N 0.45W
Lavenham England **11 F3** 52.06N 0.47E
Lawdar Yemen **95 G1** 13.53N 45.53E
Laxey I.o.M. **14 C3** 54.14N 4.24W
Laxford, Loch Scotland **18 D3** 58.25N 5.06W
Laysan I. Hawaiian Is. **108 N9** 25.46N 171.44W
Leane, L. Rep. of Ire. **20 B2** 52.03N 9.33W
Leatherhead England **11 E2** 51.18N 0.20W
Lebanon Asia **94 E5** 34.00N 36.00E
Lębork Poland **54 E6** 54.33N 17.44E
Lecce Italy **50 H4** 40.21N 18.11E
Lech *r.* Germany **48 E3** 48.45N 10.51E
Lechlade England **10 D2** 51.42N 1.40W
Ledbury England **10 C3** 52.03N 2.25W
Lee *r.* Rep. of Ire. **20 C1** 51.53N 8.25W
Leech L. U.S.A. **65 H6** 47.10N 94.30W
Leeds England **15 F2** 53.48N 1.34W
Leek England **15 E2** 53.07N 2.02W
Leeming England **15 F3** 54.18N 1.33W
Lee Moor *town* England **13 C2** 50.25N 4.01W
Leer Germany **48 C5** 53.14N 7.27E
Leeuwarden Neth. **42 E5** 53.12N 5.48E
Leeuwin, C. Australia **110 A2** 34.00S 115.00E
Leeward Is. C. America **71 L4** 18.00N 61.00W
Lefkada *i.* Greece **56 E3** 38.44N 20.37E
Legaspi Phil. **105 G6** 13.10N 123.45E
Legnica Poland **54 E4** 51.12N 16.10E
Leh Jammu & Kashmir **97 F7** 34.09N 77.35E
Le Havre France **44 D6** 49.30N 0.06E
Leiah Pakistan **95 L5** 30.58N 70.56E
Leicester England **10 D3** 52.39N 1.09W
Leicestershire *d.* England **9 E3** 52.30N 1.00W
Leichhardt *r.* Australia **110 C4** 17.35S 139.48E
Leiden Neth. **42 D4** 52.10N 4.30E
Leigh England **15 E2** 53.30N 2.33W
Leighton Buzzard England **10 E2** 51.55N 0.39W
Leinster, Mt. Rep. of Ire. **20 E2** 52.37N 6.47W
Leipzig Germany **48 F4** 51.20N 12.20E
Leiston England **11 G3** 52.13N 1.35E
Leith Scotland **17 F3** 55.59N 2.09W
Leith Hill England **11 E2** 51.11N 0.21W
Leitrim *d.* Rep. of Ire. **20 D4** 54.08N 8.00W
Leixlip Rep. of Ire. **20 E3** 53.22N 6.30W
Leizhou Pen. China **103 J2** 20.40N 109.30E
Lek *r.* Neth. **42 D3** 51.55N 4.29E
Leksozero, Ozero *l.* Russian Fed. **43 G3** 63.40N 30.52E
Lelystad Neth. **42 E4** 52.32N 5.29E
Le Mans France **44 D5** 48.01N 0.10E
Lena *r.* Russian Fed. **59 O5** 72.00N 127.10E
Leninsk-Kuznetskiy Russian Fed. **58 K3** 54.44N 86.13E
Lens France **44 E7** 50.26N 2.50E
Leoben Austria **54 D2** 47.23N 15.06E
Leominster England **10 C3** 52.15N 2.43W
León Mexico **70 D5** 21.10N 101.42W
León Spain **46 C5** 42.35N 5.34W
Leova Moldova **55 K2** 46.29N 28.12E
Le-Puy-en-Velay France **44 E4** 45.03N 3.54E
Lerum Sweden **43 C2** 57.46N 12.12E
Lerwick Scotland **19 Y9** 60.09N 1.09W
Lesatima *mtn.* Kenya **87 B2** 0.17S 36.37E
Les Cayes Haiti **71 J4** 18.15N 73.46W
Leshan China **103 I3** 29.34N 103.42E
Leskovac Yugo. **56 E5** 43.00N 21.56E
Lesotho Africa **86 B2** 29.30S 28.00E
Les Sables-d'Olonne France **44 C5** 46.30N 1.47W
Lesser Antilles *is.* C. America **71 K3** 13.00N 67.00W
Lesser Slave L. Canada **62 G3** 55.30N 115.00W
Lestijärvi *l.* Finland **43 F3** 63.32N 24.40E
Lesvos *i.* Greece **56 G3** 39.10N 26.16E
Leszno Poland **54 E4** 51.51N 16.35E
Letchworth England **11 E2** 51.58N 0.13W
Lethbridge Canada **62 G2** 49.43N 112.48W
Leti Is. Indonesia **105 H2** 8.20S 128.00E
Le Touquet-Paris-Plage France **42 A2** 50.31N 1.36E
Letterkenny Rep. of Ire. **20 D4** 54.56N 7.45W
Leuchars Scotland **17 G4** 56.23N 2.53W
Leuser, Gunung *mtn.* Indonesia **104 B4** 3.50N 97.10E
Leuven Belgium **42 D2** 50.53N 4.45E
Leven England **15 G2** 53.54N 0.18W
Leven Scotland **17 G4** 56.12N 3.00W
Leven, Loch A. and B. Scotland **18 D1** 56.43N 5.05W
Leven, Loch P. and K. Scotland **17 F4** 56.13N 3.23W
Levens England **14 E3** 54.16N 2.47W
Leverkusen Germany **48 C4** 51.02N 6.59E
Lévis Canada **65 L6** 46.47N 71.12W
Lewes England **11 F1** 50.53N 0.02E
Lewis *i.* Scotland **18 C3** 58.10N 6.40W
Lexington U.S.A. **65 J4** 38.02N 84.30W
Leyburn England **15 F3** 54.19N 1.50W
Leyland England **14 E2** 53.41N 2.42W
Leyte *i.* Phil. **105 G6** 10.40N 124.50E
L'gov Russian Fed. **55 N4** 51.41N 35.16E
Lhasa China **102 G3** 29.41N 91.10E
Lhazê China **102 F3** 29.08N 87.43E
Lhokseumawe Indonesia **104 B5** 5.09N 97.09E
Lianyungang China **103 L4** 34.37N 119.10E
Liaoning *d.* China **103 M6** 41.30N 123.00E
Liaoyuan China **103 M6** 42.53N 125.10E
Liard *r.* Canada **62 F4** 61.56N 120.35W
Libenge Dem. Rep. of Congo **86 A5** 3.39N 18.39E
Liberal U.S.A. **64 F4** 37.03N 100.56W
Liberec Czech Rep. **54 D4** 50.48N 15.05E
Liberia Africa **84 D2** 6.30N 9.30W
Libin Belgium **42 E1** 49.58N 5.15E
Liboi Kenya **87 C3** 0.23N 40.50E
Libourne France **44 C4** 44.55N 0.14W
Libreville Gabon **84 E2** 0.30N 9.25E
Libya Africa **84 F4** 26.30N 17.00E
Libyan Desert Africa **85 G4** 23.00N 26.10E
Libyan Plateau *f.* Africa **94 C5** 30.45N 26.00E
Lichfield England **10 D3** 52.40N 1.50W
Lichinga Mozambique **86 C3** 13.19S 35.13E
Lida Belarus **55 I5** 53.50N 25.19E
Liddel *r.* England/Scotland **17 G3** 55.04N 2.57W
Lidköping Sweden **43 C2** 58.30N 13.10E
Liechtenstein Europe **44 H5** 47.08N 9.35E
Liège Belgium **42 E2** 50.38N 5.35E
Liège *d.* Belgium **42 E2** 50.32N 5.35E
Lieksa Finland **43 G3** 63.13N 30.01E
Lienz Austria **54 C2** 46.50N 12.47E
Liepaja Latvia **43 E2** 56.30N 21.00E
Liévin France **42 B2** 50.27N 2.49E
Liffey *r.* Rep. of Ire. **20 E3** 53.21N 6.14W
Lifford Rep. of Ire. **20 D4** 54.50N 7.29W
Ligurian Sea Med. Sea **50 C5** 43.10N 9.00E
Likasi Dem. Rep. of Congo **86 B3** 10.58S 26.47E
Lilla Edet Sweden **43 C2** 58.10N 12.25E
Lille Belgium **42 D3** 51.14N 4.50E
Lille France **44 E7** 50.39N 3.05E
Lillehammer Norway **43 B3** 61.06N 10.27E
Lilleshall England **10 C3** 52.44N 2.24W
Lillestrøm Norway **43 B2** 59.58N 11.05E
Lilongwe Malawi **86 C3** 13.58S 33.49E
Lima Peru **76 C3** 12.06S 77.03W
Limassol Cyprus **57 K1** 34.40N 33.03E
Limavady N. Ireland **16 C3** 55.03N 6.57W
Limburg *d.* Belgium **42 E3** 51.00N 5.30E
Limburg *d.* Neth. **42 E3** 51.15N 5.45E
Limeira Brazil **77 F2** 22.34S 47.25W
Limerick Rep. of Ire. **20 C2** 52.40N 8.37W
Limerick *d.* Rep. of Ire. **20 C2** 52.30N 8.37W
Limingen *l.* Norway **43 C4** 64.50N 13.40E
Limnos *i.* Greece **56 G3** 39.55N 25.14E
Limoges France **44 D4** 45.50N 1.15E
Limousin *f.* France **44 D4** 45.30N 1.30E
Limoux France **44 E3** 43.04N 2.14E
Limpopo *r.* Mozambique **86 C2** 25.14S 33.33E
Linares Spain **46 D3** 38.05N 3.38W
Lincoln England **15 G2** 53.14N 0.32W
Lincoln U.S.A. **64 G5** 40.49N 96.41W
Lincolnshire *d.* England **9 E3** 53.10N 0.32W
Lincolnshire Wolds *hills* England **15 G2** 53.22N 0.08W
Lindesnes *c.* Norway **43 A2** 58.00N 7.05E
Lindfield England **11 E2** 51.01N 0.05W
Lindi Tanzania **86 C4** 10.00S 39.41E
Lindos Greece **57 I2** 36.05N 28.05E
Line Is. Pacific Oc. **108 P6** 3.00S 155.00W
Linfen China **103 K5** 36.07N 111.34E
Lingen Germany **48 C5** 52.32N 7.19E
Lingfield England **11 E2** 51.11N 0.01W
Linhares Brazil **77 F3** 19.22S 40.04W
Linköping Sweden **43 C2** 58.25N 15.35E
Linkou China **106 B5** 45.18N 130.17E
Linlithgow Scotland **17 F3** 55.58N 3.36W
Linney Head Wales **12 B3** 51.37N 5.05W
Linnhe, Loch Scotland **16 D4** 56.35N 5.25W
Linosa *i.* Italy **50 E1** 35.52N 12.50E
Linxia China **103 I5** 35.31N 103.08E
Linz Austria **54 D3** 48.19N 14.18E
Lions, G. of France **44 F3** 43.00N 4.15E
Lipari Is. Italy **50 F3** 38.35N 14.45E
Lipetsk Russian Fed. **55 P5** 52.37N 39.36E
Liphook England **10 E2** 51.05N 0.49W
Lipova Romania **54 G2** 46.05N 21.40E
Lippe *r.* Germany **48 C4** 51.38N 6.37E
Lisala Dem. Rep. of Congo **86 B5** 2.08N 21.37E
Lisbon Portugal **46 A3** 38.44N 9.08W
Lisburn N. Ireland **16 C2** 54.31N 6.03W
Liscannor B. Rep. of Ire. **20 B2** 52.55N 9.25W
Liskeard England **13 C2** 50.27N 4.29W
Lismore Australia **110 E3** 28.48S 153.17E
Lismore Rep. of Ire. **20 D2** 52.08N 7.59W
Lisnaskea N. Ireland **16 B2** 54.15N 7.28W
Liss England **10 E2** 51.03N 0.53W
Listowel Rep. of Ire. **20 B2** 52.27N 9.30W
Lithuania Europe **55 H6** 55.00N 24.00E
Little Andaman *i.* India **97 I3** 10.50N 92.38E
Littleborough England **15 E2** 53.39N 2.05W
Little Cayman *i.* Cayman Is. **71 H4** 19.40N 80.00W
Little Dart *r.* England **13 D2** 50.54N 3.53W
Little Falls *town* U.S.A. **65 H6** 45.59N 94.21W
Littlehampton England **11 E1** 50.48N 0.32W
Little Inagua *i.* The Bahamas **71 J5** 21.30N 73.00W
Little Karoo *f.* R.S.A. **86 B1** 33.40S 21.40E
Little Minch *str.* Scotland **18 C2** 57.40N 6.45W
Little Ouse *r.* England **11 F3** 52.34N 0.20E
Littleport England **11 F3** 52.27N 0.18E
Little Rock *town* U.S.A. **65 H3** 34.42N 92.17W
Liupanshui China **103 I3** 26.50N 104.45E
Liuzhou China **103 J2** 24.17N 109.15E
Līvāni Latvia **55 J7** 56.20N 26.12E
Liverpool England **14 E2** 53.25N 3.00W
Livingston Scotland **17 F3** 55.54N 3.31W
Livingstone Zambia **86 B3** 17.50S 25.53E
Livny Russian Fed. **55 O5** 52.25N 37.35E
Livorno Italy **50 D5** 43.33N 10.18E
Lizard England **13 B1** 49.58N 5.12W
Lizard Pt. England **13 B1** 49.57N 5.15W
Ljubljana Slovenia **54 D2** 46.04N 14.28E
Ljungan *r.* Sweden **43 D3** 62.20N 17.19E
Ljungby Sweden **43 C2** 56.49N 13.55E
Ljusdal Sweden **43 D3** 61.49N 16.09E
Llanarth Wales **12 C4** 52.12N 4.18W
Llanbadarn Fawr Wales **12 C4** 52.24N 4.05W
Llanberis Wales **12 C5** 53.07N 4.07W

Llandeilo Wales 12 D3 51.54N 4.00W
Llandovery Wales 12 D3 51.59N 3.49W
Llandrindod Wells Wales 12 D4 52.15N 3.23W
Llandudno Wales 12 D5 53.19N 3.49W
Llandwrog Wales 12 C5 53.04N 4.20W
Llandysul Wales 12 C4 52.03N 4.20W
Llanegwad Wales 12 C3 51.53N 4.09W
Llanelli Wales 12 C3 51.41N 4.11W
Llanes Spain 46 C5 43.25N 4.45W
Llanfair Caereinion Wales 12 D4 52.39N 3.20W
Llanfairfechan Wales 12 D5 53.15N 3.58W
Llanfairpwllgwyngyll Wales 12 C5 53.12N 4.13W
Llanfair-yn-Neubwll Wales 12 C5 53.15N 4.33W
Llanfyllin Wales 12 D4 52.47N 3.17W
Llangadog Wales 12 D3 51.56N 3.53W
Llangefni Wales 12 C5 53.15N 4.20W
Llangeler Wales 12 C4 52.01N 4.24W
Llangelynin Wales 12 C4 52.38N 4.06W
Llangoed Wales 12 C5 53.18N 4.05W
Llangollen Wales 12 D4 52.58N 3.10W
Llanidloes Wales 12 D4 52.28N 3.31W
Llanllwchaiarn Wales 12 D4 52.33N 3.17W
Llanllyfni Wales 12 C5 53.03N 4.17W
Llanos *f.* Colombia/Venezuela 74 B7 5.30N 72.00W
Llanos de Mojos *f.* Bolivia 76 D3 15.00S 65.00W
Llanrumney Wales 13 D3 51.32N 3.07W
Llanrwst Wales 12 D5 53.08N 3.48W
Llansannan Wales 12 D5 53.11N 3.35W
Llantrisant Wales 12 D3 51.33N 3.23W
Llantwit Major Wales 13 D3 51.24N 3.29W
Llanwnda Wales 12 C5 53.05N 4.18W
Llanybydder Wales 12 C4 52.04N 4.10W
Llawnog Wales 12 D4 52.32N 3.27W
Llay Wales 12 D5 53.06N 2.58W
Lleida Spain 46 F4 41.37N 0.38E
Llerena Spain 46 C3 38.14N 6.00W
Lleyn Pen. Wales 12 C4 52.50N 4.35W
Lloydminster Canada 62 H3 53.18N 110.00W
Llullaillaco *mtn.* Chile/Argentina 76 D2 24.43S 68.30W
Llyn Brianne Resr. Wales 12 D4 52.09N 3.44W
Llyn Celyn *l.* Wales 12 D4 52.57N 3.40W
Llyn Trawsfynydd *l.* Wales 12 D4 52.55N 3.55W
Lobito Angola 86 A3 12.20S 13.34E
Lochaline Scotland 16 D4 56.32N 5.47W
Lochboisdale *town* Scotland 18 B2 57.09N 7.19W
Lochgelly Scotland 17 F4 56.08N 3.19W
Lochgilphead Scotland 16 D4 56.02N 5.26W
Lochinver *town* Scotland 18 D3 58.09N 5.13W
Lochmaben Scotland 17 F3 55.08N 3.27W
Lochmaddy *town* Scotland 18 B2 57.36N 7.10W
Lochnagar *mtn.* Scotland 19 F1 56.57N 3.15W
Lochranza Scotland 16 D3 55.42N 5.18W
Lochy, Loch Scotland 18 E1 56.58N 4.57W
Lockerbie Scotland 17 F3 55.07N 3.21W
Loddon England 11 G3 52.30N 1.29E
Lodwar Kenya 87 B3 3.07N 35.38E
Łódź Poland 54 F4 51.49N 19.28E
Lofoten *is.* Norway 43 C5 68.15N 13.50E
Lofusa Sudan 87 A3 3.44N 33.18E
Logan, Mt. Canada 62 D4 60.45N 140.00W
Loggerheads England 10 C3 52.55N 2.23W
Logone *r.* Cameroon/Chad 82 E6 12.10N 15.00E
Logroño Spain 46 D5 42.28N 2.26W
Loire *r.* France 44 C5 47.18N 2.00W
Loja Ecuador 76 C4 3.59S 79.16W
Loja Spain 46 C2 37.10N 4.09W
Lokan tekojärvi *resr.* Finland 43 F4 67.55N 27.40E
Lokeren Belgium 42 C3 51.06N 3.59E
Lokichar Kenya 87 B3 2.23N 35.40E
Lokichokio Kenya 87 A3 4.19N 34.16E
Løkken Norway 43 B3 63.06N 9.43E
Lokoja Nigeria 84 E2 7.49N 6.44E
Lokwa Kangole Kenya 87 B3 3.32N 35.50E
Loliondo Tanzania 87 B2 2.04S 35.38E
Lolland *i.* Denmark 43 B1 54.50N 11.30E
Lom Bulgaria 56 F5 43.49N 23.13E
Loman *r.* Dem. Rep. of Congo 86 B5 0.45N 24.10E
Lombok *i.* Indonesia 104 F2 8.30S 116.20E
Lomé Togo 84 E2 6.10N 1.21E
Lommel Belgium 42 E3 51.15N 5.18E
Lomond, Loch Scotland 16 E4 56.07N 4.36W
Łomża Poland 54 H5 53.11N 22.04E
London Canada 63 J2 42.58N 81.15W
London England 11 E2 51.32N 0.06W
Londonderry N. Ireland 16 B2 55.00N 7.20W
Londonderry *d.* N. Ireland 20 E4 55.00N 7.00W
Londonderry, C. Australia 110 B4 13.58S 126.55E
Long Ashton England 10 C2 51.26N 2.37W
Long Beach *town* U.S.A. 64 C3 33.57N 118.15W
Long Bennington England 15 G1 52.59N 0.45W
Long Eaton England 10 D3 52.54N 1.16W
Longford Rep. of Ire. 20 D3 53.44N 7.48W
Longford *d.* Rep. of Ire. 20 D3 53.42N 7.45W
Longhorsley England 15 F4 55.15N 1.46W
Longhoughton England 15 F4 55.26N 1.36W
Long I. The Bahamas 71 J5 23.00N 75.00W
Long I. U.S.A. 65 L5 40.50N 73.00W
Longido Tanzania 87 B2 2.43S 36.41E
Longlac *town* Canada 63 J2 49.47N 86.34W
Long, Loch Scotland 16 E4 56.05N 4.52W
Longmount U.S.A. 64 F5 40.10N 105.06W
Longridge England 14 E2 53.50N 2.37W
Long Stratton England 11 G3 52.29N 1.14E
Longton England 14 E2 53.44N 2.48W
Longtown England 14 E4 55.01N 2.58W
Longuyon France 42 E1 49.26N 5.36E
Long Xuyên Vietnam 104 D6 10.23N 105.25E
Lons-le-Saunier France 44 F5 46.40N 5.33E
Lookout, C. U.S.A. 65 K3 34.34N 76.34W
Loolmalasin *mtn.* Tanzania 87 B2 3.06S 35.46E
Loop Head Rep. of Ire. 20 B2 52.33N 9.56W
Lop Nur *l.* China 102 G6 40.30N 90.30E
Lopphavet *est.* Norway 43 E5 70.30N 21.00E
Loralai Pakistan 95 K5 30.22N 68.36E
Lorca Spain 46 E2 37.40N 1.41W
Lord Howe I. Pacific Oc. 111 E2 31.33S 159.05E
Lordsburg U.S.A. 64 E3 32.22N 108.43W
Lorient France 44 B5 47.45N 3.21W
Losai Nat. Res. Kenya 87 B3 1.30N 37.15E
Los Angeles Chile 75 B3 37.28S 72.21W
Los Angeles U.S.A. 64 C3 34.00N 118.17W
Los Canarios Canary Is. 46 X2 28.28N 17.52W
Los Canarreos, Archipiélago de Cuba 71 H5 21.40N 82.30W
Los Chonos, Archipielago de *is.* Chile 75 B2 45.00S 74.00W
Los Estados, I. de *i.* Argentina 75 C1 54.45S 64.00W
Los Llanos de Aridane Canary Is. 46 X2 28.39N 17.54W
Los Mochis Mexico 70 C6 25.45N 108.57W
Los Roques *is.* Venezuela 71 K3 12.00N 67.00W
Lossie *r.* Scotland 19 F2 57.43N 3.18W
Lossiemouth Scotland 19 F2 57.43N 3.18W
Los Taques Venezuela 71 J3 11.50N 70.16W
Los Teques Venezuela 71 K3 10.25N 67.01W
Los Testigos *i.* Lesser Antilles 71 L3 11.24N 63.07W
Lostwithiel England 13 C2 50.24N 4.41W
Lot *r.* France 44 D4 44.17N 0.22E
Lotagipi Swamp Kenya/Sudan 87 A3 4.40N 34.30E
Lotikipi Plain *f.* Kenya 87 A3 4.25N 34.30E
Lotta *r.* Russian Fed. 43 H5 68.36N 31.06E
Louang Namtha Laos 104 C8 21.01N 101.27E
Louangphrabang Laos 104 C7 19.53N 102.10E
Loughborough England 10 D3 52.47N 1.11W
Loughor *r.* Wales 12 C3 51.43N 4.04W
Loughrea Rep. of Ire. 20 C3 53.11N 8.36W
Loughton England 11 F2 51.39N 0.03E
Louisiana *d.* U.S.A. 65 H3 31.00N 92.30W
Louisville U.S.A. 65 I4 38.13N 85.45W
Loukhi Russian Fed. 43 H4 66.05N 33.04E
Lourdes France 44 C3 43.06N 0.02W
Louth England 15 H2 53.23N 0.00
Louth *d.* Rep. of Ire. 20 E3 53.45N 6.30W
Lověch Bulgaria 56 G5 43.08N 24.44E
Lovozero Russian Fed. 43 H5 68.01N 35.08E
Lowell U.S.A. 65 L5 42.39N 71.18W
Lower California *pen.* Mexico 70 B6 27.00N 113.00W
Lower Lough Erne N. Ireland 16 B2 54.28N 7.48W
Lower Tunguska *r.* Russian Fed. 59 K4 65.50N 88.00E
Lowestoft England 11 G3 52.29N 1.44E
Łowicz Poland 54 F5 52.06N 19.55E
Loyal, Loch Scotland 19 E3 58.23N 4.21W
Loyauté, Îles N. Cal. 111 F3 21.00S 167.00E
Loyne, Loch Scotland 18 D2 57.06N 5.01W
Lozova Ukraine 55 O3 48.54N 36.20E
Luanda Angola 86 A4 8.50S 13.20E
Luarca Spain 46 B5 43.33N 6.31W
Lubango Angola 86 A3 14.55S 13.30E
Lubbock U.S.A. 64 F3 33.35N 101.53W
Lübeck Germany 48 E5 53.52N 10.40E
Lübeck B. Germany 48 E6 54.05N 11.00E
Lublin Poland 54 H4 51.18N 22.31E
Lubnaig, Loch Scotland 16 E4 56.17N 4.18W
Lubny Ukraine 55 M4 50.01N 33.00E
Lubuklinggau Indonesia 104 C3 3.24S 102.56E
Lubumbashi Dem. Rep. of Congo 86 B3 11.41S 27.29E
Luce B. Scotland 16 E2 54.45N 4.47W
Lucena Phil. 105 G6 13.56N 121.37E
Lucena Spain 46 C2 37.25N 4.29W
Lučenec Slovakia 54 F3 48.20N 19.40E
Lucknow India 97 G6 26.50N 80.54E
Lüderitz Namibia 86 A2 26.38S 15.10E
Ludgershall England 10 D2 51.15N 1.38W
Ludhiana India 96 F7 30.56N 75.52E
Ludlow England 10 C3 52.23N 2.42W
Ludvika Sweden 43 C3 60.08N 15.14E
Ludwigshafen Germany 48 D3 49.29N 8.27E
Luena Angola 86 A3 11.46S 19.55E
Lufkin U.S.A. 65 H3 31.21N 94.47W
Luga Russian Fed. 43 G2 58.42N 29.49E
Lugano Switz. 44 H5 46.01N 8.57E
Lugg *r.* England 10 C3 52.01N 2.38W
Lugo Spain 46 B5 43.00N 7.33W
Lugoj Romania 54 G1 45.42N 21.56E
Luhan'sk Ukraine 58 F2 48.35N 39.20E
Luing *i.* Scotland 16 D4 56.14N 5.38W
Lule *r.* Sweden 43 E4 65.40N 21.48E
Luleå Sweden 43 E4 65.35N 22.10E
Lüleburgaz Turkey 56 H4 41.25N 27.23E
Lumberton U.S.A. 65 K3 34.37N 79.03W
Lunan B. Scotland 19 G1 56.38N 2.30W
Lund Sweden 43 C1 55.42N 13.10E
Lundy I. England 13 C3 51.10N 4.41W
Lune *r.* England 14 E3 54.03N 2.49W
Lüneburg Germany 48 E5 53.15N 10.24E
Lunéville France 44 G6 48.36N 6.30E
Luninyets Belarus 55 J5 52.18N 26.50E
Luoyang China 103 K4 34.48N 112.25E
Lure France 44 G5 47.42N 6.30E
Lurgainn, Loch Scotland 18 D3 58.01N 5.12W
Lurgan N. Ireland 16 C2 54.28N 6.20W
Lusaka Zambia 86 B3 15.26S 28.20E
Lushoto Tanzania 87 B2 4.51S 38.19E
Luton England 11 E2 51.53N 0.25W
Luton *d.* England 9 E2 51.53N 0.25W
Luts'k Ukraine 55 I4 50.42N 25.15E
Lutterworth England 10 D3 52.28N 1.12W
Luxembourg Europe 42 F1 49.50N 6.15E
Luxembourg *town* Lux. 42 F1 49.37N 6.08E
Luxembourg *d.* Belgium 42 E1 49.58N 5.30E
Luxor Egypt 94 D4 25.41N 32.24E
Luzern Switz. 44 H5 47.03N 8.17E
Luziânia Brazil 77 F3 16.18S 47.57W
Luzon *i.* Phil. 105 G7 17.50N 121.00E
Luzon Str. Pacific Oc. 104 G8 20.20N 122.00E
L'viv Ukraine 55 I3 49.50N 24.03E
Lycksele Sweden 43 D4 64.34N 18.40E
Lydd England 11 F1 50.57N 0.56E
Lydford England 13 C2 50.39N 4.06W
Lydney England 10 C2 51.43N 2.32W
Lyepyel' Belarus 55 K6 54.48N 28.40E
Lyme B. England 10 C1 50.40N 2.55W
Lyme Regis England 10 C1 50.44N 2.57W
Lymington England 10 D1 50.46N 1.32W
Lynchburg U.S.A. 65 K4 37.24N 79.10W
Lyndhurst England 10 D1 50.53N 1.33W
Lynmouth England 13 D3 51.15N 3.50W
Lynn Lake *town* Canada 62 H3 56.51N 101.01W
Lynton England 13 D3 51.14N 3.50W
Lyon France 44 F4 45.46N 4.50E
Lyon *r.* Scotland 16 E4 56.37N 3.59W
Lysychans'k Ukraine 55 P3 48.53N 38.25E
Lytchett Minster England 10 C1 50.44N 2.04W
Lytham St. Anne's England 14 D2 53.45N 3.01W
Azul, Cordillera *mts.* Peru 64 B6 9.00S 76.30W

M

Ma'an Jordan 94 E5 30.11N 35.43E
Maas *r.* Neth. 42 D3 51.44N 4.42E
Maaseik Belgium 42 E3 51.08N 5.48E
Maastricht Neth. 42 E2 50.51N 5.42E
Mablethorpe England 15 H2 53.21N 0.14W
Macaé Brazil 77 F2 22.21S 41.48W
Macapá Brazil 77 E5 0.01N 51.01W
Macas Ecuador 76 C4 2.22S 78.08W
Macau Asia 103 K2 22.13N 113.36E
Macclesfield England 15 E2 53.16N 2.09W
McClintock Channel Canada 62 H5 71.20N 102.00W
McClure Str. Canada 62 G5 74.30N 116.00W
McCook U.S.A. 64 F5 40.15N 100.45W
Macdonnell Ranges *mts.* Australia 110 C3 23.30S 132.00E
Macduff Scotland 19 G2 57.40N 2.30W
Macedo de Cavaleiros Portugal 46 B4 41.32N 6.58W
Macedonia Europe 56 E4 41.15N 21.15E
Maceió Brazil 77 G4 9.34S 35.47W
Macgillycuddy's Reeks *mts.* Rep. of Ire. 20 B2 52.00N 9.45W
McGrath U.S.A. 62 C4 62.58N 155.40W
Mach Pakistan 95 K4 29.52N 67.20E
Machakos Kenya 87 B2 1.31S 37.15E
Machala Ecuador 76 C4 3.20S 79.57W
Machilipatnam India 97 G4 16.13N 81.12E
Machrihanish Scotland 16 D3 55.25N 5.44W
Machynlleth Wales 12 D4 52.35N 3.51W
Mackay Australia 110 D3 21.10S 149.10E
Mackay, L. Australia 110 B3 22.30S 128.58E
Mackenzie *r.* Canada 62 E4 69.20N 134.00W
Mackenzie King I. Canada 62 G5 77.30N 112.00W
Mackenzie Mts. Canada 62 E4 64.00N 130.00W
McKinley, Mt. U.S.A. 62 C4 63.00N 151.00W
MacLeod, L. Australia 110 A3 24.10S 113.35E
Macomer Italy 50 C4 40.16N 8.45E
Mâcon France 44 F5 46.18N 4.50E
Macon U.S.A. 65 J3 32.47N 83.37W
Macquarie I. Pacific Oc. 108 K1 54.29S 158.58E
Macroom Rep. of Ire. 20 C1 51.53N 8.59W
Madadeni R.S.A. 86 C2 27.43S 30.02E
Madagascar Africa 86 D2 20.00S 46.30E
Madeira *r.* Brazil 77 E4 3.50S 58.30W
Madeira *i.* Atlantic Oc. 84 C5 32.45N 17.00W
Madeley England 10 C3 52.39N 2.28W
Madhya Pradesh *d.* India 97 F5 23.00N 79.30E
Madīnat ath Thawrah Syria 57 N1 35.50N 38.35E
Madini *r.* Bolivia 76 D3 12.32S 66.50W
Madison U.S.A. 65 I5 43.04N 89.22W
Mado Gashi Kenya 87 B3 0.40N 39.11E
Madras *see* Chennai India 97
Madre de Dios *r.* Bolivia 76 D3 11.00S 66.30W
Madre del Sur, Sierra *mts.* Mexico 70 E4 17.00N 100.00W
Madre Lagoon Mexico 70 E5 25.00N 97.30W
Madre Occidental, Sierra *mts.* Mexico 70 C6 25.00N 105.00W
Madre Oriental, Sierra *mts.* Mexico 70 D5 24.00N 101.00W
Madre, Sierra *mts.* C. America 60 K3 15.00N 90.00W
Madrid Spain 46 D4 40.25N 3.43W
Madukani Tanzania 87 B2 3.52S 35.46E
Madura *i.* Indonesia 104 E2 7.00S 113.30E
Madurai India 97 F2 9.55N 78.07E
Maebashi Japan 106 C3 36.30N 139.04E
Maesteg Wales 12 D3 51.36N 3.40W
Maestra, Sierra *mts.* Cuba 71 I5 20.10N 76.30W
Mafia I. Tanzania 86 C4 7.50S 39.50E
Mafraq Jordan 94 E5 32.20N 36.12E
Magadan Russian Fed. 59 R3 59.38N 150.50E
Magadi Kenya 87 B2 1.52S 36.18E
Magdalena Mexico 70 B7 30.38N 110.59W
Magdalena *r.* Colombia 74 B7 10.56N 74.58W
Magdalena, B. Mexico 70 B5 24.30N 112.00W
Magdeburg Germany 48 E5 52.08N 11.36E
Magellan, Str. of Chile 75 B1 53.00S 71.00W
Magerøya *i.* Norway 43 F5 71.00N 25.50E
Maggiore, L. Italy 50 C7 46.00N 8.37E
Maghâgha Egypt 53 J2 28.39N 30.50E
Maghera N. Ireland 16 C2 54.51N 6.42W
Magherafelt N. Ireland 16 C2 54.45N 6.38W
Maghull England 14 E2 53.31N 2.56W
Magilligan Pt. N. Ireland 16 C3 55.11N 6.58W
Magnitogorsk Russian Fed. 58 H3 53.28N 59.06E
Magny-en-Vexin France 42 A1 49.09N 1.47E
Magu Tanzania 87 A2 2.35S 33.27E
Maguarinho, Cabo *c.* Brazil 77 F4 0.15S 48.23W
Magwe Myanmar 97 J5 20.10N 95.00E
Mahabad Iran 95 G6 36.44N 45.44E
Mahagi Dem. Rep. of Congo 85 H2 2.16N 30.59E
Mahajanga Madagascar 86 D3 15.40S 46.20E
Maharashtra *d.* India 96 E4 20.00N 75.00E
Mahé I. Seychelles 85 J1 4.41S 55.30E
Mahilyow Belarus 55 L5 53.54N 30.20E
Mahón Spain 46 H3 39.55N 4.18E
Maicao Colombia 71 J3 11.25N 72.10W
Maidenhead England 10 E2 51.32N 0.44W
Maidstone England 11 F2 51.17N 0.32E
Maiduguri Nigeria 84 F3 11.53N 13.16E
Main *r.* N. Ireland 16 C2 54.43N 6.19W
Mai-Ndombe, L. Dem. Rep. of Congo 86 A4 2.00S 18.20E
Maine *d.* U.S.A. 65 M6 45.00N 69.00W
Mainland *i.* Orkney Is. Scotland 19 F3 59.00N 3.10W
Mainland *i.* Shetland Is. Scotland 19 Y9 60.15N 1.22W
Mainz Germany 48 D3 50.00N 8.16E
Maio *i.* Cape Verde 84 B3 15.10N 23.10W
Maiquetía Venezuela 71 K3 10.03N 66.57W
Maitland Australia 110 E2 32.33S 151.33E
Maíz, Is. del Nicaragua 71 H3 12.12N 83.00W
Maizuru Japan 106 C3 35.30N 135.20E
Majene Indonesia 104 F3 3.33S 118.59E
Maji Ethiopia 87 B4 6.11N 35.38E
Makale Indonesia 104 F3 3.06S 119.53E
Makassar Str. Indonesia 104 F3 3.00S 118.00E
Makgadikgadi *f.* Botswana 86 B2 20.50S 25.45E
Makhachkala Russian Fed. 58 G2 42.59N 47.30E
Makindu Kenya 87 B2 2.18S 37.50E
Makiyivka Ukraine 55 P3 48.01N 38.00E
Makran *f.* Asia 95 J4 25.40N 62.00E
Makurdi Nigeria 84 E2 7.44N 8.35E
Malabar Coast *f.* India 96 F3 11.00N 75.00E
Malabo Equat. Guinea 84 E2 3.45N 8.48E
Malacca, Str. of Indian Oc. 104 C4 3.00N 100.30E
Maladzyechna Belarus 55 J6 54.16N 26.50E
Málaga Spain 46 C2 36.43N 4.25W
Malahide Rep. of Ire. 20 E3 53.27N 6.09W
Malaita *i.* Solomon Is. 111 F5 9.00S 161.00E
Malakal Sudan 85 H2 9.31N 31.40E
Malakula *i.* Vanuatu 111 F4 16.15S 167.30E
Malang Indonesia 104 E2 7.59S 112.45E
Malanje Angola 86 A4 9.36S 16.21E
Mala Pt. *c.* Panama 71 I2 7.30N 80.00W
Mälaren *l.* Sweden 43 D2 59.30N 17.00E
Malatya Turkey 57 N3 38.22N 38.18E
Malawi Africa 86 C3 13.00S 34.00E
Malaya *f.* Malaysia 104 C4 5.00N 102.00E
Malayer Iran 95 G5 34.19N 48.51E
Malaysia Asia 104 D5 5.00N 110.00E
Malbork Poland 54 F6 54.02N 19.01E
Malden I. Kiribati 108 P6 4.03S 154.49W
Maldives Indian Oc. 96 E2 6.20N 73.00E
Maldon England 11 F2 51.43N 0.41E
Maléa, C. Greece 56 F2 36.27N 23.11E
Malgomaj *l.* Sweden 43 D4 64.45N 16.00E
Mali Africa 84 D3 16.00N 3.00W
Malili Indonesia 104 G3 2.38S 121.06E
Malindi Kenya 87 A2 3.14S 40.07E
Malin Head Rep. of Ire. 20 D5 55.22N 7.24W
Mallaig Scotland 18 D1 57.00N 5.50W
Mallawi Egypt 53 J2 27.44N 30.50E
Mallorca *i.* Spain 46 G3 39.35N 3.00E
Mallow Rep. of Ire. 20 C2 52.08N 8.39W
Malmédy Belgium 42 F2 50.25N 6.02E
Malmesbury England 10 C2 51.35N 2.05W
Malmö Sweden 43 C1 55.35N 13.00E
Måløy Norway 43 A3 61.57N 5.06E
Malpas England 14 E2 53.01N 2.46W
Malpelo, I. de Pacific Oc. 60 K2 4.00N 81.35W
Malta Europe 50 F1 35.55N 14.25E
Maltby England 15 F2 53.25N 1.12W
Malton England 15 G3 54.09N 0.48W
Malvern Hills England 10 C3 52.07N 2.19W
Mamelodi R.S.A. 86 B2 25.41S 28.22E
Mamoré *r.* Brazil/Bolivia 76 D3 9.41S 65.20W
Mamuju Indonesia 104 F3 2.41S 118.55E
Manacapuru Brazil 77 D4 3.16S 60.37W
Manacor Spain 46 G3 39.32N 3.12E
Manadao Indonesia 105 G4 1.30N 124.58E
Managua Nicaragua 70 G3 12.06N 86.18W
Managua, L. Nicaragua 70 G3 12.10N 86.30W
Manama Bahrain 95 H4 26.12N 50.36E
Mananjary Madagascar 86 D2 21.13S 48.20E
Manaus Brazil 77 E4 3.06S 60.00W
Manavgat Turkey 57 J2 36.47N 31.28E
Manchester England 15 E2 53.30N 2.15W
Manchester U.S.A. 65 L5 42.59N 71.28W
Manchuria *f.* Asia 90 O7 45.00N 125.00E
Mand *r.* Iran 95 H4 28.09N 51.16E
Mandala, Peak Indonesia 105 K3 4.45S 140.15E
Mandalay Myanmar 97 J5 21.57N 96.04E
Mandalgovĭ Mongolia 103 J7 45.40N 106.10E
Mandera Kenya 87 C3 3.55N 41.50E
Mandioré, Lagoa *l.* Brazil/Bolivia 77 E3 18.05S 57.30W
Manfredonia Italy 50 F4 41.38N 15.54E
Mangaia *i.* Cook Is. 108 P4 21.56S 157.56W
Mangalia Romania 57 I5 43.50N 28.35E
Mangalore India 96 E3 12.54N 74.51E
Mangaung R.S.A. 86 B2 29.30S 26.30E
Mangoky *r.* Madagascar 86 D2 21.20S 43.30E
Mangotsfield England 10 C2 51.29N 2.29W
Mangu Kenya 87 B2 0.58S 36.57E
Manihiki *i.* Cook Is. 108 O5 10.24S 161.01W
Maniitsoq Greenland 63 M4 65.40N 53.00W
Manila Phil. 104 G6 14.36N 120.59E
Manipur *d.* India 97 I5 25.00N 93.40E
Manisa Turkey 56 H3 38.37N 27.28E
Man, Isle of U.K. 14 C3 54.15N 4.30W
Manitoba *d.* Canada 62 I3 54.00N 96.00W
Manitoba, L. Canada 62 I3 51.35N 99.00W
Manizales Colombia 74 B7 5.03N 75.32W
Manmad India 96 E5 20.15N 74.27E
Mannheim Germany 48 D3 49.30N 8.28E
Manningtree England 11 G2 51.56N 1.03E
Manokwari Indonesia 105 I3 0.53S 134.05E
Manorbier Wales 12 C3 51.39N 4.48W
Manorhamilton Rep. of Ire. 20 C4 54.18N 8.14W
Manra *i.* Kiribati 108 N6 4.29S 172.10W
Manresa Spain 46 F4 41.43N 1.50E
Mansa Zambia 86 B3 11.10S 28.52E
Mansel I. Canada 63 J4 62.00N 80.00W
Mansfield England 15 F2 53.09N 1.12W
Mansfield U.S.A. 65 J5 40.46N 82.31W
Manston England 10 C1 50.57N 2.16W
Manta Ecuador 76 B4 0.59S 80.44W
Mantua Italy 50 D6 45.09N 10.47E
Manukau New Zealand 111 G2 36.59S 174.53E
Manyara, L. Tanzania 87 B2 3.36S 35.44E
Manzanares Spain 46 D3 39.00N 3.23W
Manzhouli China 103 L7 49.36N 117.28E
Maoke Range *mts.* Indonesia 105 J3 4.00S 137.30E
Mapeura *r.* Brazil 77 E4 1.10S 57.00W
Maputo Mozambique 86 C2 25.58S 32.35E
Maraba Brazil 77 F4 5.23S 49.10W
Maracaibo Venezuela 71 J3 10.44N 71.37W
Maracaibo, L. Venezuela 71 J2 10.00N 71.30W
Maraca, Ilha de *i.* Brazil 77 E5 2.00N 50.30W
Maracaju, Serra de *mts.* Brazil 77 E2 21.38S 55.10W
Maracay Venezuela 71 K3 10.20N 67.28W
Maradah Libya 85 F4 29.14N 19.13E
Maradi Niger 84 E3 13.29N 7.10E
Maragheh Iran 95 G6 37.25N 46.13E
Marajó, I. de Brazil 77 F4 1.00S 49.30W
Maralal Kenya 87 B3 1.04N 36.41E
Marand Iran 95 G6 38.25N 45.50E
Maranhão *d.* Brazil 77 F4 6.00S 45.30W
Marañón *r.* Peru 77 C4 4.00S 73.30W
Marathonas Greece 56 G3 38.10N 23.59E
Marazion England 13 B2 50.08N 5.29W
Marbella Spain 46 C2 36.31N 4.53W
Marburg Germany 48 D4 50.49N 8.36E
March England 11 F3 52.33N 0.05E
Marche-en-Famenne Belgium 42 E2 50.13N 5.21E
Mar Chiquita, L. Argentina 77 D1 30.42S 62.36W
Mardan Pakistan 95 L5 34.14N 72.05E
Mar del Plata Argentina 75 D3 38.00S 57.32W
Mardin Turkey 94 F6 37.19N 40.43E
Maree, Loch Scotland 18 D2 57.41N 5.28W
Maresfield England 11 F1 51.00N 0.05E
Margarita *i.* Venezuela 71 L3 11.00N 64.00W
Margate England 11 G2 51.23N 1.24E
Margery Hill England 15 F2 53.26N 1.42W
Marhanets' Ukraine 55 N2 47.37N 34.40E
Marianas Trench *f.* Pacific Oc. 117 Q6

Marianna U.S.A. 65 I3 30.45N 85.15W
Marías, I. *is.* Mexico 70 C5 21.25N 106.30W
Mariato Pt. *c.* Panama 71 H2 7.12N 80.52W
Maria van Diemen, C. New Zealand 111 G2 34.29S 172.39E
Ma'rib Yemen 94 G2 15.01N 45.30E
Maribor Slovenia 54 D2 46.35N 15.40E
Marie Galante *i.* Guadeloupe 71 L4 15.54N 61.11W
Mariehamn Finland 43 D3 60.05N 19.55E
Mariental Namibia 86 A2 24.38S 17.58E
Mariestad Sweden 43 C2 58.44N 13.50E
Marijampole Lithuania 54 H6 54.31N 23.20E
Marília Brazil 77 E2 22.13S 50.20W
Maringá Brazil 77 E2 23.36S 52.02W
Maritsa *r.* Bulgaria 56 H4 41.40N 26.25E
Mariupol' Ukraine 55 O2 47.05N 37.34E
Marka Somalia 86 D5 1.42N 44.47E
Markermeer *l.* Neth. 42 E4 52.30N 5.15E
Market Deeping England 11 E3 52.40N 0.20W
Market Drayton England 10 C3 52.55N 2.30W
Market Harborough England 10 E3 52.29N 0.55W
Markethill N. Ireland 16 C2 54.17N 6.31W
Market Rasen England 15 G2 53.24N 0.20W
Market Weighton England 15 G2 53.52N 0.40W
Marlborough England 10 D2 51.26N 1.44W
Marlborough Downs *hills* England 10 D2 51.28N 1.48W
Marlow England 10 E2 51.35N 0.48W
Marmande France 44 D4 44.30N 0.10E
Marmara *i.* Turkey 57 H4 40.38N 27.37E
Marmara, Sea of Turkey 57 I4 40.45N 28.15E
Marmaris Turkey 57 I2 36.52N 28.17E
Marne *r.* France 44 E6 48.50N 2.25E
Marne-la-Vallée France 44 E6 48.50N 2.30E
Marondera Zimbabwe 86 C3 18.11S 31.31E
Maroni *r.* French Guiana 74 D7 5.30N 54.00W
Marotiri *i.* French Polynesia 109 Q4 27.55S 143.26W
Maroua Cameroon 84 F3 10.35N 14.20E
Marple England 15 E2 53.23N 2.05W
Marquesas Is. Pacific Oc. 109 R6 9.00S 139.30W
Marquette U.S.A. 65 I6 46.33N 87.23W
Marrakesh Morocco 84 D5 31.49N 8.00W
Marra Plateau Sudan 94 B1 13.00N 23.50E
Marsa Alam Egypt 94 D4 25.03N 34.44E
Marsa al Burayqah Libya 85 G4 30.25N 19.35E
Marsabit Kenya 87 B3 2.20N 37.59E
Marsabit Nat. Res. Kenya 87 B3 2.20N 37.58E
Marsala Italy 50 E2 37.48N 12.27E
Marsa Matrûh Egypt 94 C5 31.21N 27.15E
Marseille France 44 F3 43.18N 5.22E
Mar, Serra do *mts.* Brazil 77 F2 28.00S 49.40W
Marsfjället *mtn.* Sweden 43 C4 65.10N 15.20E
Marshall Is. Pacific Oc. 108 L8 10.00N 167.00E
Märsta Sweden 43 D2 59.37N 17.52E
Martaban Myanmar 97 J4 16.32N 97.35E
Martaban, G. of Indian Oc. 97 J4 15.10N 96.30E
Martapura Indonesia 104 C3 4.20S 104.22E
Martin Slovakia 54 F3 49.05N 18.55E
Martinique *i.* Windward Is. 71 L3 14.40N 61.00W
Martinsville U.S.A. 65 K4 36.43N 79.53W
Martin Vaz Is. Atlantic Oc. 72 H4 20.30S 28.51W
Martock England 13 E2 50.59N 2.46W
Martre, Lac La *l.* Canada 62 G4 63.15N 116.55W
Mary Turkmenistan 95 J6 37.42N 61.54E
Maryborough Australia 110 E3 25.32S 152.36E
Mary Byrd Land Antarctica 112 80.00S 130.00W
Maryland *d.* U.S.A. 65 K4 39.00N 76.30W
Maryport England 14 D3 54.43N 3.30W
Masai Mara Nat. Res. Kenya 87 A2 1.25S 35.05E
Masai Steppe *f.* Tanzania 87 B2 5.00S 37.00E
Masan S. Korea 103 N5 35.10N 128.35E
Masbate Phil. 105 G6 12.21N 123.36E
Masbate *i.* Phil. 105 G6 12.00N 123.30E
Mascara Algeria 52 D4 35.24N 0.08E
Maseno Kenya 87 A3 0.00 34.32E
Maseru Lesotho 86 B2 29.19S 27.29E
Mashhad Iran 95 I6 36.16N 59.34E
Masinga Resr. Kenya 87 B2 1.10S 37.40E
Masirah *i.* Oman 95 I3 20.30N 58.50E
Masirah, G. of Oman 95 I2 20.00N 58.10E
Masjed Soleyman Iran 95 G5 31.59N 49.18E
Mask, Lough Rep. of Ire. 20 B3 53.38N 9.22W
Mason City U.S.A. 65 H5 43.10N 93.10W
Massachusetts *d.* U.S.A. 65 L5 42.20N 72.00W
Massakory Chad 84 F3 13.02N 15.43E
Massawa Eritrea 85 H3 15.36N 39.29E
Massif Central *mts.* France 44 E4 45.00N 3.30E
Massif Ennedi *mts.* Chad 94 B2 17.00N 23.00E
Mastung Pakistan 95 K4 29.48N 66.51E
Masuda Japan 106 B2 34.42N 131.51E
Masurian Lakes Poland 54 G5 53.50N 21.40E
Masvingo Zimbabwe 86 C2 20.10S 30.49E
Maswe Game Res. Tanzania 87 A2 2.30S 34.00E
Matadi Dem. Rep. of Congo 86 A4 5.50S 13.32E
Matam Senegal 84 C3 15.40N 13.18W
Matamoros Mexico 70 E6 25.50N 97.31W
Matanzas Cuba 71 H5 23.04N 81.35W
Matapan, C. Greece 56 F2 36.22N 22.28E
Matara Sri Lanka 97 G2 5.57N 80.32E
Mataram Indonesia 104 F2 8.36S 116.07E
Mataró Spain 46 G4 41.32N 2.27E
Matā'utu Wallis and Futuna Is. 108 N5 13.17S 176.07W
Matera Italy 50 G4 40.41N 16.36E
Matheniko Game Res. Uganda 87 A3 2.55N 34.25E
Mathura India 97 F6 27.30N 77.30E
Matlock England 15 F2 53.08N 1.32W
Mato Grosso *d.* Brazil 77 E3 13.00S 55.00W
Mato Grosso do Sul *d.* Brazil 77 E2 20.00S 54.30W
Mato Grosso, Planalto do *f.* Brazil 77 E3 15.00S 55.00W
Matrah Oman 95 I3 23.37N 58.33E
Matsue Japan 106 B3 35.29N 133.00E
Matsu Is. Taiwan 103 M3 26.12N 120.00E
Matsumoto Japan 106 C3 36.18N 137.58E
Matsusaka Japan 106 C2 34.34N 136.32E
Matsuyama Japan 106 B2 33.50N 132.47E
Matterhorn *mtn.* Switz. 44 G4 45.58N 7.38E
Maturín Venezuela 71 L2 9.45N 63.16W
Mau *mtn.* Kenya 87 B2 0.30S 35.50E
Maubeuge France 42 C2 50.17N 3.58E
Mauchline Scotland 16 E3 55.31N 4.23W
Maughold Head I.o.M. 14 C3 54.18N 4.19W
Maug Is. N. Mariana Is. 105 L8 20.02N 14.19E
Maui *i.* Hawaiian Is. 108 P9 20.45N 156.15W
Mauke *i.* Cook Is. 108 P5 19.49S 157.41W
Maumere Indonesia 105 G2 8.35S 122.13E
Maun Botswana 86 B2 19.52S 23.40E
Mauritania Africa 84 C3 20.00N 10.00W
Mauritius Indian Oc. 115 L4 20.10S 58.00E
Maya *r.* Russian Fed. 59 P4 60.25N 134.28E
Mayaguana *i.* The Bahamas 71 J5 22.30N 73.00W
Mayagüez Puerto Rico 71 K4 18.13N 67.09W
Mayamey Iran 95 I6 36.27N 55.40E
Maya Mts. Belize 70 G4 16.30N 89.00W
Mayar *mtn.* Scotland 19 F1 56.50N 3.16W
Maybole Scotland 16 E3 55.21N 4.41W
Mayen Germany 42 G2 50.19N 7.14E
Mayenne *r.* France 44 C5 47.30N 0.37W
Mayo, Isle of Scotland 17 G4 56.12N 2.32W
Maykop Russian Fed. 53 L5 44.37N 40.48E
Mayo Canada 62 E4 63.45N 135.45W
Mayo *d.* Rep. of Ire. 20 B3 53.47N 9.07W
Mayotte *i.* Indian Oc. 86 D3 12.50S 45.10E
Mazar-e Sharif Afghan. 95 K6 36.43N 67.05E
Mazatlán Mexico 70 C5 23.11N 106.25W
Mažeikiai Lithuania 54 H7 56.06N 23.06E
Mazyr Belarus 55 K5 52.02N 29.10E
Mbabane Swaziland 86 C2 26.20S 31.08E
Mbale Uganda 87 A3 1.02N 34.11E
Mbandaka Dem. Rep. of Congo 86 A5 0.03N 18.28E
M'banza Congo Angola 86 A4 6.18S 14.16E
Mbarara Uganda 86 C4 0.36S 30.40E
Mbeya Tanzania 86 C4 8.54S 33.29E
Mbuji-Mayi Dem. Rep. of Congo 86 B4 6.10S 23.39E
Mbulu Tanzania 87 B2 3.51S 35.32E
Mdantsane R.S.A. 86 B1 32.54S 27.54E
Mead, L. U.S.A. 64 D4 36.10N 114.25W
Mealasta I. Scotland 18 B3 58.05N 7.07W
Meall a'Bhuiridh *mtn.* Scotland 16 E4 56.37N 4.50W
Meath *d.* Rep. of Ire. 20 E3 53.30N 6.30W
Meaux France 44 E6 48.58N 2.54E
Mecca Saudi Arabia 94 E3 21.26N 39.49E
Mechelen Belgium 42 D3 51.01N 4.28E
Mecheria Algeria 52 C3 33.33N 0.17W
Meckenheim Germany 42 G2 50.37N 7.02E
Medan Indonesia 104 B4 3.35N 98.39E
Medanosa, Punta *c.* Argentina 75 C2 48.08S 65.55W
Medellín Colombia 71 I2 6.15N 75.36W
Meden *r.* England 15 G2 53.19N 0.55W
Medenine Tunisia 84 F5 33.24N 10.25E
Medetsiz *mtn.* Turkey 57 L2 37.33N 34.38E
Medicine Hat Canada 62 G3 50.03N 110.41W
Medina Saudi Arabia 94 E3 24.30N 39.35E
Medina del Campo Spain 46 C4 41.20N 4.55W
Mediterranean Sea 52-53
Medway *r.* England 11 F2 51.24N 0.31E
Medway Towns *d.* England 9 F2 51.24N 0.33E
Meerut India 97 F6 29.00N 77.42E
Mēga Ethiopia 87 B3 4.02N 38.19E
Mēga Escarpment *f.* Ethiopia 87 B3 4.00N 38.10E
Megara Greece 56 F3 38.00N 23.21E
Meghalaya *d.* India 97 I6 25.30N 91.00E
Meiktila Myanmar 97 J5 20.53N 95.54E
Meiningen Germany 48 E4 50.34N 10.25E
Meissen Germany 48 F4 51.10N 13.28E
Meizhou China 103 L2 24.20N 116.15E
Mejicana *mtn.* Argentina 76 D2 29.00S 67.50W
Mek'ele Ethiopia 94 E1 13.33N 39.30E
Meknès Morocco 84 D5 33.53N 5.37W
Mekong *r.* Asia 104 D6 10.00N 106.20E
Mekong, Mouths of the Vietnam 104 D6 10.00N 106.20E
Melaka Malaysia 104 C4 2.11N 102.16E
Melbourn England 11 F3 52.05N 0.01E
Melbourne Australia 110 D2 37.45S 144.58E
Melby Scotland 19 Y9 60.19N 1.40W
Melilla Spain 84 D5 35.17N 2.57W
Melitopol' Ukraine 55 N2 46.51N 35.22E
Melka Guba Ethiopia 87 B3 4.52N 39.18E
Melksham England 10 C2 51.22N 2.09W
Melrose Scotland 17 G3 55.36N 2.43W
Meltham England 15 F2 53.36N 1.52W
Melton Mowbray England 10 E3 52.46N 0.53W
Melun France 44 E6 48.32N 2.40E
Melvich Scotland 19 F3 58.32N 3.55W
Melville Canada 62 H3 50.57N 102.49W
Melville B. Greenland 63 L5 75.00N 63.00W
Melville I. Australia 110 C4 11.30S 131.00E
Melville I. Canada 62 G5 75.30N 110.00W
Melville Pen. Canada 63 J4 68.00N 84.00W
Melvin, Lough Rep. of Ire./N. Ireland 16 A2 54.25N 8.10W
Memberamo *r.* Indonesia 105 J3 1.45S 137.25E
Memmingen Germany 48 E2 47.59N 10.11E
Memphis U.S.A. 65 I4 35.05N 90.00W
Menai Bridge *town* Wales 12 C5 53.14N 4.11W
Menai Str. Wales 12 C5 53.17N 4.20W
Mendawai *r.* Indonesia 104 E3 3.17S 113.20E
Mende France 44 E4 44.32N 3.30E
Menderes *r.* Turkey 56 H2 37.30N 27.05E
Mendip Hills England 13 E3 51.15N 2.40W
Mendoza Argentina 75 C3 33.00S 68.52W
Menongue Angola 86 A3 14.40S 17.41E
Menorca *i.* Spain 46 H3 40.00N 4.00E
Mentawai Is. Indonesia 104 B3 2.50S 99.00E
Mentok Indonesia 104 D3 2.04S 105.12E
Menzel Bourguiba Tunisia 52 E4 37.10N 9.48E
Meppel Neth. 42 F4 52.42N 6.12E
Meppen Germany 42 G4 52.42N 7.17E
Merano Italy 50 D7 46.41N 11.10E
Merauke Indonesia 105 K2 8.30S 140.22E
Merced U.S.A. 64 B4 37.17N 120.29W
Mere England 10 C2 51.05N 2.16W
Mergui Myanmar 97 J3 12.26N 98.34E
Mergui Archipelago *is.* Myanmar 97 J3 11.30N 98.30E
Mérida Mexico 70 G5 20.59N 89.39W
Mérida Spain 46 B3 38.55N 6.20W
Mérida Venezuela 71 J2 8.24N 71.08W
Mérida, Cordillera de *mts.* Venezuela 71 J2 8.00N 71.30W
Meridian U.S.A. 65 I3 32.21N 88.42W
Merkys *r.* Lithuania 55 I6 54.05N 24.03E
Merowe Sudan 85 H3 18.30N 31.49E
Merrick *mtn.* Scotland 16 E3 55.08N 4.29W
Mersch Lux. 42 F1 49.44N 6.05E
Mersey *r.* England 14 E2 53.22N 2.37W
Merseyside *d.* England 9 D3 53.28N 3.00W
Mersin Turkey 57 L2 36.47N 34.37E
Merthyr Tydfil Wales 12 D3 51.45N 3.23W
Merthyr Tydfil *d.* Wales 9 D2 51.45N 3.23W
Merti Kenya 87 B3 1.07N 38.39E
Méru France 42 B1 49.14N 2.08E
Meru Kenya 87 B3 0.03N 37.38E
Meru *mtn.* Tanzania 87 B2 3.16S 36.53E
Merzifon Turkey 57 L4 40.52N 35.28E
Merzig Germany 42 F1 49.26N 6.39E
Mesolóngion Greece 56 E3 38.23N 21.23E
Mesopotamia *f.* Iraq 85 I5 33.30N 44.30E
Messina Italy 50 F3 38.13N 15.34E
Messina, G. of Greece 56 F2 36.50N 22.05E
Mesta *r.* see Nestos *r.* Bulgaria 56
Meta *r.* Venezuela 74 C7 6.10N 67.30W
Metheringham England 15 G2 53.09N 0.22W
Methwold England 11 F3 52.30N 0.33E
Metković Croatia 56 C5 43.03N 17.38E
Metlika Slovenia 54 D1 45.39N 15.19E
Metz France 44 G6 49.07N 6.11E
Meuse *r.* see Maas *r.* Belgium 42
Mevagissey England 13 C2 50.16N 4.48W
Mexborough England 15 F2 53.29N 1.18W
Mexicali Mexico 70 A7 32.26N 115.30W
Mexico C. America 70 D5 20.00N 100.00W
México *d.* Mexico 70 E4 19.45N 99.30W
Mexico City Mexico 70 E4 19.25N 99.10W
Mexico, G. of N. America 70 F5 25.00N 90.00W
Meymaneh Afghan. 95 J6 35.54N 64.43E
Mezen Russian Fed. 58 G4 65.50N 44.20E
Mezen *r.* Russian Fed. 35 H4 65.50N 44.18E
Mezha *r.* Russian Fed. 55 L6 55.47N 31.24E
Mezitli Turkey 57 L2 36.45N 34.30E
Mezquital *r.* Mexico 70 C5 21.58N 105.30W
Mfanganu I. Kenya 87 A2 0.27S 34.00E
Miami-Fort Lauderdale U.S.A. 65 J2 25.45N 80.10W
Miandowab Iran 95 G6 36.57N 46.06E
Mīāneh Iran 95 G6 37.23N 47.45E
Mianwali Pakistan 95 L5 32.32N 71.33E
Miass Russian Fed. 58 I3 55.00N 60.00E
Michigan *d.* U.S.A. 65 I5 45.00N 85.00W
Michigan, L. U.S.A. 65 I5 44.00N 87.00W
Michipicoten I. Canada 65 I6 47.40N 85.50W
Michipicoten River *town* Canada 65 J6 47.57N 84.55W
Michoacán *d.* Mexico 70 D4 19.20N 101.00W
Michurinsk Russian Fed. 58 G3 52.54N 40.30E
Middlesbrough England 15 F3 54.34N 1.13W
Middlesbrough *d.* England 9 E4 54.33N 1.13W
Middleton England 15 E2 53.33N 2.12W
Middleton-in-Teesdale England 15 E3 54.38N 2.05W
Middleton St. George England 15 F3 54.30N 1.28W
Middlewich England 15 E2 53.12N 2.28W
Midhurst England 10 E1 50.59N 0.44W
Midi, Canal du France 44 D3 43.27N 1.45E
Midland U.S.A. 64 F3 32.00N 102.05W
Midleton Rep. of Ire. 20 C1 51.55N 8.12W
Midlothian *d.* Scotland 8 D4 55.50N 3.10W
Midway Is. Hawaiian Is. 108 N9 28.15N 177.25W
Miekojärvi *l.* Finland 43 F4 66.35N 24.20E
Mikhaylovskiy Russian Fed. 102 D8 51.41N 79.47E
Mikkeli Finland 43 F3 61.44N 27.15E
Milan Italy 50 C6 45.28N 9.16E
Milas Turkey 57 H2 37.18N 27.48E
Milborne Port England 13 E2 50.58N 2.28W
Mildenhall England 11 F3 52.20N 0.30E
Mildura Australia 110 D2 34.14S 142.13E
Miles City U.S.A. 64 E6 46.24N 105.48W
Milford Haven *town* Wales 12 B3 51.43N 5.02W
Milk *r.* U.S.A. 64 E6 47.55N 106.15W
Millárs *r.* Spain 46 E3 39.58N 0.01W
Millau France 44 E4 44.06N 3.05E
Mille Lacs *l.* U.S.A. 65 H6 46.15N 93.40W
Millennium I. Kiribati 109 Q6 10.00S 150.30W
Milleur Pt. Scotland 16 D3 55.01N 5.07W
Millom England 14 D3 54.13N 3.16W
Millport Scotland 16 E3 55.45N 4.56W
Milnthorpe England 14 E3 54.14N 2.47W
Milos *i.* Greece 56 G2 36.40N 24.26E
Milton Keynes England 10 E3 52.03N 0.42W
Milton Keynes *d.* England 9 E3 52.03N 0.42W
Milwaukee U.S.A. 65 I5 43.03N 87.56W
Mimizan France 44 C4 44.12N 1.14W
Mīnab Iran 95 I4 27.07N 57.05E
Minas Gerais *d.* Brazil 77 F3 18.00S 45.00W
Minatitlán Mexico 70 F4 17.59N 94.32W
Mindanao *i.* Phil. 105 H5 7.30N 125.00E
Mindelo Cape Verde 84 B3 16.54N 25.00W
Mindoro *i.* Phil. 104 G6 13.00N 121.00E
Mindoro Str. Pacific Oc. 104 G6 12.30N 120.10E
Mindra, Mt. Romania 56 F6 45.20N 23.32E
Minehead England 13 D3 51.12N 3.29W
Mingulay *i.* Scotland 18 B1 56.48N 7.37W
Minna Nigeria 84 E2 9.39N 6.32E
Minneapolis-St. Paul U.S.A. 65 H5 45.00N 93.15W
Minnesota *d.* U.S.A. 65 H5 46.00N 95.00W
Miño *r.* Spain 46 A4 41.50N 8.52W
Minot U.S.A. 64 F6 48.16N 101.19W
Minsk Belarus 55 J5 53.51N 27.30E
Minsterly England 10 C3 52.38N 2.56W
Mintlaw Scotland 19 G2 57.31N 2.00W
Miranda *r.* Brazil 77 E2 19.22S 57.15W
Miranda de Ebro Spain 46 D5 42.41N 2.57W
Mirandela Portugal 46 B4 41.28N 7.10W
Mirbat Oman 95 H2 17.00N 54.45E
Miri Malaysia 104 E4 4.28N 114.00E
Mirim, Lagoa *l.* Brazil 75 D3 33.10S 53.30W
Mirpur Khas Pakistan 96 D6 25.33N 69.05E
Mirzapur India 97 G6 25.09N 82.34E
Miskolc Hungary 54 G3 48.07N 20.47E
Misoöl *i.* Indonesia 105 I3 1.50S 130.10E
Misratah Libya 84 F5 32.24N 15.04E
Missinaibi *r.* Canada 63 J3 50.44N 81.29W
Mississippi *r.* U.S.A. 65 I2 28.55N 89.05W
Mississippi *d.* U.S.A. 65 I3 33.00N 90.00W
Mississippi Delta U.S.A. 65 I2 29.00N 89.10W
Missoula U.S.A. 64 D6 46.52N 114.00W
Missouri *r.* U.S.A. 65 H4 38.40N 90.20W
Missouri *d.* U.S.A. 65 H4 39.00N 93.00W
Mistassini, L. Canada 63 K3 50.45N 73.40W
Mistissini Canada 63 K3 50.20N 73.30W
Mitchell U.S.A. 64 G5 43.40N 98.01W
Mitchell *r.* Australia 110 D4 15.12S 141.40E
Mitchell, Mt. U.S.A. 65 J4 35.57N 82.16W
Mitchelstown Rep. of Ire. 20 C2 52.16N 8.19W
Mito Japan 106 D3 36.30N 140.29E
Mittimatalik *see* Pond Inlet Canada 63
Mitumba, Chaine des *mts.* Dem. Rep. of Congo 86 B4 8.00S 28.00E
Miyako Japan 106 D3 39.40N 141.59E
Miyazaki Japan 106 B2 31.58N 131.50E
Mizdah Libya 52 F3 31.26N 12.59E
Mizen Head Cork Rep. of Ire. 20 B1 51.27N 9.50W
Mizen Head Wicklow Rep. of Ire. 20 E2 52.52N 6.03W
Mizoram *d.* India 97 I5 23.40N 92.40E
Mjölby Sweden 43 C2 58.19N 15.10E
Mkomazi Tanzania 87 B2 4.41S 38.02E
Mkomazi Game Res. Tanzania 87 B2 4.00S 38.00E
Mljet *i.* Croatia 56 C5 42.45N 17.30E
Mmabatho R.S.A. 86 B2 25.46S 25.37E
Mobile U.S.A. 65 I3 30.40N 88.05W
Mobile B. U.S.A. 65 I3 30.30N 87.50W
Mobridge U.S.A. 64 F6 45.31N 100.25W
Moçambique *town* Mozambique 86 D3 15.00S 40.55E
Mochudi Botswana 86 B2 24.26S 26.07E
Modbury England 13 D2 50.21N 3.53W
Modena Italy 50 D6 44.39N 10.55E
Modica Italy 50 F2 36.51N 14.51E
Moelfre Wales 12 C5 53.21N 4.15W
Moel Sych *mtn.* Wales 12 D4 52.52N 3.22W
Moffat Scotland 17 F3 55.20N 3.27W
Mogadishu Somalia 85 I2 2.02N 45.21E
Mohyliv Podil's'kyy Ukraine 55 J3 48.29N 27.49E
Mo i Rana Norway 43 C4 66.20N 14.12E
Moji das Cruzes Brazil 77 F2 23.33S 46.14W
Mold Wales 12 D5 53.10N 3.08W
Molde Norway 43 A3 62.44N 7.08E
Moldova Europe 55 K2 47.30N 28.30E
Mole *r.* England 13 D2 50.58N 3.53W
Molopo *r.* R.S.A. 86 B2 28.30S 20.07E
Moluccas *is.* Indonesia 108 H6 2.00S 128.00E
Molucca Sea Pacific Oc. 105 H4 2.00N 126.30E
Mombasa Kenya 87 B2 4.04S 39.40E
Møn *i.* Denmark 43 C1 54.58N 12.20E
Monach Is. Scotland 18 B2 57.32N 7.38W
Monach, Sd. of Scotland 18 B2 57.34N 7.35W
Monaco Europe 44 G3 43.40N 7.25E
Monadhliath Mts. Scotland 19 E2 57.09N 4.08W
Monaghan Rep. of Ire. 20 E4 54.15N 6.58W
Monaghan *d.* Rep. of Ire. 20 E4 54.10N 7.00W
Mona I. Puerto Rico 71 K4 18.06N 67.54W
Mona Passage Dom. Rep. 71 K4 18.10N 68.00W
Monar, Loch Scotland 18 D2 57.25N 5.05W
Monbetsu Japan 106 D4 44.21N 143.18E
Monchegorsk Russian Fed. 43 H4 67.55N 33.01E
Mönchengladbach Germany 48 C4 51.12N 6.25E
Monclova Mexico 70 D6 26.55N 101.20W
Moncton Canada 63 L2 46.06N 64.50W
Mondego *r.* Portugal 46 A4 40.09N 8.52W
Mondovì Italy 50 B6 44.24N 7.50E
Moneymore N. Ireland 16 C2 54.42N 6.41W
Monforte Spain 46 B5 42.32N 7.30W
Monga Dem. Rep. of Congo 88 B5 4.05N 22.56E
Mongolia Asia 103 I7 46.30N 104.00E
Mongora Pakistan 95 L5 34.47N 72.22E
Mongu Zambia 86 B3 15.10S 23.09E
Moniaive Scotland 17 F3 55.12N 3.55W
Monmouth Wales 12 E3 51.48N 2.43W
Monmouthshire *d.* Wales 9 D2 51.44N 2.50W
Monnow *r.* England/Wales 12 E3 51.49N 2.42W
Monroe U.S.A. 65 H3 32.31N 92.06W
Monrovia Liberia 84 C2 6.20N 10.46W
Mons Belgium 42 C2 50.27N 3.57E
Montana *d.* U.S.A. 64 D6 47.00N 110.00W
Montargis France 44 E5 48.00N 2.44E
Montauban France 44 D4 44.01N 1.20E
Montbéliard France 44 G5 47.31N 6.48E
Mont-de-Marsan *town* France 44 C3 43.54N 0.30W
Montdidier France 42 B1 49.39N 2.34E
Monte Alegre *town* Brazil 77 E4 2.01S 54.04W
Monte Carlo Monaco 44 G3 43.44N 7.25E
Montecristo *i.* Italy 50 D5 42.20N 10.19E
Montego Bay *town* Jamaica 71 I4 18.27N 77.56W
Montélimar France 44 F4 44.33N 4.45E
Monte Lindo *r.* Paraguay 77 E2 23.57S 57.10W
Montemorelos Mexico 70 E6 25.12N 99.50W
Montenegro *d.* Yugo 56 D5 42.49N 19.24E
Montería Colombia 71 I2 8.45N 75.54W
Monterrey Mexico 70 D6 25.40N 100.20W
Monte Santu, C. Italy 50 C4 40.05N 9.44E
Montes Claros Brazil 77 F3 16.45S 43.52W
Montevideo Uruguay 75 D3 34.55S 56.10W
Montgomery U.S.A. 65 I3 32.22N 86.20W
Montgomery Wales 12 D4 52.34N 3.09W
Montluçon France 44 E5 46.20N 2.36E
Montmagny Canada 63 L2 46.59N 70.33W
Montpelier U.S.A. 65 L5 44.16N 72.34W
Montpellier France 44 E3 43.36N 3.53E
Montréal Canada 63 K2 45.30N 73.36W
Montreuil France 42 A2 50.28N 1.46E
Montrose Scotland 19 G1 56.43N 2.29W
Montrose U.S.A. 64 E4 38.29N 107.53W
Montserrat *i.* Leeward Is. 71 L4 16.45N 62.14W
Monywa Myanmar 97 I5 22.07N 95.11E
Monza Italy 50 C6 45.35N 9.16E
Monzón Spain 46 F4 41.52N 0.10E
Moore, L. Australia 110 A3 29.30S 117.30E
Moorhead U.S.A. 64 G6 46.51N 96.44W
Moose Jaw Canada 62 H3 50.23N 105.35W
Moosonee Canada 63 J3 51.18N 80.40W
Mopti Mali 84 D3 14.29N 4.10W
Mora Sweden 43 C3 61.00N 14.30E
Morar, Loch Scotland 18 D1 56.56N 5.40W
Morava *r.* Yugo. 56 E6 44.43N 21.02E
Moray *d.* Scotland 8 D5 57.30N 3.20W
Moray Firth *est.* Scotland 19 F2 57.45N 3.50W
Morbach Germany 42 G1 49.48N 7.07E
Morden Canada 62 I2 49.15N 98.10W
Morecambe England 14 E3 54.03N 2.52W
Morecambe B. England 14 E3 54.05N 3.00W
Moree Australia 110 D3 29.29S 149.53E
Morehead P.N.G. 105 K2 8.55S 141.33E
Morelia Mexico 70 D4 19.40N 101.11W
More, Loch Scotland 18 E3 58.16N 4.51W
Morelos *d.* Mexico 70 E4 18.40N 99.00W
Morena, Sierra *mts.* Spain 46 C3 38.10N 5.00W
Moretonhampstead England 13 D2 50.39N 3.45W
Morgan City U.S.A. 65 H2 29.41N 91.13W
Mori Japan 106 D4 42.07N 140.30E
Morioka Japan 106 D3 39.43N 141.10E
Moriston *r.* Scotland 18 E2 57.12N 4.37W
Morlaix France 44 B6 48.35N 3.50W
Morley England 15 F2 53.45N 1.36W
Morocco Africa 84 D5 31.00N 5.00W
Moro G. Phil. 105 G5 6.30N 123.20E
Morogoro Tanzania 86 C4 6.49S 37.40E
Mörön Mongolia 103 I7 49.36N 100.08E
Morondava Madagascar 86 D2 20.17S 44.17E

Morón de la Frontera Spain 46 C2 37.08N 5.27W
Moroni Comoros 86 D3 11.40S 43.19E
Morotai *i.* Indonesia 105 H4 2.10N 128.30E
Moroto Uganda 87 A3 2.30N 34.40E
Moroto, Mt. Uganda 87 A3 2.30N 34.46E
Morpeth England 15 F4 55.10N 1.40W
Morriston Wales 12 D3 51.40N 3.55W
Morro, Punta *c.* Chile 76 C2 27.06S 71.00W
Morrosquillo, G. of Colombia 71 I2 9.30N 75.50W
Morte B. England 13 C3 51.10N 4.14W
Mortehoe England 13 C3 51.12N 4.12W
Mortes *r.* Brazil 77 E3 11.59S 50.25W
Morungole *mtn.* Uganda 87 A3 3.50N 34.02E
Morvern *f.* Scotland 16 D4 56.37N 5.45W
Mosborough England 15 F2 53.19N 1.21W
Moscow Russian Fed. 58 F3 55.45N 37.42E
Moselle *r.* Germany 48 C4 50.23N 7.37E
Moshi Tanzania 87 B2 3.20S 37.21E
Mosjöen Norway 43 C4 65.50N 13.10E
Mosquitos Coast *f.* Nicaragua 71 H3 13.00N 84.00W
Mosquitos, G. of Panama 71 H2 9.00N 81.00W
Moss Norway 43 B2 59.26N 10.41E
Mossoró Brazil 77 G4 5.10S 37.20W
Mostaganem Algeria 52 D4 35.56N 0.05E
Mostar Bosnia. 56 C5 43.20N 17.50E
Mosul Iraq 94 F6 36.21N 43.08E
Møsvatnet *l.* Norway 43 A2 59.55N 8.00E
Motala Sweden 43 C2 58.34N 15.05E
Motherwell Scotland 17 F3 55.48N 4.00W
Motril Spain 46 D2 36.45N 3.31W
Moulins France 44 E5 46.34N 3.20E
Moulmein Myanmar 97 J4 16.20N 97.50E
Moundou Chad 84 F2 8.36N 16.02E
Mt. Elgon Nat. Park Uganda 87 A3 1.05N 34.20E
Mount Gambier *town* Australia 110 D2 37.51S 140.50E
Mount Isa *town* Australia 110 C3 20.50S 139.29E
Mountmellick Rep. of Ire. 20 D3 53.08N 7.21W
Mountrath Rep. of Ire. 20 D3 53.00N 7.28W
Mount's B. England 13 B2 50.05N 5.25W
Mountsorrel *town* England 10 D3 52.44N 1.07W
Mourdi, Dépression du *f.* Chad 94 B2 18.00N 23.30E
Mourne *r.* N. Ireland 16 B2 54.50N 7.29W
Mourne Mts. N. Ireland 16 C2 54.10N 6.02W
Mousa *i.* Scotland 19 Y8 60.00N 1.10W
Mouscron Belgium 42 C2 50.46N 3.10E
Mouzon France 42 E1 49.36N 5.05E
Moville Rep. of Ire. 16 B3 55.11N 7.03W
Moy *r.* Rep. of Ire. 20 B4 54.10N 9.09W
Moyale Kenya 87 B3 3.31N 39.01E
Moyen Atlas *mts.* Morocco 52 B3 33.30N 5.00W
Mozambique Africa 86 C3 18.00S 35.00E
Mozambique Channel Indian Oc. 86 D3 16.00S 42.30E
M'Saken Tunisia 52 F4 35.42N 10.33E
Msambweni Kenya 87 B2 4.27S 39.28E
Mtelo *mtn.* Kenya 87 B3 1.40N 35.23E
Mtsensk Russian Fed. 55 O5 53.18N 36.35E
Mtwara Tanzania 86 D3 10.17S 40.11E
Muar Malaysia 104 C4 2.01N 102.35E
Muarabungo Indonesia 104 C3 1.29S 102.06E
Muchinga Mts. Zambia 82 G3 12.00S 31.00E
Much Wenlock England 10 C3 52.36N 2.34W
Muck *i.* Scotland 18 C1 56.50N 6.14W
Muckish Mtn. Rep. of Ire. 16 A3 55.06N 7.59W
Muckle Roe *i.* Scotland 19 Y9 60.22N 1.26W
Mudanjiang China 103 N6 44.36N 129.42E
Muğla Turkey 57 I2 37.12N 28.22E
Muhammad Qol Sudan 94 E3 20.53N 37.09E
Mühlhausen Germany 48 E4 51.12N 10.27E
Muhos Finland 43 F4 64.49N 26.00E
Mui Ethiopia 87 B4 5.59N 35.29E
Mui Ca Mau *c.* Vietnam 104 C5 8.30N 104.35E
Muine Bheag Rep. of Ire. 20 E2 52.41N 6.59W
Muirkirk Scotland 16 E3 55.31N 4.04W
Muirneag *mtn.* Scotland 18 C3 58.24N 6.21W
Mukacheve Ukraine 54 H3 48.26N 22.45E
Mukalla Yemen 95 G1 14.34N 49.09E
Mukono Uganda 87 A3 0.21N 32.27E
Mulanje, Mt. Malawi 86 C3 15.57S 35.33E
Mulhacén *mtn.* Spain 46 D2 37.04N 3.22W
Mulhouse France 44 G5 47.45N 7.21E
Muling *r.* China 106 B5 45.53N 133.40E
Mull *i.* Scotland 16 D4 56.28N 5.56W
Mull, Sd. of *str.* Scotland 16 D4 56.32N 5.55W
Mullaghareirk Mts. Rep. of Ire. 20 B2 52.20N 9.10W
Mull Head Scotland 19 G4 59.23N 2.53W
Mullingar Rep. of Ire. 20 D3 53.31N 7.21W
Mull of Galloway *c.* Scotland 16 E2 54.39N 4.52W
Mull of Kintyre *c.* Scotland 16 D3 55.17N 5.45W
Mull of Oa *c.* Scotland 16 C3 55.36N 6.20W
Multan Pakistan 96 E7 30.10N 71.36E
Mumbai India 96 E4 10.56N 72.51E
Muna *i.* Indonesia 105 G2 5.00S 122.30E
Mundesley England 11 G3 52.53N 1.24E
Mundford England 11 F3 52.31N 0.39E
Munger India 97 H6 25.24N 86.29E
Munich Germany 48 E3 48.08N 11.35E
Munim *r.* Brazil 77 F4 2.51S 44.05W
Münster Germany 48 C4 51.58N 7.37E
Muojärvi *l.* Finland 43 G4 65.56N 29.40E
Muonio Finland 43 E4 67.52N 23.45E
Muonio *r.* Sweden/Finland 43 E4 67.13N 23.30E
Murallón *mtn.* Argentina/Chile 75 B2 49.48S 73.26W
Muranga Kenya 87 B2 0.43S 37.10E
Murchison *r.* Australia 110 A3 27.30S 114.10E
Murcia Spain 46 E2 37.59N 1.08W
Mureş *r.* Romania 54 G2 46.16N 20.10E
Muret France 44 D3 43.28N 1.19E
Müritz, L. Germany 48 F5 53.25N 12.45E
Murmansk Russian Fed. 58 F4 68.59N 33.08E
Murom Russian Fed. 58 G3 55.04N 42.04E
Muroran Japan 106 D4 42.21N 140.59E
Murray *r.* Australia 110 C2 35.23S 139.20E
Murray Bridge *town* Australia 110 C2 35.10S 139.17E
Murrumbidgee *r.* Australia 110 D2 34.38S 143.10E
Mururoa *i.* French Polynesia 109 Q4 22.00S 140.00W
Murwara India 97 G5 23.49N 80.28E
Murzuq Libya 84 F4 25.56N 13.57E
Muscat Oman 95 I3 23.36N 58.37E
Musgrave Ranges *mts.* Australia 110 C3 26.30S 131.10E
Muskegon U.S.A. 65 I5 43.13N 86.10W
Muskogee U.S.A. 65 G4 35.45N 95.21W
Musmar Sudan 94 E2 18.13N 35.38E
Musoma Tanzania 87 A2 1.29S 33.48E
Musselburgh Scotland 17 F3 55.57N 3.04W
Mut Egypt 94 C4 25.29N 28.59E
Mut Turkey 57 K2 36.38N 33.27E
Mutare Zimbabwe 86 C3 18.58S 32.38E
Mutis *mtn.* Indonesia 105 G2 9.35S 124.15E
Mutsu Japan 106 D4 41.16N 141.12E
Muzaffargarh Pakistan 95 L5 30.04N 71.12E
Muzaffarpur India 97 H6 26.07N 85.23E
Mwanza Tanzania 87 A2 2.30S 32.54E
Mwene-Ditu Dem. Rep. of Congo 86 B4 7.01S 23.27E
Mweru, L. Zambia/Dem. Rep. of Congo 86 B4 9.00S 28.40E
Myanmar Asia 97 J5 21.00N 95.00E
Myingyan Myanmar 97 J5 21.25N 95.20E
Mykolayiv Ukraine 55 M2 46.57N 32.00E
Mynydd Eppynt *mts.* Wales 12 D4 52.06N 3.30W
Mysore India 96 F3 12.18N 76.37E
My Tho Vietnam 104 D6 10.21N 106.21E
Mytilini Greece 56 H3 39.06N 26.34E
Mytishchi Russian Fed. 55 O6 55.54N 37.47E
Mzuzu Malawi 86 C3 11.26S 34.02E

N

Naas Rep. of Ire. 20 E3 53.13N 6.41W
Naberera Tanzania 87 B2 4.10S 36.57E
Naberezhnyye Chelny Russian Fed. 58 H3 55.42N 52.20E
Nabeul Tunisia 52 F4 36.28N 10.44E
Nacala Mozambique 86 D3 14.30S 40.37E
Nador Morocco 52 C4 35.12N 2.55W
Næstved Denmark 43 B1 55.14N 11.47E
Nafplio Greece 56 F2 37.33N 22.47E
Naga Phil. 105 G6 13.36N 123.12E
Nagaland *d.* India 97 I6 26.10N 94.30E
Nagano Japan 106 C3 36.39N 138.10E
Nagaoka Japan 106 C3 37.30N 138.50E
Nagaon India 97 I6 26.20N 92.41E
Nagasaki Japan 106 A2 32.45N 129.52E
Nagercoil India 96 F2 8.11N 77.30E
Nagha Kalat Pakistan 95 K4 27.24N 65.08E
Nagichot Sudan 87 A3 4.16N 33.34E
Nagoya Japan 106 C3 35.08N 136.53E
Nagpur India 97 F5 21.10N 79.12E
Nagykanizsa Hungary 54 E2 46.27N 17.01E
Naha Japan 103 N3 26.10N 127.40E
Nahanni *r.* Canada 62 F4 61.00N 123.20W
Nahavand Iran 95 G5 34.13N 48.23E
Nailsworth England 10 C2 51.41N 2.12W
Nain Canada 63 L3 56.30N 61.45W
Na'īn Iran 95 H5 32.52N 53.05E
Nairn Scotland 19 F2 57.35N 3.52W
Nairn *r.* Scotland 19 F2 57.35N 3.52W
Nairobi Kenya 87 B2 1.17S 36.50E
Nairobi *d.* Kenya 87 B2 1.15S 36.50E
Naivasha Kenya 87 B2 0.44S 36.26E
Naivasha, L. Kenya 87 B2 0.45S 36.22E
Najafabad Iran 95 H5 32.38N 51.23E
Najd *d.* Saudi Arabia 94 F3 25.00N 43.00E
Najin N. Korea 106 B4 42.10N 130.20E
Najran Saudi Arabia 94 F2 17.28N 44.06E
Nakhodka Russian Fed. 59 P2 42.53N 132.54E
Nakhon Pathom Thailand 97 J3 13.50N 100.01E
Nakhon Ratchasima Thailand 104 C7 15.02N 102.12E
Nakhon Sawan Thailand 104 C7 15.35N 100.10E
Nakhon Si Thammarat Thailand 104 B5 8.29N 99.55E
Naknek U.S.A. 62 C3 58.45N 157.00W
Nakskov Denmark 43 B1 54.50N 11.10E
Nakuru Kenya 87 B2 0.16S 36.04E
Nalut Libya 84 F5 31.53N 10.59E
Namakzar-e Shadad *f.* Iran 95 I5 30.00N 59.00E
Namanga Kenya 87 B2 2.31S 36.47E
Namangan Uzbekistan 102 C6 40.59N 71.41E
Namaqualand *f.* Namibia 86 A2 25.30S 17.00E
Nam Co *l.* China 102 G4 30.40N 90.30E
Nam Dinh Vietnam 104 D8 20.25N 106.12E
Namib Desert Namibia 86 A2 22.50S 14.40E
Namibe Angola 86 A3 15.10S 12.10E
Namibia Africa 86 A2 22.00S 17.00E
Namlea Indonesia 105 H3 3.15S 127.07E
Nampo N. Korea 103 N5 38.40N 125.30E
Nampula Mozambique 86 C3 15.09S 39.14E
Namsos Norway 43 B4 64.28N 11.30E
Namur Belgium 42 D2 50.28N 4.52E
Namur *d.* Belgium 42 D2 50.20N 4.45E
Nan Thailand 104 C7 18.45N 100.42E
Nanaimo Canada 64 B6 49.08N 123.58W
Nanao Japan 106 C3 37.03N 136.58E
Nanchang China 103 L3 28.38N 115.56E
Nanchong China 103 J4 30.54N 106.06E
Nancy France 44 G6 48.42N 6.12E
Nandurbar India 96 E5 21.22N 74.15E
Nanjing China 103 L4 32.00N 118.40E
Nan Ling *mts.* China 103 K3 25.20N 112.30E
Nanning China 103 J2 22.50N 108.19E
Nanortalik Greenland 63 N4 60.09N 45.15W
Nanping China 103 L3 26.40N 118.07E
Nansio Tanzania 87 A2 2.07S 33.03E
Nantes France 44 C5 47.14N 1.35W
Nantong China 103 M4 32.05N 120.59E
Nantucket I. U.S.A. 65 M5 41.16N 70.00W
Nantwich England 14 E2 53.05N 2.31W
Nant-y-moch Resr. Wales 12 D4 52.28N 3.50W
Nanumea *i.* Tuvalu 108 M6 5.40S 176.10E
Nanyuki Kenya 87 B2 0.01N 37.08E
Napamute U.S.A. 62 C4 61.31N 158.45W
Napier New Zealand 111 G2 39.30S 176.54E
Naples Italy 50 F4 40.50N 14.14E
Naples U.S.A. 65 J2 26.09N 81.48W
Napo *r.* Peru 76 C4 3.30S 73.10W
Narberth Wales 12 C3 51.48N 4.45W
Narbonne France 44 E3 43.11N 3.00E
Narborough Leics. England 10 D3 52.35N 1.11W
Narborough Norfolk England 11 F3 52.42N 0.35E
Nares Str. Canada 63 K5 78.30N 72.00W
Narmada *r.* India 96 E5 21.40N 73.00E
Narodnaya *mtn.* Russian Fed. 58 I4 65.00N 61.00E
Narok Kenya 87 B2 1.05S 35.55E
Närpes Finland 43 E3 62.28N 21.19E
Narva Estonia 43 G2 59.22N 28.17E
Narvik Norway 43 D5 68.26N 17.25E
Naryan Mar Russian Fed. 58 H4 67.37N 53.02E
Naryn Kyrgyzstan 102 D6 41.24N 76.00E
Nashville U.S.A. 65 I4 36.10N 86.50W
Näsijärvi *l.* Finland 43 E3 61.30N 23.50E
Nasik India 96 E5 20.00N 73.52E
Nassau The Bahamas 71 I6 25.03N 77.20W
Nassau *i.* Cook Is. 108 O5 11.33S 165.25W
Nasser, L. Egypt 94 D3 22.40N 32.00E
Nässjö Sweden 43 C2 57.39N 14.40E
Nata Tanzania 87 A2 2.00S 34.28E
Natal Brazil 77 G4 5.46S 35.15W
Natchez U.S.A. 65 H3 31.22N 91.24W
Natron, L. Tanzania 87 B2 2.18S 36.05E
Natuna Besar *i.* Indonesia 104 D4 4.00N 108.20E
Natuna Is. Indonesia 104 D4 3.00N 108.50E
Nauru Pacific Oc. 108 L6 0.32S 166.55E
Navalmoral de la Mata Spain 46 C3 39.54N 5.33W
Navan Rep. of Ire. 20 E3 53.39N 6.42W
Navapolatsk Belarus 55 K6 55.34N 28.40E
Naver *r.* Scotland 19 E3 58.29N 4.12W
Naver, Loch Scotland 19 E3 58.17N 4.20W
Navlya Russian Fed. 55 N5 52.51N 34.30E
Navrongo Ghana 84 D3 10.51N 1.03W
Nawabshah Pakistan 96 D6 26.15N 68.26E
Naxçivan Azerbaijan 95 G6 39.12N 45.22E
Naxos *i.* Greece 56 G2 37.03N 25.30E
Nayarit *d.* Mexico 70 D5 21.30N 105.00W
Nazareth Israel 94 E5 32.41N 35.16E
Nazas *r.* Mexico 70 D6 25.34N 103.25W
Nazca Peru 76 C3 14.53S 74.54W
Nazilli Turkey 57 I2 37.55N 28.20E
Nazret Ethiopia 85 H2 8.32N 39.22E
Nazwá Oman 95 I3 22.56N 57.33E
N'dalatando Angola 86 A4 9.12S 14.54E
Ndéle C.A.R. 85 G2 8.24N 20.39E
Ndeni *i.* Solomon Is. 111 F4 10.30S 166.00E
Ndjamena Chad 84 F3 12.10N 14.59E
Ndola Zambia 86 B3 13.00S 28.39E
Ndoto *mtn.* Kenya 87 B3 1.42N 37.10E
Neagh, Lough N. Ireland 16 C2 54.36N 6.26W
Neath Wales 12 D3 51.39N 3.49W
Neath *r.* Wales 12 D3 51.39N 3.50W
Neath Port Talbot *d.* Wales 9 D2 51.42N 3.47W
Nebitdag Turkmenistan 95 H6 39.31N 54.24E
Neblina, Pico da *mtn.* Colombia/Brazil 76 D5 0.50N 66.00W
Nebraska *d.* U.S.A. 64 F5 41.30N 100.00W
Nebrodi Mts. Italy 50 F2 38.00N 14.50E
Nechisar Nat. Park Ethiopia 87 B4 6.00N 37.50E
Neckar *r.* Germany 48 D3 49.32N 8.26E
Necker I. Hawaiian Is. 108 O9 23.35N 164.42W
Needham Market England 11 G3 52.09N 1.02E
Needles U.S.A. 64 D3 34.51N 114.36W
Neftekumsk Russian Fed. 58 G2 44.46N 44.10E
Nefyn Wales 12 C4 52.55N 4.31W
Negēlē Ethiopia 87 B4 5.20N 39.36E
Negev *des.* Israel 94 D5 30.42N 34.55E
Negotin Yugo. 56 F6 44.14N 22.33E
Negra, Cordillera *mts.* Peru 76 C4 10.00S 78.00W
Negra, Punta *c.* Peru 76 B4 6.06S 81.09W
Negro *r.* Argentina 75 C2 41.00S 62.48W
Negro *r.* Amazonas Brazil 77 D4 3.30S 60.00W
Negro *r.* Mato Grosso do Sul Brazil 77 E3 19.15S 57.15W
Negro, C. Morocco 46 C1 35.41N 5.17W
Negros *i.* Phil. 105 G5 10.00N 123.00E
Neijiang China 103 J3 29.32N 105.03E
Neiva Colombia 74 B7 2.58N 75.15W
Nek'emte Ethiopia 85 H2 9.02N 36.31E
Neksø Denmark 54 D6 55.04N 15.09E
Nelkan Russian Fed. 59 P3 57.40N 136.04E
Nellore India 97 G3 14.29N 80.00E
Nelson Canada 62 G2 49.29N 117.17W
Nelson England 15 E2 53.50N 2.14W
Nelson New Zealand 111 G1 41.16S 173.15E
Nelson *r.* Canada 63 I3 57.00N 93.20W
Nelspruit R.S.A. 86 C2 25.27S 30.58E
Neman Russian Fed. 54 H6 55.02N 22.02E
Neman *r.* Europe 54 G6 55.23N 21.15E
Nementcha, Mts. de Algeria/Tunisia 52 E4 35.00N 7.00E
Nenagh Rep. of Ire. 20 C2 52.52N 8.13W
Nene *r.* England 11 F3 52.49N 0.12E
Nenjiang China 103 N7 49.10N 125.15E
Nepal Asia 97 G6 28.00N 84.00E
Nephin *mtn.* Rep. of Ire. 20 B4 54.00N 9.25W
Neris *r.* Lithuania 55 H6 54.52N 23.55E
Ness *r.* Scotland 19 E2 57.27N 4.15W
Ness, Loch Scotland 19 E2 57.16N 4.30W
Neston England 14 D2 53.17N 3.03W
Nestos *r.* Greece 56 G4 40.51N 24.48E
Netherlands Europe 42 E4 52.00N 5.30E
Netherlands Antilles *is.* S. America 71 K3 12.30N 69.00W
Netley England 10 D1 50.52N 1.19W
Nettilling L. Canada 63 K4 66.30N 70.40W
Neubrandenburg Germany 48 F5 53.33N 13.16E
Neuchâtel Switz. 44 G5 47.00N 6.56E
Neuchâtel, Lac de *l.* Switz. 44 G5 46.55N 6.55E
Neufchâteau Belgium 42 E1 49.51N 5.26E
Neumünster Germany 48 E6 54.05N 10.01E
Neunkirchen Germany 42 G1 49.21N 7.12E
Neuquén Argentina 75 C3 38.55S 68.55W
Neustrelitz Germany 48 F5 53.22N 13.05E
Neuwied Germany 48 C4 50.26N 7.28E
Nevada *d.* U.S.A. 64 C4 39.00N 117.00W
Nevada, Sierra *mts.* Spain 46 D2 37.04N 3.20W
Nevada, Sierra *mts.* U.S.A. 64 C4 37.30N 119.00W
Nevers France 44 E5 47.00N 3.09E
Nevėžis *r.* Lithuania 55 H6 54.52N 23.55E
Nevis, Loch Scotland 18 D1 57.00N 5.40W
Nevşehir Turkey 57 L3 38.38N 34.43E
New Addington England 11 E2 51.21N 0.01W
New Alresford England 10 D2 51.06N 1.10W
Newark U.S.A. 65 L5 40.44N 74.11W
Newark-on-Trent England 15 G2 53.06N 0.48W
New Bedford U.S.A. 65 L5 41.38N 70.55W
New Bern U.S.A. 65 K4 35.05N 77.04W
Newbiggin-by-the-Sea England 15 F4 55.11N 1.30W
Newbridge Rep. of Ire. 20 E3 53.11N 6.48W
Newbridge Wales 12 D3 51.41N 3.09W
New Britain *i.* P.N.G. 110 D5 6.00S 143.00E
New Brunswick *d.* Canada 63 L2 47.00N 66.00W
Newburgh Scotland 17 F4 56.21N 3.15W
Newbury England 10 D2 51.24N 1.19W
New Caledonia *i.* Pacific Oc. 111 F3 22.00S 165.00E
Newcastle Australia 110 E2 32.55S 151.46E
Newcastle N. Ireland 16 D2 54.13N 5.54W
Newcastle Emlyn Wales 12 C4 52.02N 4.29W
Newcastle-under-Lyme England 10 C3 53.02N 2.15W
Newcastle upon Tyne England 15 F3 54.58N 1.36W
Newcastle West Rep. of Ire. 20 B2 52.27N 9.04W
New Cumnock Scotland 16 E3 55.24N 4.11W
New Delhi India 96 F6 28.37N 77.13E
Newent England 10 C2 51.56N 2.24W
New Forest *f.* England 10 D1 50.50N 1.35W
Newfoundland *i.* Canada 63 M2 48.30N 56.00W
Newfoundland *d.* Canada 63 L3 55.00N 60.00W
New Galloway Scotland 16 E3 55.05N 4.09W
New Guinea *i.* Austa. 110 D5 5.00S 140.00E
New Hampshire *d.* U.S.A. 65 L5 44.00N 71.30W
New Haven U.S.A. 65 L5 41.14N 72.50W
New Ireland *i.* P.N.G. 110 E5 2.30S 151.30E
New Jersey *d.* U.S.A. 65 L5 40.00N 74.30W
New Liskeard Canada 65 K6 47.31N 79.41W
Newmarket England 11 F3 52.15N 0.23E
Newmarket on-Fergus Rep. of Ire. 20 C2 52.46N 8.55W
New Mexico *d.* U.S.A. 64 E3 34.00N 106.00W
New Milton England 10 D1 50.45N 1.39W
Newnham England 10 C2 51.48N 2.27W
New Orleans U.S.A. 65 H2 30.00N 90.03W
New Pitsligo Scotland 19 G2 57.35N 2.12W
Newport Essex England 11 F2 51.58N 0.13E
Newport Hants. England 10 D1 50.43N 1.18W
Newport Shrops. England 10 C3 52.47N 2.22W
Newport Newport Wales 12 E3 51.34N 2.59W
Newport Pem. Wales 12 C4 52.01N 4.51W
Newport *d.* Wales 9 D2 51.33N 3.00W
Newport B. Wales 12 C4 52.03N 4.53W
Newport News U.S.A. 65 K4 36.59N 76.26W
Newport Pagnell England 10 E3 52.05N 0.42W
New Providence *i.* The Bahamas 71 I6 25.03N 77.25W
Newquay England 13 B2 50.24N 5.06W
New Quay Wales 12 C4 52.13N 4.22W
New Romney England 11 F1 50.59N 0.58E
New Ross Rep. of Ire. 20 E2 52.23N 6.59W
Newry N. Ireland 16 C2 54.11N 6.20W
Newry Canal N. Ireland 16 C2 54.15N 6.22W
New Scone Scotland 17 F4 56.25N 3.25W
New Siberian Is. Russian Fed. 59 Q5 76.00N 144.00E
New South Wales *d.* Australia 110 D2 33.45S 147.00E
Newton Abbot England 13 D2 50.32N 3.37W
Newton Aycliffe England 15 F3 54.36N 1.34W
Newtonhill Scotland 19 G2 57.02N 2.08W
Newton-le-Willows England 14 E2 53.28N 2.38W
Newton Mearns Scotland 16 E3 55.46N 4.18W
Newtonmore Scotland 19 E2 57.03N 4.10W
Newton Stewart Scotland 16 E2 54.57N 4.29W
Newtown Wales 12 D4 52.31N 3.19W
Newtownabbey N. Ireland 16 D2 54.40N 5.57W
Newtownards N. Ireland 16 D2 54.35N 5.42W
Newtown St. Boswells Scotland 17 G3 55.35N 2.40W
Newtownstewart N. Ireland 16 B2 54.43N 7.25W
New York U.S.A. 65 L5 40.40N 73.50W
New York *d.* U.S.A. 65 K5 43.00N 75.00W
New Zealand Austa. 111 G1 41.00S 175.00E
Neyriz Iran 95 H4 29.12N 54.17E
Neyshabur Iran 95 I6 36.13N 58.49E
Ngaoundéré Cameroon 84 F2 7.20N 13.35E
Ngorongoro Conservation Area Tanzania 87 B2 3.00S 35.30E
Nguigmi Niger 84 F3 14.00N 13.11E
Ngulu *i.* Fed. States of Micronesia 105 J5 8.30N 137.30E
Nha Trang Vietnam 104 D6 12.15N 109.10E
Niamey Niger 84 E3 13.32N 2.05E
Niangara Dem. Rep. of Congo 85 G2 3.45N 27.54E
Nias *i.* Indonesia 104 B4 1.05N 97.30E
Nicaragua C. America 71 H3 13.00N 85.00W
Nicaragua, L. Nicaragua 71 G3 11.30N 85.30W
Nice France 44 G3 43.42N 7.16E
Nicobar Is. India 97 I2 8.00N 94.00E
Nicosia Cyprus 57 K1 35.11N 33.23E
Nicoya, G. of Costa Rica 71 H2 9.30N 85.00W
Nidd *r.* England 15 F3 54.01N 1.12W
Nidzica Poland 54 G5 53.22N 20.26E
Niers *r.* Neth. 42 E3 51.43N 5.56E
Nieuwpoort Belgium 42 B3 51.08N 2.45E
Niğde Turkey 57 L2 37.58N 34.42E
Niger Africa 84 E3 17.00N 10.00E
Niger *r.* Nigeria 84 E2 4.15N 6.05E
Nigeria Africa 84 E2 9.00N 9.00E
Nigg B. Scotland 19 E2 57.42N 4.01W
Niigata Japan 106 C3 37.58N 139.02E
Nijmegen Neth. 42 E3 51.50N 5.52E
Nikel' Russian Fed. 43 G5 69.20N 29.44E
Nikolayevsk-na-Amure Russian Fed. 59 Q3 53.20N 140.44E
Nikopol' Ukraine 55 N2 47.34N 34.25E
Niksar Turkey 57 M4 40.35N 36.59E
Nikšić Yugo. 56 D5 42.48N 18.56E
Nikumaroro *i.* Kiribati 108 N6 4.40S 174.32W
Nile *r.* Egypt 94 D5 31.30N 30.25E
Nilgiri Hills India 96 F3 11.30N 77.30E
Nîmes France 44 F3 43.50N 4.21E
Ningbo China 103 M3 29.54N 121.33E
Ningxia Huizu Zizhiqu *d.* China 103 J5 37.00N 106.00E
Ninigo Group *is.* P.N.G. 105 K3 2.00S 143.00E
Nioro Mali 84 D3 15.12N 9.35W
Niort France 44 C5 46.19N 0.27W
Nipigon Canada 63 J2 49.02N 88.26W
Nipigon, L. Canada 63 J2 49.50N 88.30W
Niš Yugo. 56 E5 43.20N 21.54E
Niterói Brazil 77 F2 22.45S 43.06W
Nith *r.* Scotland 17 F3 55.00N 3.35W
Nitra Slovakia 54 F3 48.20N 18.05E
Niue *i.* Cook Is. 108 O5 19.02S 169.52W
Nivelles Belgium 42 D2 50.36N 4.20E
Nizamabad India 97 F4 18.40N 78.05E
Nizhneudinsk Russian Fed. 59 L3 54.55N 99.00E
Nizhnevartovsk Russian Fed. 58 J3 60.57N 76.40E
Nizhniy Novgorod Russian Fed. 58 G3 56.20N 44.00E
Nizhniy Tagil Russian Fed. 58 I3 58.00N 60.00E
Nizhyn Ukraine 55 L4 51.03N 31.54E
Nizip Turkey 57 M2 37.02N 37.47E
Nkongsamba Cameroon 84 E2 4.59N 9.53E
Nobeoka Japan 106 B2 32.36N 131.40E
Nogales Mexico 70 B7 31.20N 111.00W
Nogent-le-Rotrou France 44 D6 48.19N 0.50E
Nogent-sur-Oise France 42 B1 49.17N 2.28E
Nogwak-san *mtn.* S. Korea 106 A3 37.20N 128.50E
Nohfelden Germany 42 G1 49.35N 7.09E
Noirmoutier, Île de *i.* France 44 B5 47.00N 2.15W
Nok Kundi Pakistan 95 J4 28.46N 62.46E
Nome U.S.A. 62 B4 64.30N 165.30W
Nomoi Is. Fed. States of Micronesia 108 K7 5.21N 153.42E

Pangkalpinang Indonesia **104 D3** 2.05S 106.09E
Pangnirtung Canada **63 L4** 66.05N 65.45W
Panjgur Pakistan **95 J4** 26.58N 64.06E
Pantelleria *i.* Italy **50 E2** 36.48N 12.00E
Paola Italy **50 G3** 39.21N 16.03E
Papa Stour *i.* Scotland **19 Y9** 60.20N 1.42W
Papa Westray *i.* Scotland **19 G4** 59.22N 2.54W
Papeete Fr. Polynesia **109 Q5** 17.30S 149.31W
Papenburg Germany **42 G5** 53.05N 7.25E
Paphos Cyprus **57 K1** 34.45N 32.25E
Papua, G. of P.N.G. **110 D5** 8.50S 145.00E
Papua New Guinea Austa. **110 D5** 6.00S 148.00E
Par England **13 C2** 50.21N 4.43W
Pará *d.* Brazil **77 E4** 4.00S 53.00W
Paracel Is. S. China Sea **104 E7** 16.20N 112.00E
Paragua *r.* Bolivia **77 D3** 13.28S 61.50W
Paraguai *r. see* Paraguay *r.* Brazil **77**
Paraguay S. America **77 E2** 23.00S 58.00W
Paraguay *r.* Argentina **77 E2** 27.30S 58.50W
Paraíba *r.* Brazil **75 E4** 21.45S 41.10W
Paraíba *d.* Brazil **77 G4** 7.30S 36.30W
Parakow Benin **84 E2** 9.23N 2.40E
Paramaribo Suriname **74 D7** 5.52N 55.14W
Paraná Argentina **77 D1** 31.45S 60.30W
Paraná *r.* Argentina **75 D3** 34.00S 58.30W
Paraná *r.* Brazil **77 F3** 12.30S 48.10W
Paraná *d.* Brazil **77 E2** 24.30S 52.00W
Paranaguá Brazil **77 F2** 25.32S 48.36W
Paranaíba *r.* Brazil **77 E3** 20.00S 51.00W
Paranapanema *r.* Brazil **77 E2** 22.30S 53.03W
Paranapiacaba, Serra *mts.* Brazil **77 F2** 24.30S 49.15W
Parczew Poland **54 H4** 51.39N 22.54E
Pardo *r.* Bahia Brazil **77 G3** 15.40S 39.38W
Pardo *r.* Mato Grosso do Sul Brazil **77 E2** 21.56S 52.07W
Pardubice Czech Rep. **54 D4** 50.03N 15.45E
Parecis, Serra dos *mts.* Brazil **77 E3** 13.30S 60.00W
Parepare Indonesia **104 F3** 4.03S 119.40E
Paria Pen. Venezuela **71 L3** 10.45N 62.30W
Pariñas, Punta *c.* Peru **76 B4** 4.45S 81.22W
Parintins Brazil **77 E4** 2.36S 56.44W
Paris France **44 E6** 48.52N 2.20E
Parma Italy **50 D6** 44.48N 10.18E
Parnaíba Brazil **77 F4** 2.58S 41.46W
Parnaíba *r.* Brazil **77 F4** 3.00S 42.00W
Parnassos *mtn.* Greece **56 F3** 38.33N 22.35E
Pärnu Estonia **43 F2** 58.28N 24.30E
Paropamisus Afghan. **95 J5** 35.00N 63.00E
Paros *i.* Greece **56 G2** 37.04N 25.11E
Parrett *r.* England **13 E3** 51.10N 3.00W
Parry, C. Canada **62 F5** 70.10N 124.33W
Parry, C. Greenland **63 K5** 76.50N 71.00W
Parry Is. Canada **62 H5** 76.00N 106.00W
Parsęta *r.* Poland **54 D6** 54.12N 15.33E
Parthenay France **44 C5** 46.39N 0.14W
Partry Mts. Rep. of Ire. **20 B3** 53.40N 9.30W
Paru *r.* Brazil **77 E4** 1.33S 52.38W
Pasadena U.S.A. **64 C3** 34.10N 118.09W
Pasado, C. Ecuador **76 B4** 0.22S 80.30W
Pashkovskiy Russian Fed. **57 N6** 45.01N 39.06E
Pasig Phil. **104 G6** 14.30N 120.54E
Pasni Pakistan **95 J4** 25.16N 63.28E
Passau Germany **48 F3** 48.35N 13.28E
Passero, C. Italy **50 F2** 36.40N 15.08E
Passo Fundo Brazil **77 E2** 28.16S 52.20W
Pastavy Belarus **55 J6** 55.07N 26.50E
Pastaza *r.* Peru **76 C4** 4.50S 76.25W
Pasto Colombia **74 B7** 1.12N 77.17W
Patagonia *f.* Argentina **75 C2** 45.00S 68.00W
Patchway England **10 C2** 51.30N 2.34W
Pate I. Kenya **87 A2** 2.07S 41.08E
Paterson U.S.A. **65 L5** 40.55N 74.10W
Pathfinder Resr. U.S.A. **64 E5** 42.25N 106.55W
Patna India **97 H6** 25.37N 85.12E
Patos de Minas Brazil **77 F3** 18.35S 46.32W
Patos, Lagoa dos *l.* Brazil **77 E1** 31.00S 51.10W
Patras Greece **56 E3** 38.15N 21.45E
Patras, G. of Med. Sea **56 E3** 38.15N 21.35E
Patrington England **15 G2** 53.41N 0.02W
Pau France **44 C3** 43.18N 0.22W
Paulo Afonso Brazil **77 G4** 9.25S 38.15W
Pavia Italy **50 C6** 45.10N 9.10E
Pavlodar Kazakhstan **102 D8** 52.21N 76.59E
Pavlohrad Ukraine **55 N3** 48.34N 35.50E
Pavlovskaya Russian Fed. **53 K6** 46.18N 39.48E
Pazardzhik Bulgaria **56 G5** 42.12N 24.20E
Peace *r.* Canada **62 G3** 59.00N 111.26W
Peacehaven England **11 E1** 50.45N 0.01W
Peace River *town* Canada **62 G3** 56.15N 117.18W
Peale, Mt. U.S.A. **64 E4** 38.26N 109.14W
Pearl *r.* U.S.A. **65 I3** 30.15N 89.25W
Peć Yugo. **56 E5** 42.40N 20.17E
Pechora *r.* Russian Fed. **58 H4** 68.10N 54.00E
Pechora G. Russian Fed. **58 H4** 69.00N 56.00E
Pechory Russian Fed. **43 F2** 57.49N 27.37E
Pecos U.S.A. **64 F3** 31.25N 103.30W
Pecos *r.* U.S.A. **64 F2** 29.45N 101.25W
Pécs Hungary **54 F2** 46.05N 18.14E
Pedro Juan Caballero Paraguay **77 E2** 22.30S 55.44W
Peebles Scotland **17 F3** 55.39N 3.12W
Peel I.o.M. **14 C3** 54.14N 4.42W
Peel *r.* Canada **62 E4** 68.13N 135.00W
Pegasus B. New Zealand **111 G1** 43.15S 173.00E
Pegu Myanmar **97 J4** 17.18N 96.31E
Pegwell B. England **11 G2** 51.18N 1.25E
Peipholsfjöll *mtn.* Iceland **43 X2** 65.40N 22.25W
Peipus, L. Europe **43 F2** 58.30N 27.30E
Pekalongan Indonesia **104 D2** 6.54S 109.37E
Pelat, Mont *mtn.* France **44 G4** 44.17N 6.41E
Peleng *i.* Indonesia **105 G3** 1.30S 123.10E
Pello Finland **43 E4** 66.47N 23.55E
Pelotas Brazil **77 E1** 31.45S 52.20W
Pemba Mozambique **86 D3** 13.02S 40.30E
Pemba Channel Tanzania **87 B2** 5.00S 39.20E
Pemba I. Tanzania **87 B1** 5.10S 39.45E
Pembrey Wales **12 C3** 51.42N 4.16W
Pembroke Wales **12 C3** 51.41N 4.57W
Pembroke Dock Wales **12 C3** 51.42N 4.56W
Pembrokeshire *d.* Wales **9 C2** 51.50N 5.00W
Pembury England **11 F2** 51.07N 0.19E
Peña Nevada, Cerro *mtn.* Mexico **70 E5** 23.49N 99.51W
Peñaranda de Bracamonte Spain **46 C4** 40.54N 5.13W
Penarth Wales **13 D3** 51.26N 3.11W
Penas, Golfo de *g.* Chile **75 B2** 47.20S 75.00W
Pendine Wales **12 C3** 51.44N 4.33W
Pendlebury England **15 E2** 53.32N 2.21W
Pendle Hill England **15 E2** 53.52N 2.18W
Penicuik Scotland **17 F3** 55.49N 3.13W
Peninsular Malaysia *d.* Malaysia **104 C4** 5.00N 102.00E
Penitente, Serra do *mts.* Brazil **77 F4** 9.00S 46.15W
Penkridge England **10 C3** 54.44N 2.07W
Pennsylvania *d.* U.S.A. **65 K5** 41.00N 78.00W
Penny Icecap *f.* Canada **63 L4** 67.10N 66.50W
Penrhyn Mawr *c.* Wales **12 C4** 52.51N 4.46W
Penrith England **14 E3** 54.40N 2.45W
Penryn England **13 B2** 50.10N 5.07W
Pensacola U.S.A. **65 I3** 30.30N 87.12W
Penticton Canada **64 C6** 49.29N 119.38W
Pentire Pt. England **13 C2** 50.35N 4.55W
Pentland Firth *str.* Scotland **19 F3** 58.40N 3.00W
Pentland Hills Scotland **17 F3** 55.50N 3.20W
Pentland Skerries Scotland **19 G3** 58.41N 2.55W
Penygadair *mtn.* Wales **12 D4** 52.40N 3.55W
Pen-y-Ghent *mtn.* England **15 E3** 54.10N 2.14W
Penza Russian Fed. **58 G3** 53.11N 45.00E
Penzance England **13 B2** 50.07N 5.32W
Penzhina, G. of Russian Fed. **59 S4** 61.00N 163.00E
Peoria U.S.A. **65 I5** 40.43N 89.38W
Pereira Colombia **74 B7** 4.47N 75.46W
Périgueux France **44 D4** 45.12N 0.44E
Perija, Sierra de *mts.* Venezuela **71 J2** 9.00N 73.00W
Perito Moreno Argentina **75 B2** 46.35S 71.00W
Perlas Pt. *c.* Nicaragua **71 H3** 12.23N 83.30W
Perm Russian Fed. **58 H3** 58.01N 56.10E
Pernambuco *d.* Brazil **77 G4** 8.00S 39.00W
Pernik Bulgaria **56 F5** 42.35N 23.03E
Péronne France **42 B1** 49.56N 2.57E
Perpignan France **44 E3** 42.42N 2.54E
Perranporth England **13 B2** 50.21N 5.09W
Pershore England **10 C3** 52.07N 2.04W
Perth Australia **110 A2** 31.58S 115.49E
Perth Scotland **17 F4** 56.24N 3.28W
Perth and Kinross *d.* Scotland **8 D5** 56.30N 3.40W
Peru S. America **76 C4** 10.00S 75.00W
Perugia Italy **50 E5** 43.06N 12.24E
Pervouralsk Russian Fed. **58 H3** 56.59N 59.58E
Pesaro Italy **50 E5** 43.54N 12.54E
Pescara Italy **50 F5** 42.27N 14.13E
Pescara *r.* Italy **50 F5** 42.28N 14.13E
Peshawar Pakistan **96 E7** 34.01N 71.40E
Peshkopi Albania **56 E4** 41.41N 20.25E
Peski Karakumy *f.* Turkmenistan **95 I6** 37.45N 60.00E
Peterborough Canada **63 K2** 44.18N 78.19W
Peterborough England **11 E3** 52.35N 0.14W
Peterborough *d.* England **9 E3** 52.35N 0.14W
Peterhead Scotland **19 H2** 57.30N 1.46W
Peterlee England **15 F3** 54.45N 1.18W
Petersfield England **10 E2** 51.00N 0.56W
Petersville U.S.A. **62 C4** 62.30N 150.48W
Petrich Bulgaria **56 F4** 41.25N 23.13E
Petrolina Brazil **77 F4** 9.22S 40.30W
Petropavlovsk Kazakhstan **58 I3** 54.53N 69.13E
Petropavlovsk-Kamchatskiy Russian Fed. **59 R3** 53.03N 158.43E
Petroşani Romania **56 F6** 45.25N 23.22E
Petrozavodsk Russian Fed. **58 F4** 61.46N 34.19E
Petworth England **11 E1** 50.59N 0.37W
Pevensey England **11 F1** 50.49N 0.20E
Pewsey England **10 D2** 51.20N 1.46W
Pforzheim Germany **48 D3** 48.53N 8.41E
Phan Thiêt Vietnam **104 D6** 10.56N 108.06E
Phatthalung Thailand **104 C5** 7.38N 100.05E
Phayao Thailand **104 B7** 19.10N 99.55E
Phet Buri Thailand **104 B6** 13.01N 99.55E
Philadelphia U.S.A. **65 K5** 39.55N 75.10W
Philippeville Belgium **42 D2** 50.12N 4.32E
Philippines Asia **105 G6** 13.00N 123.00E
Philippine Trench *f.* Pacific Oc. **117 P6**
Phitsanulok Thailand **104 C7** 16.50N 100.15E
Phnom Penh Cambodia **104 C6** 11.35N 104.55E
Phoenix U.S.A. **64 D3** 33.30N 111.55W
Phoenix Is. Kiribati **108 N6** 4.00S 172.00W
Phôngsali Laos **104 C8** 21.40N 102.06E
Phrae Thailand **104 C7** 18.07N 100.09E
Phuket Thailand **104 B5** 8.00N 98.28E
Piacenza Italy **50 C6** 45.03N 9.42E
Pianosa *i.* Italy **50 D5** 42.35N 10.05E
Pian-Upe Game Res. Uganda **87 A3** 1.30N 34.40E
Piatra-Neamţ Romania **55 J2** 46.56N 26.22E
Piauí *d.* Brazil **77 F4** 7.45S 42.30W
Piave *r.* Italy **50 E6** 45.33N 12.45E
Picardie *f.* France **44 D6** 49.47N 2.00E
Pickering England **15 G3** 54.15N 0.46W
Piedras Negras Mexico **70 D6** 28.40N 100.32W
Pieksämäki Finland **43 F3** 62.18N 27.10E
Pielinen *l.* Finland **43 G3** 63.20N 29.50E
Pierre U.S.A. **64 F5** 44.23N 100.20W
Pietermaritzburg R.S.A. **86 C2** 29.36S 30.24E
Pietersburg R.S.A. **86 B2** 23.54S 29.23E
Pihlajavesi *l.* Finland **43 G3** 61.45N 29.10E
Piła Poland **54 E5** 53.09N 16.44E
Pilar Paraguay **77 E2** 26.52S 58.23W
Pilcomayo *r.* Argentina/Paraguay **77 E2** 25.15S 57.43W
Pilica *r.* Poland **54 G4** 51.52N 21.17E
Pimperne England **10 C1** 50.53N 2.09W
Pinang *i.* Malaysia **104 C5** 5.30N 100.10E
Pınarbaşı Turkey **57 M3** 38.43N 36.23E
Pinar del Rio Cuba **71 H5** 22.24N 83.42W
Pinatubo, Mt. Philippines **104 G7** 15.08N 120.35E
Pindaré *r.* Brazil **77 F4** 3.10S 44.40W
Pindus Mts. Albania/Greece **56 E4** 39.40N 21.00E
Pine Bluff *town* U.S.A. **65 H3** 34.13N 92.00W
Pinega Russian Fed. **58 G4** 64.42N 43.28E
Pineios *r.* Greece **56 F3** 39.51N 22.37E
Pinerolo Italy **50 B6** 44.53N 7.21E
Pines, I. of Cuba **71 H5** 21.40N 82.40W
Pingdingshan China **103 K4** 33.38N 113.30E
Pingxiang Guangxi China **103 J2** 22.05N 106.46E
Pingxiang Jiangxi China **103 K3** 27.36N 113.48E
Pinsk Belarus **55 J5** 52.08N 26.01E
Piombino Italy **50 D5** 42.56N 10.30E
Piotrków Trybunalski Poland **54 F4** 51.25N 19.42E
Pipa Dingzi *mtn.* China **106 A4** 44.00N 128.10E
Piracicaba Brazil **77 F2** 22.20S 47.40W
Piraeus Greece **56 F2** 37.56N 23.38E
Pirot Yugo. **56 F5** 43.10N 22.32E
Pisa Italy **50 D5** 43.43N 10.24E
Pisco Peru **76 C3** 13.46S 76.12W
Písek Czech Rep. **54 D3** 49.19N 14.10E
Pisuerga *r.* Spain **46 C4** 41.35N 5.40W
Pisz Poland **54 G5** 53.38N 21.49E
Pitcairn I. Pacific Oc. **109 R4** 25.04S 130.06W
Piteå Sweden **43 E4** 65.19N 21.30E
Piteşti Romania **56 G6** 44.52N 24.51E
Pitlochry Scotland **19 F1** 56.44N 3.47W
Pittsburgh U.S.A. **65 K5** 40.26N 79.58W
Piura Peru **76 B4** 5.15S 80.38W
Plasencia Spain **46 B4** 40.02N 6.05W
Platí, C. Greece **56 F4** 40.26N 23.59E
Platinum U.S.A. **62 B3** 59.00N 161.50W
Plauen Germany **48 F4** 50.29N 12.08E
Playa Blanca Canary Is. **46 Z2** 28.51N 13.49W
Plenty, B. of New Zealand **111 G2** 37.40S 176.50E
Pleven Bulgaria **56 G5** 43.25N 24.39E
Pljevlja Yugo. **56 D5** 43.22N 19.22E
Płock Poland **54 F5** 52.33N 19.43E
Ploieşti Romania **56 H6** 44.57N 26.02E
Plomb du Cantal *mtn.* France **44 E4** 45.04N 2.45E
Plovdiv Bulgaria **56 G5** 42.09N 24.45E
Plungė Lithuania **54 G6** 55.52N 21.49E
Plymouth England **13 C2** 50.23N 4.09W
Plymouth *d.* England **9 C2** 50.23N 4.09W
Plympton England **13 C2** 50.24N 4.02W
Plymstock England **13 C2** 50.21N 4.05W
Plynlimon *mtn.* Wales **12 D4** 52.28N 3.47W
Plzeň Czech Rep. **54 C3** 49.45N 13.22E
Po *r.* Italy **50 E6** 44.51N 12.30E
Pobeda, Mt. Russian Fed. **59 Q4** 65.20N 145.50E
Pocatello U.S.A. **64 D5** 42.53N 112.26W
Pochep Russian Fed. **55 M5** 52.55N 33.29E
Pocklington England **15 G2** 53.56N 0.48W
Poços de Caldas Brazil **77 F2** 21.48S 46.33W
Podgorica Yugo. **56 D5** 42.30N 19.16E
Podol'sk Russian Fed. **55 O6** 55.23N 37.32E
P'ohang S. Korea **106 A3** 36.00N 129.26E
Pohnpei *i.* Fed. States of Micronesia **108 K7** 6.55N 158.15E
Pointe-à-Pitre Guadeloupe **71 L4** 16.14N 61.32W
Pointe Noire *town* Congo **84 F1** 4.46S 11.53E
Poitiers France **44 D5** 46.35N 0.20E
Pokaran India **96 E6** 26.55N 71.55E
Poland Europe **54 F5** 52.30N 19.00E
Polatlı Turkey **57 K3** 39.34N 32.08E
Polatsk Belarus **55 B5** 55.30N 28.43E
Polegate England **11 F1** 50.49N 0.15E
Pol-e-Khomrī Afghan. **95 K6** 35.55N 68.45E
Policastro, G. of Med. Sea **50 F3** 40.00N 15.35E
Pollaphuca Resr. Rep. of Ire. **20 E3** 53.08N 6.30W
Pollino, Monte *mtn.* Italy **50 G3** 39.53N 16.11E
Polperro England **13 C2** 50.19N 4.31W
Poltava Ukraine **55 N3** 49.35N 34.35E
Pombal Portugal **46 A3** 39.55N 8.38W
Ponce Puerto Rico **71 K4** 18.00N 66.40W
Pondicherry India **97 F3** 11.59N 79.50E
Pond Inlet *town* Canada **63 K5** 72.40N 77.59W
Ponferrada Spain **46 B5** 42.32N 6.31W
Ponta do Sol Cape Verde **84 B3** 17.12N 25.03W
Ponta Grossa Brazil **77 E2** 25.07S 50.00W
Pontardawe Wales **12 D3** 51.44N 3.51W
Pontefract England **15 F2** 53.42N 1.19W
Ponteland England **15 F4** 55.03N 1.43W
Pontevedra Spain **46 A5** 42.25N 8.39W
Pontianak Indonesia **104 D3** 0.05S 109.16E
Pontine Is. Italy **50 E4** 40.56N 12.58E
Pontine Mts. Turkey **57 M4** 40.32N 38.00E
Pontoise France **44 E6** 49.03N 2.05E
Pont-Ste.-Maxence France **42 B1** 49.18N 2.37E
Pontycymer Wales **12 D3** 51.36N 3.35W
Pontypool Wales **12 D3** 51.42N 3.01W
Pontypridd Wales **12 D3** 51.36N 3.21W
Poole England **10 D1** 50.42N 1.58W
Poole *d.* England **9 E2** 50.42N 1.58W
Poole B. England **10 D1** 50.40N 1.55W
Pooley Bridge *town* England **14 E3** 54.37N 2.49W
Popoó, L. Bolivia **76 D3** 19.00S 67.00W
Popayán Colombia **76 C5** 2.27N 76.32W
Poplar Bluff *town* U.S.A. **65 H4** 36.40N 90.25W
Popocatépetl *mtn.* Mexico **70 E5** 19.02N 98.38W
Poprad Slovakia **54 G3** 49.03N 20.18E
Porbandar India **96 D5** 21.40N 69.40E
Porcupine *r.* U.S.A. **62 D4** 66.25N 145.20W
Pori Finland **43 E3** 61.28N 21.45E
Porlock England **13 D3** 51.14N 3.36W
Poronaysk Russian Fed. **59 Q2** 49.13N 142.55E
Porsangen *est.* Norway **43 F5** 70.30N 25.45E
Porsgrunn Norway **43 B2** 59.10N 9.40E
Porsuk *r.* Turkey **57 J3** 39.41N 31.56E
Portadown N. Ireland **16 C2** 54.25N 6.27W
Portaferry N. Ireland **16 D2** 54.23N 5.33W
Portage la Prairie *town* Canada **62 I2** 49.58N 98.20W
Portalegre Portugal **46 B3** 39.17N 7.25W
Port Angeles U.S.A. **64 B6** 48.06N 123.26W
Port Arthur U.S.A. **65 H2** 29.55N 93.56W
Port Askaig Scotland **16 C3** 55.51N 6.07W
Port Augusta Australia **110 C2** 32.30S 137.46E
Port-au-Prince Haiti **71 J4** 18.33N 72.20W
Portavogie N. Ireland **16 D2** 54.26N 5.27W
Port Blair India **97 I3** 11.40N 92.30E
Port-de-Paix Haiti **71 J4** 19.57N 72.50W
Port Elizabeth R.S.A. **86 B1** 33.58S 25.36E
Port Ellen Scotland **16 C3** 55.38N 6.12W
Port Erin I.o.M. **14 C3** 54.05N 4.45W
Port-Gentil Gabon **84 E1** 0.40S 8.50E
Port Glasgow Scotland **16 E3** 55.56N 4.40W
Port Harcourt Nigeria **84 E2** 4.43N 7.05E
Port Hawkesbury Canada **65 N6** 45.37N 61.21W
Porthcawl Wales **13 D3** 51.28N 3.42W
Port Hedland Australia **110 A3** 20.24S 118.36E
Porthleven England **13 B2** 50.05N 5.20W
Porthmadog Wales **12 C4** 52.55N 4.08W
Port Hope Simpson Canada **63 M3** 52.18N 55.51W
Portimão Portugal **46 A2** 37.08N 8.32W
Port Isaac B. England **13 C2** 50.36N 4.50W
Portishead England **10 C2** 51.29N 2.46W
Portknockie Scotland **19 G2** 57.42N 2.52W
Portland Australia **110 D2** 38.21S 141.38E
Portland Maine U.S.A. **65 L5** 43.41N 70.18W
Portland Oreg. U.S.A. **64 B6** 45.32N 122.40W
Portlaoise Rep. of Ire. **20 D3** 53.03N 7.20W
Portlethen Scotland **19 G2** 57.03N 2.07W
Port Lincoln Australia **110 C2** 34.43S 135.49E
Port Macquarie Australia **110 E2** 31.28S 152.25E
Port Moresby P.N.G. **110 D5** 9.30S 147.07E
Portnahaven Scotland **16 C3** 55.41N 6.31W
Pôrto Alegre Brazil **77 E1** 30.03S 51.10W
Port of Ness Scotland **18 C3** 58.29N 6.14W
Port of Spain Trinidad **71 L3** 10.38N 61.31W
Porto-Novo Benin **84 E2** 6.30N 2.47E
Porto Torres Italy **50 C4** 40.49N 8.24E
Porto-Vecchio France **44 H2** 41.35N 9.16E
Pôrto Velho Brazil **76 D4** 8.45S 63.54W
Portoviejo Ecuador **76 B4** 1.07S 80.28W
Portpatrick Scotland **16 D2** 54.51N 5.07W
Port Pirie Australia **110 C2** 33.11S 138.01E
Portree Scotland **18 C2** 57.24N 6.12W
Portrush N. Ireland **16 C3** 55.12N 6.40W
Port Said Egypt **94 D5** 31.17N 32.18E
Portsmouth England **10 D1** 50.48N 1.06W
Portsmouth *d.* England **9 E2** 50.48N 1.06W
Portsoy Scotland **19 G2** 57.42N 2.42W
Portstewart N. Ireland **16 C3** 55.10N 6.43W
Port Sudan Sudan **85 H3** 19.39N 37.01E
Port Talbot Wales **12 D3** 51.35N 3.48W
Portugal Europe **46 A3** 39.30N 8.05W
Portumna Rep. of Ire. **20 C3** 53.06N 8.13W
Port Vila Vanuatu **111 F4** 17.44S 168.19E
Port William Scotland **16 E2** 54.46N 4.35W
Posadas Argentina **77 E2** 27.25S 55.48W
Poso Indonesia **104 G3** 1.23S 120.45E
Potenza Italy **50 F4** 40.40N 15.47E
Potosí Bolivia **76 D3** 19.34S 65.45W
Potsdam Germany **48 F5** 52.24N 13.04E
Poulton-le-Fylde England **14 E2** 53.51N 3.00W
Pouthisat Cambodia **104 C6** 12.33N 103.55E
Powell, L. U.S.A. **64 D4** 37.30N 110.45W
Powick England **10 C3** 52.10N 2.14W
Powys *d.* Wales **9 D3** 52.26N 3.26W
Poyang Hu *l.* China **103 L3** 29.05N 116.20E
Požarevac Yugo. **56 E6** 44.38N 21.12E
Poza Rica Mexico **70 E5** 20.34N 97.26W
Poznań Poland **54 E5** 52.25N 16.53E
Pozoblanco Spain **46 C3** 38.23N 4.51W
Prabumulih Indonesia **104 C3** 3.29S 104.14E
Prachuap Khiri Khan Thailand **104 B6** 11.50N 99.49E
Prague Czech Rep. **54 D4** 50.05N 14.25E
Praia Cape Verde **84 B3** 14.53N 23.30W
Prapat Indonesia **104 B4** 2.42N 98.56E
Prato Italy **50 D5** 43.52N 11.06E
Prawle Pt. England **13 D2** 50.12N 3.43W
Pregel *r.* Russian Fed. **54 G6** 54.41N 20.22E
Preparis I. Myanmar **97 I3** 14.40N 93.40E
Prescott U.S.A. **64 D3** 34.34N 112.28W
Presidente Prudente Brazil **77 E2** 22.09S 51.24W
Prespa, L. Europe **56 E4** 40.53N 21.02E
Presque Isle *town* U.S.A. **65 M6** 46.42N 68.01W
Prestatyn Wales **12 D5** 53.20N 3.24W
Presteigne Wales **12 E4** 52.17N 3.00W
Preston Dorset England **10 C1** 50.39N 2.25W
Preston Lancs. England **14 E2** 53.46N 2.42W
Prestonpans Scotland **17 G3** 55.57N 3.00W
Prestwick Scotland **16 E3** 55.30N 4.36W
Pretoria R.S.A. **86 B2** 25.45S 28.12E
Preveza Greece **56 E3** 38.58N 20.43E
Pribilof Is. U.S.A. **62 A3** 57.00N 170.00W
Prievidza Slovakia **54 F3** 48.47N 18.35E
Prilep Macedonia **56 E4** 41.20N 21.32E
Prince Albert Canada **62 H3** 53.13N 105.45W
Prince Alfred, C. Canada **62 F5** 74.30N 125.00W
Prince Charles I. Canada **63 K4** 67.50N 76.00W
Prince Edward Is. Indian Oc. **117 K3** 47.00S 37.00E
Prince Edward Island *d.* Canada **63 L2** 46.15N 63.10W
Prince George Canada **62 F3** 53.55N 122.49W
Prince of Wales I. Australia **110 D4** 10.55S 142.05E
Prince of Wales I. Canada **63 I5** 73.00N 99.00W
Prince of Wales I. U.S.A. **62 E3** 55.00N 132.30W
Prince Patrick I. Canada **62 G5** 77.00N 120.00W
Prince Rupert Canada **62 E3** 54.09N 130.20W
Princes Risborough England **10 E2** 51.43N 0.50W
Príncipe *i.* São Tomé & Príncipe **84 E2** 1.37N 7.27E
Pripet *r.* Europe **55 L4** 51.08N 30.30E
Pripet Marshes *f.* Belarus **55 J5** 52.15N 29.00E
Priština Yugo. **56 E5** 42.39N 21.10E
Privas France **44 F4** 44.44N 4.36E
Prizren Yugo. **56 E5** 42.13N 20.42E
Probolinggo Indonesia **104 E2** 7.45S 113.09E
Probus England **13 C2** 50.17N 4.57W
Providence U.S.A. **65 L5** 41.50N 71.30W
Providence, C. New Zealand **111 F1** 46.01S 166.28E
Providencia, I. de Colombia **71 H3** 13.21N 81.22W
Provins France **44 E6** 48.34N 3.18E
Provo U.S.A. **64 D5** 40.15N 111.40W
Prudhoe Bay *town* U.S.A. **62 D5** 70.20N 148.25W
Prüm Germany **42 F2** 50.12N 6.25E
Pruszków Poland **54 G5** 52.11N 20.48E
Prut *r.* Europe **55 K1** 45.29N 28.14E
Pruzhany Belarus **55 I5** 52.23N 24.28E
Prydz B. Antarctica **112** 68.30S 74.00E
Pryluky Ukraine **55 M4** 50.35N 32.24E
Przemyśl Poland **54 H3** 49.48N 22.48E
Psara *i.* Greece **56 G3** 38.34N 25.35E
Psel *r.* Ukraine/Russian Fed. **55 M3** 49.00N 33.30E
Pskov Russian Fed. **58 E3** 57.48N 28.00E
Ptsich *r.* Belarus **55 K5** 52.09N 28.52E
Pucallpa Peru **76 C4** 8.21S 74.33E
Pudsey England **15 F2** 53.47N 1.40W
Puebla Mexico **70 E4** 19.03N 98.10W
Puebla *d.* Mexico **70 E4** 18.30N 98.00W
Pueblo U.S.A. **64 F4** 38.17N 104.38W
Puente-Genil Spain **46 C2** 37.24N 4.46W
Puerto de la Cruz Canary Is. **46 X2** 28.24N 16.33W
Puerto del Rosario Canary Is. **46 Z2** 28.29N 13.52W
Puertollano Spain **46 C3** 38.41N 4.07W
Puerto Maldonado Peru **76 D3** 12.37S 69.11W
Puerto Montt Chile **75 B2** 41.28S 73.00W
Puerto Natales Chile **75 B1** 51.41S 72.15W
Puerto Peñasco Mexico **70 B7** 31.20N 113.35W
Puerto Princesa Phil. **104 F5** 9.46N 118.45E
Puerto Rico C. America **71 K4** 18.20N 66.30W
Pula Croatia **56 A6** 44.52N 13.53E
Pulborough England **11 E1** 50.58N 0.30W
Pułtusk Poland **54 G5** 52.42N 21.02E
Pune India **96 E4** 18.34N 73.58E
Punjab *d.* India **96 F7** 30.30N 75.15E
Punta Arenas *town* Chile **75 B1** 53.10S 70.56W
Puntjak Jaya *mtn.* Indonesia **105 J3** 4.00S 137.15E
Pur *r.* Russian Fed. **58 J4** 67.30N 75.30E
Purari *r.* P.N.G. **110 D5** 7.50S 145.10E
Puri India **97 H4** 19.49N 85.54E
Purmerend Neth. **42 D4** 52.30N 4.56E
Purus *r.* Brazil **77 D4** 3.15S 61.30W
Pusan S. Korea **103 N5** 35.05N 129.02E
Puting, Tanjung *c.* Indonesia **104 E3** 3.35S 111.52E

St. George's Channel U.K./Rep. of Ire. **20 E1** 52.00N 6.00W
St. Germans England **13 C2** 50.24N 4.18W
St. Govan's Head Wales **12 C3** 51.36N 4.55W
St. Helena *i.* Atlantic Oc. **116 I5** 16.00S 6.00W
St. Helena B. R.S.A. **86 A1** 32.35S 18.00E
St. Helens England **14 E2** 53.28N 2.43W
St. Helens, Mt. U.S.A. **64 B6** 46.12N 122.11W
St. Helier Channel Is. **13 Z8** 49.12N 2.07W
St. Ives Cambs. England **11 E3** 52.20N 0.05W
St. Ives Cornwall England **13 B2** 50.13N 5.29W
St. Ives B. England **13 B2** 50.14N 5.26W
St.-Jean, L. Canada **65 L6** 48.35N 72.00W
St. John Canada **63 L2** 45.16N 66.03W
St. John Channel Is. **13 Z8** 49.15N 2.08W
St. John *r.* Canada **63 L2** 45.30N 66.05W
St. John *i.* U.S.V. Is. **71 K4** 18.21N 64.48W
St. John's Antigua **71 L4** 17.07N 61.51W
St. John's Canada **63 M2** 47.34N 52.41W
St. John's Pt. N. Ireland **16 D2** 54.13N 5.39W
St. Jordi, G. of Spain **46 F4** 40.50N 1.10E
St. Joseph U.S.A. **65 H4** 39.45N 94.51W
St. Joseph, Lac *l.* Canada **63 I3** 51.05N 90.35W
St. Just England **13 B2** 50.07N 5.41W
St. Keverne England **13 B2** 50.03N 5.05W
St. Kilda *i.* Scotland **18 A2** 57.49N 8.34W
St. Kitts-Nevis Leeward Is. **71 L4** 17.20N 62.45W
St. Lawrence *r.* Canada/U.S.A. **63 L2** 48.45N 68.30W
St. Lawrence, G. of Canada **63 L2** 48.00N 62.00W
St. Lawrence I. U.S.A. **62 A4** 63.00N 170.00W
St.-Lô France **44 C6** 49.07N 1.05W
St. Louis Senegal **84 C3** 16.01N 16.30W
St. Louis U.S.A. **65 H4** 38.40N 90.15W
St. Lucia Windward Is. **71 L3** 14.05N 61.00W
St. Magnus B. Scotland **19 Y9** 60.25N 1.35W
St.-Malo France **44 B6** 48.39N 2.00W
St.-Malo, Golfe de *g.* France **44 B6** 49.00N 2.00W
St. Margaret's Hope Scotland **19 G3** 58.50N 2.57W
St. Martin Guernsey Channel Is. **13 Y9** 49.27N 2.34W
St. Martin Jersey Channel Is. **13 Z8** 49.13N 2.03W
St. Martin *i.* Leeward Is. **71 L4** 18.05N 63.05W
St. Martin's *i.* England **13 A1** 49.57N 6.16W
St. Mary's *i.* England **13 A1** 49.55N 6.16W
St. Matthew I. U.S.A. **62 A4** 60.30N 172.45W
St. Maurice *r.* Canada **65 L6** 46.21N 72.31W
St. Mawes England **13 B2** 50.10N 5.01W
St. Moritz Switz. **44 H5** 46.30N 9.51E
St.-Nazaire France **44 B5** 47.17N 2.12W
St. Neots England **11 E3** 52.14N 0.16W
St.-Niklaas Belgium **42 D3** 51.10N 4.09E
St.-Omer France **42 B2** 50.45N 2.15E
St. Ouen Channel Is. **13 Z8** 49.13N 2.14W
St. Peter Port Channel Is. **13 Y9** 49.27N 2.32W
St. Petersburg Russian Fed. **58 F3** 59.55N 30.25E
St. Pierre and Miquelon *is.* N. America **63 M2** 47.00N 56.15W
St. Pölten Austria **54 D3** 48.13N 15.37E
St.-Quentin France **44 E6** 49.51N 3.17E
St. Sampson Channel Is. **13 Y9** 49.29N 2.31W
St. Vincent and the Grenadines C. America **71 L3** 13.10N 61.15W
St. Vincent, C. Portugal **46 A2** 37.01N 8.59W
St.-Vith Belgium **42 F2** 50.15N 6.08E
St. Wendel Germany **42 G1** 49.27N 7.10E
Saipan *i.* N. Mariana Is. **105 L7** 15.12N 145.43E
Sajama *mtn.* Bolivia **76 D3** 18.06S 69.00W
Sakai Japan **106 C3** 34.37N 135.28E
Sakakah Saudi Arabia **94 F5** 29.59N 40.12E
Sakakawea, L. U.S.A. **64 F6** 47.30N 102.20W
Sakarya *r.* Turkey **57 J4** 41.08N 30.36E
Sakata Japan **106 C3** 38.55N 139.51E
Sakhalin *i.* Russian Fed. **59 Q3** 50.00N 143.00E
Sakura Japan **106 D3** 35.43N 140.13E
Sal *i.* Cape Verde **84 B3** 16.45N 23.00W
Sala Sweden **43 D2** 59.55N 16.38E
Salado *r.* Argentina **77 D1** 31.40S 60.41W
Salado *r.* Mexico **70 E6** 26.46N 98.55W
Şalalah Oman **95 I2** 17.00N 54.04E
Salamanca Spain **46 C4** 40.58N 5.40W
Salar de Arizaro *f.* Argentina **76 D2** 24.50S 67.40W
Salar de Atacama *f.* Chile **76 D2** 23.30S 68.46W
Salar de Coipasa *l.* Bolivia **76 D3** 19.20S 68.00W
Salar de Uyuni *l.* Bolivia **76 D2** 20.30S 67.45W
Salavat Russian Fed. **58 H3** 53.22N 55.50E
Salayar *i.* Indonesia **104 G2** 6.07S 120.28E
Sala y Gómez, I. Pacific Oc. **109 U4** 26.28S 105.28W
Salcombe England **13 D2** 50.14N 3.47W
Saldanha R.S.A. **86 A1** 33.00S 17.56E
Sale Australia **110 D2** 38.06S 147.06E
Sale England **15 E2** 53.26N 2.19W
Salehurst England **11 F1** 50.58N 0.29E
Salekhard Russian Fed. **58 I4** 66.33N 66.35E
Salem India **97 F3** 11.38N 78.08E
Salem U.S.A. **64 B5** 44.57N 123.01W
Salerno Italy **50 F4** 40.41N 14.45E
Salerno, G. of Med. Sea **50 F4** 40.30N 14.45E
Salford England **15 E2** 53.30N 2.17W
Salgado *r.* Brazil **77 G4** 4.27S 37.46W
Salihli Turkey **57 I3** 38.29N 28.08E
Salihorsk Belarus **55 J5** 52.41N 27.29E
Salina U.S.A. **64 G4** 38.50N 97.37W
Salinas U.S.A. **64 B4** 36.40N 121.38W
Salinas Grandes *f.* Argentina **76 D2** 29.37S 64.56W
Salinosó Lachay, Punta *c.* Peru **76 C3** 11.20S 77.29W
Salisbury England **10 D2** 51.04N 1.48W
Salisbury U.S.A. **65 K4** 38.22N 75.37W
Salisbury Plain *f.* England **10 D2** 51.15N 1.55W
Salitre *r.* Brazil **77 F4** 9.23S 40.35W
Salluit Canada **63 K4** 62.10N 75.40W
Salmon *r.* U.S.A. **64 C6** 45.50N 116.50W
Salmon Arm Canada **62 G3** 50.41N 119.18W
Salmon River Mts. U.S.A. **64 D5** 44.30N 114.30W
Salo Finland **43 E3** 60.23N 23.10E
Sal'sk Russian Fed. **53 L6** 46.30N 41.33E
Salta Argentina **76 D2** 24.46S 65.28W
Saltash England **13 C2** 50.25N 4.13W
Saltcoats Scotland **16 E3** 55.37N 4.47W
Saltee Is. Rep. of Ire. **20 E2** 52.08N 6.36W
Saltillo Mexico **70 D6** 25.30N 101.00W
Salt Lake City U.S.A. **64 D5** 40.45N 111.55W
Salto Uruguay **77 E1** 31.23S 57.58W
Salton Sea *l.* U.S.A. **64 C3** 33.25N 115.45W
Salvador Brazil **77 G3** 12.58S 38.20W
Salween *r.* Myanmar **97 J4** 16.30N 97.33E
Salyan Azerbaijan **95 G6** 39.36N 48.59E
Salzburg Austria **54 C2** 47.54N 13.03E
Salzgitter Germany **48 E5** 52.02N 10.22E
Samandağı Turkey **57 L2** 36.07N 35.55E
Samani Japan **106 D4** 42.09N 142.50E
Samar *i.* Phil. **105 H6** 11.45N 125.15E
Samara Russian Fed. **58 H3** 53.10N 50.15E
Samara *r.* Ukraine **55 N3** 48.27N 35.07E
Samarinda Indonesia **104 F3** 0.30S 117.09E
Samarra' Iraq **94 F5** 34.13N 43.52E
Sambalpur India **97 G5** 21.28N 84.04E
Sambas Indonesia **104 D4** 1.20N 109.15E
Sambre *r.* Belgium **42 D2** 50.29N 4.52E
Same Tanzania **87 B2** 4.04S 37.44E
Samoa Pacific Oc. **108 N5** 13.55S 172.00W
Samos *i.* Greece **56 H2** 37.44N 26.45E
Samothraki *i.* Greece **56 G4** 40.26N 25.35E
Sampit Indonesia **104 E3** 2.34S 112.59E
Samso *i.* Denmark **54 B6** 55.52N 10.37E
Samsun Turkey **57 M4** 41.17N 36.22E
San Mali **84 D3** 13.21N 4.57W
Sana Yemen **94 F2** 15.23N 44.14E
San Ambrosio *i.* Chile **109 X4** 26.28S 79.53W
Sanandaj Iran **95 G6** 35.18N 47.01E
San Andrés, I. de Colombia **71 H3** 12.35N 81.42W
San Antonio U.S.A. **64 G2** 29.25N 98.30W
San Antonio, C. Cuba **71 H5** 21.50N 84.57W
San Benedicto, I. Mexico **70 B4** 19.10N 110.50W
San Bernardino U.S.A. **64 C3** 34.07N 117.18W
San Blas, C. U.S.A. **65 I2** 29.40N 85.25W
San Cristóbal Venezuela **71 J2** 7.46N 72.15W
San Cristóbal, I. Galapagos Is. **76 B4** 0.50S 89.30W
San Cristobal *i.* Solomon Is. **111 F4** 10.40S 162.00E
San Cristóbal de la Laguna Canary Is. **46 X2** 28.29N 16.19W
Sanda I. Scotland **16 D3** 55.17N 5.34W
Sandakan Malaysia **104 F5** 5.52N 118.04E
Sandane Norway **43 A3** 61.47N 6.14E
Sanday *i.* Scotland **19 G4** 59.15N 2.33W
Sanday Sd. Scotland **19 G4** 59.11N 2.35W
Sandbach England **15 E2** 53.09N 2.23W
Sandhurst England **10 E2** 51.21N 0.49W
San Diego U.S.A. **64 C3** 32.45N 117.10W
Sandnes Norway **43 A2** 58.51N 5.45E
Sandnessjøen Norway **43 C4** 66.01N 12.40E
Sandoway Myanmar **97 I4** 18.28N 94.20E
Sandown England **10 D1** 50.39N 1.09W
Sandpoint *town* U.S.A. **64 C6** 48.17N 116.34W
Sandviken Sweden **43 D3** 60.38N 16.50E
Sandwich England **11 G2** 51.16N 1.21E
Sandy England **11 E3** 52.08N 0.18W
Sandy C. Australia **110 E3** 24.42S 153.17E
Sandy Lake *town* Canada **63 I3** 53.00N 93.00W
San Felipe Mexico **70 B7** 31.03N 114.52W
San Félix *i.* Chile **109 W4** 26.23S 80.05W
San Fernando Phil. **104 G7** 16.39N 120.19E
San Fernando Spain **46 B2** 36.28N 6.12W
San Fernando de Apure Venezuela **71 K2** 7.53N 67.15W
San Francisco U.S.A. **64 B4** 37.45N 122.27W
San Francisco, C. de Ecuador **76 B5** 0.38N 80.08W
San Francisco de Paula, C. Argentina **75 C2** 49.44S 67.38W
Sangha *r.* Congo **82 E4** 1.10S 16.47E
Sangir Is. Indonesia **105 H4** 2.45N 125.20E
Sangkulirang Indonesia **104 F4** 1.00N 117.58E
Sangli India **96 E4** 16.55N 74.37E
Sangmélima Cameroon **84 F2** 2.57N 11.56E
Sangre de Cristo Range *mts.* U.S.A. **64 E4** 37.30N 106.00W
Sangue *r.* Brazil **77 E3** 11.00S 58.30W
San Jorge, Golfo de *g.* Argentina **75 C2** 46.00S 66.00W
San José Costa Rica **71 H3** 9.59N 84.04W
San Jose U.S.A. **64 B4** 37.20N 121.55W
San José *i.* Mexico **70 B6** 25.00N 110.38W
San Juan Argentina **76 D1** 31.33S 68.31W
San Juan Puerto Rico **71 K4** 18.29N 66.08W
San Juan *r.* Costa Rica **71 H3** 10.50N 83.40W
San Juan *r.* U.S.A. **64 D4** 37.20N 110.05W
San Juan Bautista Paraguay **77 E2** 26.37S 57.06W
San Juan Mts. U.S.A. **64 E4** 37.30N 107.00W
Şanlıurfa Turkey **57 N2** 37.08N 38.45E
Sanlúcar de Barrameda Spain **46 B2** 36.46N 6.21W
San Lucas, C. Mexico **70 B5** 22.50N 110.00W
San Luis, Lago de *l.* Bolivia **76 D3** 13.40S 64.00W
San Luis Obispo U.S.A. **64 B4** 35.16N 120.40W
San Luis Potosí Mexico **70 D5** 22.10N 101.00W
San Luis Potosí *d.* Mexico **70 D5** 23.00N 100.00W
San Marino Europe **50 E5** 43.55N 12.27E
San Marino *town* San Marino **50 E5** 43.55N 12.27E
San Martín *r.* Bolivia **76 D3** 13.05S 63.48W
San Martín, L. Argentina **75 B2** 49.00S 72.30W
San Matías, Golfo *g.* Argentina **75 C2** 41.30S 64.00W
San Miguel El Salvador **70 G3** 13.28N 88.10W
San Miguel *r.* Bolivia **76 D3** 12.25S 64.25W
San Miguel de Tucumán Argentina **76 D2** 26.47S 65.15W
Sanming China **103 L3** 26.25N 117.35E
Sanndray *i.* Scotland **18 B1** 56.53N 7.31W
San Pablo Phil. **104 G6** 13.58N 121.10E
San Pedro Paraguay **77 E2** 24.08S 57.08W
San Pedro, Sierra de *mts.* Spain **46 B3** 39.20N 6.20W
San Pedro Sula Honduras **70 G4** 15.26N 88.01W
San Pietro *i.* Italy **50 C3** 39.09N 8.16E
Sanquhar Scotland **17 F3** 55.22N 3.56W
San Quintin, C. Mexico **64 C3** 30.20N 116.00W
San Remo Italy **50 B5** 43.48N 7.46E
San Salvador El Salvador **70 G3** 13.40N 89.10W
San Salvador *i.* The Bahamas **71 J5** 24.00N 74.32W
San Salvador de Jujuy Argentina **76 D2** 24.10S 65.20W
San Sebastian de la Gomera Canary Is. **46 X2** 28.06N 17.06W
San Severo Italy **50 F4** 41.40N 15.24E
Santa Ana El Salvador **70 G3** 14.00N 89.31W
Santa Ana U.S.A. **64 C3** 33.44N 117.54W
Santa Barbara U.S.A. **64 C3** 34.25N 119.41W
Santa Catarina *d.* Brazil **77 E2** 27.00S 50.00W
Santa Clara Cuba **71 H5** 22.25N 79.58W
Santa Cruz Bolivia **76 D3** 17.58S 63.14W
Santa Cruz, I. Galapagos Is. **76 A4** 0.40S 90.20W
Santa Cruz de la Palma Canary Is. **46 X2** 28.41N 17.46W
Santa Cruz de Tenerife Canary Is. **46 X2** 28.27N 16.14W
Santa Cruz do Sul Brazil **77 E2** 29.42S 52.25W
Santa Cruz Is. Solomon Is. **111 F4** 10.30S 166.00E
Santa Elena, B. de Ecuador **76 B4** 2.10S 80.50W
Santa Elena, C. Costa Rica **70 G3** 10.54N 85.56W
Santa Fé Argentina **77 D1** 31.38S 60.43W
Santa Fe U.S.A. **64 E4** 35.41N 105.57W
Santa Isabel *i.* Solomon Is. **111 E5** 8.00S 159.00E
Santa Maria Brazil **77 E2** 29.40S 53.47W
Santa Maria U.S.A. **64 B3** 34.56N 120.25W
Santa Maria *r.* Mexico **70 C7** 31.10N 107.05W
Santa Maria di Leuca, C. Italy **50 H3** 39.47N 18.24E
Santa Marta Colombia **71 J3** 11.18N 74.10W
Santander Spain **46 D5** 43.28N 3.48W
Santarém Brazil **76 E4** 2.26S 54.41W
Santarém Portugal **46 A3** 39.14N 8.40W
Santa Rosa Argentina **75 C3** 36.00S 64.40W
Santa Rosa U.S.A. **64 B4** 38.26N 122.34W
Santa Rosalía Mexico **70 B6** 27.20N 112.20W
Santiago Chile **75 B3** 33.30S 70.40W
Santiago Dom. Rep. **71 J4** 19.30N 70.42W
Santiago Spain **46 A5** 42.52N 8.33W
Santiago de Cuba Cuba **71 I5** 20.00N 75.49W
Santo André Brazil **77 F2** 23.39S 46.29W
Santo Antão *i.* Cape Verde **84 B3** 17.00N 25.10W
Santo Domingo Dom. Rep. **71 K4** 18.30N 69.57W
Santoña Spain **46 D5** 43.27N 3.26W
Santos Brazil **77 F2** 23.56S 46.22W
San Valentín *mtn.* Chile **75 B2** 46.33S 73.20W
São Carlos Brazil **77 F2** 22.01S 47.54W
São Francisco *r.* Brazil **77 G3** 10.10S 36.40W
São José do Rio Prêto Brazil **77 F2** 20.50S 49.20W
São Luís Brazil **77 F4** 2.34S 44.16W
São Marcos, Baia de *b.* Brazil **77 F4** 2.30S 44.15W
Saône *r.* France **44 F4** 45.46N 4.52E
São Paulo Brazil **77 F2** 23.33S 46.39W
São Roque, C. de Brazil **77 G4** 5.00S 35.00W
São Sebastião, I. de Brazil **77 F2** 23.53S 45.17W
São Tiago *i.* Cape Verde **84 B3** 15.00N 23.40W
São Tomé *i.* São Tomé & Príncipe **84 E2** 0.19N 6.43E
São Tomé, Cabo de *c.* Brazil **75 E4** 21.54S 40.59W
São Tomé & Príncipe Africa **84 E2** 1.00N 7.00E
São Vicente Brazil **77 F2** 23.57S 46.23W
Sã Paulo *d.* Brazil **77 F2** 22.05S 48.00W
Sapporo Japan **106 D4** 43.05N 141.21E
Sapri Italy **50 F4** 40.04N 15.38E
Saqqaq Greenland **63 M5** 70.00N 52.00W
Saqqez Iran **95 G6** 36.14N 46.15E
Sarab Iran **95 G6** 37.56N 47.35E
Sara Buri Thailand **104 C6** 14.32N 100.53E
Sarajevo Bosnia. **56 D5** 43.52N 18.26E
Sarandë Albania **56 E3** 39.52N 20.00E
Saransk Russian Fed. **58 G3** 54.12N 45.10E
Sarasota U.S.A. **65 J2** 27.20N 82.32W
Saratov Russian Fed. **58 G3** 51.30N 45.55E
Saravan Iran **95 J4** 27.25N 62.17E
Sarawak *d.* Malaysia **104 E4** 3.00N 114.00E
Sardindida Plain *f.* Kenya **87 C3** 2.30N 40.00E
Sardinia *i.* Italy **50 C4** 40.00N 9.00E
Sarektjåkkå *mtn.* Sweden **43 D4** 67.25N 17.45E
Sar-e Pol Afghan. **95 K6** 36.13N 65.55E
Sargodha Pakistan **96 E7** 32.01N 72.40E
Sarh Chad **85 F2** 9.08N 18.22E
Sarī Iran **95 H6** 36.33N 53.06E
Sarigan *i.* N. Mariana Is. **105 L7** 16.43N 145.47E
Sarıyer Turkey **57 I4** 41.11N 29.03E
Sark *i.* Channel Is. **13 Y9** 49.26N 2.22W
Şarkişla Turkey **57 M3** 39.21N 36.27E
Särna Sweden **43 C3** 61.40N 13.10E
Sarny Ukraine **55 J4** 51.21N 26.31E
Saros, G. of Turkey **56 H4** 40.32N 26.25E
Sarpsborg Norway **43 B2** 59.17N 11.06E
Sarrebourg France **44 G6** 48.43N 7.03E
Sarria Spain **46 B5** 42.47N 7.25W
Sarthe *r.* France **44 C5** 47.29N 0.30W
Sasebo Japan **106 A2** 33.10N 129.42E
Saskatchewan *r.* Canada **62 H3** 53.25N 100.15W
Saskatchewan *d.* Canada **62 H3** 55.00N 105.00W
Saskatoon Canada **62 H3** 52.10N 106.40W
Sassandra Côte d'Ivoire **84 D2** 4.58N 6.08W
Sassari Italy **50 C4** 40.43N 8.33E
Sassnitz Germany **48 F6** 54.32N 13.40E
Satpura Range *mts.* India **96 F5** 21.50N 76.00E
Satu Mare Romania **54 H2** 47.48N 22.52E
Saudhárkrókur Iceland **43 Y2** 65.45N 19.39W
Saudi Arabia Asia **94 F3** 25.00N 44.00E
Sault Sainte Marie Canada **63 J2** 46.32N 84.20W
Sault Sainte Marie U.S.A. **65 J6** 46.30N 84.21W
Saumlakki Indonesia **105 I2** 7.59S 131.22E
Saundersfoot Wales **12 C3** 51.43N 4.42W
Saurimo Angola **86 B4** 9.38S 20.20E
Sava *r.* Yugo. **56 E6** 44.50N 20.26E
Savaii *i.* Samoa **108 N5** 13.36S 172.27W
Savannah U.S.A. **65 J3** 32.09N 81.01W
Savannah *r.* U.S.A. **65 J3** 32.10N 81.00W
Savannakhét Laos **104 C7** 16.34N 104.55E
Save *r.* Mozambique **86 C2** 21.00S 35.01E
Savona Italy **50 C6** 44.18N 8.28E
Sawel Mt. N. Ireland **16 B2** 54.49N 7.03W
Sawston England **11 F3** 52.07N 0.11E
Sawtry England **11 E3** 52.27N 0.15W
Sawu *i.* Indonesia **105 G1** 10.30S 121.50E
Sawu Sea Pacific Oc. **105 G2** 9.30S 122.30E
Saxilby England **15 G2** 53.17N 0.40W
Saxmundham England **11 G3** 52.13N 1.29E
Sayhut Yemen **95 H2** 15.12N 51.12E
Saylac Somalia **85 I3** 11.21N 43.30E
Saynshand Mongolia **103 K6** 44.58N 110.10E
Scafell Pike *mtn.* England **14 D3** 54.27N 3.12W
Scalasaig Scotland **16 C4** 56.04N 6.12W
Scalby England **15 G3** 54.18N 0.26W
Scalp *mtn.* Donegal Rep. of Ire. **16 B3** 55.05N 7.22W
Scalp *mtn.* Galway Rep. of Ire. **20 C3** 53.00N 8.29W
Scalpay *i.* High. Scotland **18 D2** 57.18N 5.58W
Scalpay *i.* W.Isles Scotland **18 C2** 57.52N 6.40W
Scandinavia *f.* Europe **34 E4** 65.00N 18.00E
Scapa Flow *str.* Scotland **19 F3** 58.53N 3.05W
Scarba *i.* Scotland **16 D4** 56.11N 5.42W
Scarborough England **15 G3** 54.17N 0.24W
Scarp *i.* Scotland **18 B3** 58.02N 7.07W
Schaffhausen Switz. **44 H5** 47.42N 8.38E
Schagen Neth. **42 D4** 52.47N 4.47E
Schefferville Canada **63 L3** 54.50N 67.00W
Schelde *r.* Belgium **42 D3** 51.13N 4.25E
Schiehallion *mtn.* Scotland **19 E1** 56.40N 4.08W
Schiermonnikoog *i.* Neth. **42 F5** 53.28N 6.15E
Schleswig Germany **48 D6** 54.32N 9.34E
Schwaner Mts. Indonesia **104 E3** 0.45S 113.20E
Schwedt Germany **48 G5** 53.04N 14.17E
Schweinfurt Germany **48 E4** 50.03N 10.16E
Schwerin Germany **48 E5** 53.38N 11.25E
Scilly, Isles of England **13 A1** 49.55N 6.20W
Scole England **11 G3** 52.22N 1.10E
Scotland U.K. **8-9**
Scottish Borders *d.* Scotland **8 D4** 55.30N 2.53W
Scottsbluff U.S.A. **64 F5** 41.52N 103.40W
Scourie Scotland **18 D3** 58.20N 5.09W
Scranton U.S.A. **65 K5** 41.25N 75.40W
Scridain, Loch Scotland **16 C4** 56.22N 6.06W
Scunthorpe England **15 G2** 53.35N 0.38W
Scuol Switz. **44 G5** 46.48N 10.18E
Seaford England **11 F1** 50.46N 0.08E
Seamer England **15 G3** 54.14N 0.27W
Seascale England **14 D3** 54.24N 3.29W
Seaton Cumbria England **14 D3** 54.41N 3.31W
Seaton Devon England **13 D2** 50.43N 3.05W
Seaton Delaval England **15 F4** 55.05N 1.31W
Seattle U.S.A. **64 B6** 47.35N 122.20W
Sebastián Vizcaíno B. Mexico **70 B6** 28.20N 114.45W
Sechura, Bahía de *b.* Peru **76 B4** 5.30S 81.00W
Secunderabad India **97 F4** 17.27N 78.27E
Sedan France **44 F6** 49.42N 4.57E
Sedbergh England **14 E3** 54.20N 2.31W
Seefin *mtn.* Rep. of Ire. **20 D2** 52.13N 7.36W
Segovia Spain **46 C4** 40.57N 4.07W
Segura, Sierra de *mts.* Spain **46 D3** 38.00N 2.50W
Seiland *i.* Norway **43 E5** 70.30N 23.00E
Seinäjoki Finland **43 E3** 62.45N 22.55E
Seine *r.* France **44 D6** 49.28N 0.25E
Sekondi-Takoradi Ghana **84 D2** 4.59N 1.43W
Selaru *i.* Indonesia **105 I2** 8.15S 131.00E
Selatan I. Indonesia **104 C3** 3.00S 100.18E
Selatan, Tanjung *c.* Indonesia **104 E3** 4.20S 114.45E
Selattyn England **10 B3** 52.54N 3.05W
Selawik U.S.A. **62 C4** 66.35N 160.10W
Selby England **15 F2** 53.47N 1.05W
Selkirk Canada **62 I3** 50.10N 96.52W
Selkirk Scotland **17 G3** 55.33N 2.51W
Sellindge England **11 G2** 51.07N 1.00E
Selsey England **10 E1** 50.44N 0.47W
Selsey Bill *c.* England **10 E1** 50.44N 0.47W
Selvas *f.* Brazil **77 D4** 8.00S 68.00W
Selwyn Mts. Canada **62 E4** 63.00N 130.00W
Selwyn Range *mts.* Australia **110 C4** 21.35S 140.35E
Seman *r.* Albania **56 D4** 40.53N 19.25E
Semarang Indonesia **104 E2** 6.58S 110.29E
Semenivka Ukraine **55 M5** 52.08N 32.36E
Seminoe Resr. U.S.A. **64 E5** 42.05N 106.50W
Semipalatinsk Kazakhstan **102 E8** 50.26N 80.16E
Semnan Iran **95 H6** 35.31N 53.24E
Semois *r.* France/Belgium **42 D1** 49.53N 4.45E
Sendai Japan **106 A3** 38.20N 140.50E
Senegal Africa **84 C3** 14.15N 14.15W
Sénégal *r.* Senegal/Mauritania **84 C3** 16.00N 16.28W
Senja *i.* Norway **43 D5** 69.20N 17.30E
Senlis France **42 B1** 49.12N 2.35E
Sennar Sudan **85 H3** 13.31N 33.38E
Sens France **44 E6** 48.12N 3.18E
Seoul S. Korea **103 N5** 37.30N 127.00E
Sepik *r.* P.N.G. **105 K3** 3.53S 144.33E
Sept-Îles *town* Canada **63 L3** 50.13N 66.22W
Seram *i.* Indonesia **105 H3** 3.10S 129.30E
Seram Sea Pacific Oc. **105 H3** 2.50S 128.00E
Serang Indonesia **104 D2** 6.07S 106.09E
Serbia *d.* Yugo. **56 E5** 43.52N 21.00E
Seremban Malaysia **104 C4** 2.42N 101.54E
Serengeti Nat. Park Tanzania **87 A2** 2.20S 34.55E
Serengeti Plains *f.* Kenya **87 B2** 3.30S 37.50E
Serere Uganda **87 A3** 1.31N 33.25E
Seret *r.* Ukraine **55 I3** 48.38N 25.52E
Sergipe *d.* Brazil **77 G3** 11.00S 37.00W
Seria Brunei **104 E4** 4.39N 114.23E
Serov Russian Fed. **58 I3** 59.42N 60.32E
Serowe Botswana **86 B2** 22.25S 26.44E
Serpukhov Russian Fed. **55 O6** 54.53N 37.25E
Serranías del Burro *mts.* Mexico **70 D6** 28.30N 102.00W
Serres Greece **56 F4** 41.04N 23.32E
Serui Indonesia **105 J3** 1.53S 136.15E
Sestroretsk Russian Fed. **43 G3** 60.09N 29.58E
Sète France **44 E3** 43.25N 3.43E
Sete Lagoas Brazil **77 F3** 19.29S 44.15W
Setesdal *f.* Norway **43 A2** 59.20N 7.25E
Sétif Algeria **84 E5** 36.09N 5.26E
Seto-naikai *str.* Japan **106 B2** 34.00N 132.30E
Settat Morocco **84 D5** 33.04N 7.37W
Settle England **15 E3** 54.05N 2.18W
Setúbal Portugal **46 A3** 38.31N 8.54W
Setúbal, B. of Portugal **46 A3** 38.20N 9.00W
Seul, Lac *l.* Canada **65 H7** 50.20N 92.30W
Sevastopol' Ukraine **55 M1** 44.36N 33.31E
Seven Heads Rep. of Ire. **20 C1** 51.34N 8.42W
Sevenoaks England **11 F2** 51.16N 0.12E
Severn *r.* Canada **63 J3** 56.00N 87.40W
Severn *r.* England **10 C2** 51.50N 2.21W
Severnaya Zemlya *is.* Russian Fed. **59 L5** 80.00N 96.00E
Severo-Kurilsk Russian Fed. **59 R3** 50.40N 156.01E
Seville Spain **46 C2** 37.24N 5.59W
Seward U.S.A. **62 D4** 60.05N 149.34W
Seward Pen. U.S.A. **62 B4** 65.00N 164.10W
Seychelles Indian Oc. **85 J1** 5.00S 55.00E
Seydhisfjördhur Iceland **43 Z2** 65.16N 14.02W
Seydişehir Turkey **57 J2** 37.25N 31.51E
Seym *r.* Russian Fed. **55 M4** 51.30N 32.30E
Sfântu Gheorghe Romania **56 G6** 45.52N 25.50E
Sfax Tunisia **84 F5** 34.45N 10.43E
Sgurr Dhomhnuill *mtn.* Scotland **18 D1** 56.47N 5.25W
Sgurr Mor *mtn.* Scotland **18 D2** 57.41N 5.01W
Shaanxi *d.* China **103 J4** 34.00N 109.00E
Shabeelle *r.* Somalia **85 I2** 0.30N 43.10E
Shaftesbury England **10 C2** 51.00N 2.12W
Shahr-e Kord Iran **95 H5** 32.40N 50.52E
Shandong *d.* China **103 L5** 35.45N 117.30E
Shandong Pen. China **103 M5** 37.00N 121.30E
Shanghai China **103 M4** 31.13N 121.25E
Shangqiu China **103 L4** 34.21N 115.40E
Shangzhi China **106 A5** 45.13N 127.59E
Shanklin England **10 D1** 50.38N 1.10W
Shannon *r.* Rep. of Ire. **20 C2** 52.39N 8.43W
Shannon, Mouth of the *est.* Rep. of Ire. **20 B2** 52.29N 9.57W
Shantar Is. Russian Fed. **59 P3** 55.00N 138.00E
Shanxi *d.* China **103 K5** 36.45N 112.00E
Shaoguan China **103 K2** 24.54N 113.33E
Shaoxing China **103 M3** 30.02N 120.35E